The Multiverse

The Multiverse

Special Issue Editors

Mariusz P. Dąbrowski
Ana Alonso-Serrano
Thomas Naumann

MDPI • Basel • Beijing • Wuhan • Barcelona • Belgrade • Manchester • Tokyo • Cluj • Tianjin

Special Issue Editors
Mariusz P. Dąbrowski
Institute of Physics,
University of Szczecin
Poland

Ana Alonso-Serrano
Max-Planck Institute for
Gravitational Physics,
Albert Einstein Institute
Germany

Thomas Naumann
Deutsches Elektronen-Synchrotron
DESY
Germany

Editorial Office
MDPI
St. Alban-Anlage 66
4052 Basel, Switzerland

This is a reprint of articles from the Special Issue published online in the open access journal *Universe* (ISSN 2218-1997) (available at: https://www.mdpi.com/journal/universe/special_issues/the_multiverse).

For citation purposes, cite each article independently as indicated on the article page online and as indicated below:

LastName, A.A.; LastName, B.B.; LastName, C.C. Article Title. *Journal Name* **Year**, *Article Number*, Page Range.

ISBN 978-3-03928-867-0 (Hbk)
ISBN 978-3-03928-868-7 (PDF)

© 2020 by the authors. Articles in this book are Open Access and distributed under the Creative Commons Attribution (CC BY) license, which allows users to download, copy and build upon published articles, as long as the author and publisher are properly credited, which ensures maximum dissemination and a wider impact of our publications.

The book as a whole is distributed by MDPI under the terms and conditions of the Creative Commons license CC BY-NC-ND.

Contents

About the Special Issue Editors vii

Ana Alonso-Serrano, Mariusz P. Dąbrowski and Thomas Naumann
Post-Editorial of "The Multiverse" Special Volume
Reprinted from: *Universe* **2020**, 6, 17, doi:10.3390/universe6010017 1

Michael Heller
Multiverse—Too Much or Not Enough?
Reprinted from: *Universe* **2019**, 5, 113, doi:10.3390/universe5050113 5

Ana Alonso-Serrano and Gil Jannes
Conceptual Challenges on the Road to the Multiverse
Reprinted from: *Universe* **2019**, 5, 212, doi:10.3390/universe5100212 15

Mariusz P. Dąbrowski
Anthropic Selection of Physical Constants, Quantum Entanglement, and the Multiverse Falsifiability
Reprinted from: *Universe* **2019**, 5, 172, doi:10.3390/universe5070172 37

Michael Douglas
The String Theory Landscape
Reprinted from: *Universe* **2019**, 5, 176, doi:10.3390/universe5070176 59

Salvador J. Robles-Perez
Time Reversal Symmetry in Cosmology and the Creation of a Universe–Antiuniverse Pair
Reprinted from: *Universe* **2019**, 5, 150, doi:10.3390/universe5060150 75

Eckhard Rebhan
Possible Origins and Properties of an Expanding, Dark Energy Providing *Dark Multiverse*
Reprinted from: *Universe* **2019**, 5, 178, doi:10.3390/universe5080178 87

McCullen Sandora
Multiverse Predictions for Habitability: The Number of Stars and Their Properties
Reprinted from: *Universe* **2019**, 5, 149, doi:10.3390/universe5060149 109

McCullen Sandora
Multiverse Predictions for Habitability: Number of Potentially Habitable Planets
Reprinted from: *Universe* **2019**, 5, 157, doi:10.3390/universe5060157 135

McCullen Sandora
Multiverse Predictions for Habitability: Fraction of Planets that Develop Life
Reprinted from: *Universe* **2019**, 5, 171, doi:10.3390/universe5070171 169

McCullen Sandora
Multiverse Predictions for Habitability: Fraction of Life That Develops Intelligence
Reprinted from: *Universe* **2019**, 5, 175, doi:10.3390/universe5070175 199

About the Special Issue Editors

Mariusz P. Dąbrowski is a theoretical physicist specializing in cosmology, nuclear physics, and particle physics. He is Full Professor at the Institute of Physics, University of Szczecin, and Head of the Szczecin Cosmology Group, which he created. He received his Ph.D. in theoretical physics at the Institute of Theoretical Physics, University of Wrocław, Poland, in 1989 and was awarded the title of Full Professor by the President of Poland in 2012. He has authored about 100 research papers in cosmology and related fields and has led many grants, including a recent project (2013–2018) dealing with the variability of fundamental constants in physics and cosmology funded by the Polish National Science Centre. His main research accomplishments are the proposal of some new inhomogeneous models of the universe (Stephani–Dąbrowski models), the exploration of exotic singularities in the universe, and discovering a new type of barotropic index w-singularity. Most recently he suggested a novel test for astronomical measurement of the variability of the speed of light in the universe, formulated cyclic models of the multiverse, including calculations of inter-universal entanglement entropy, and worked on some aspects of generalized and extended uncertainty principles in quantum mechanics with consequences for black holes and the universe. Prof. Dąbrowski is also active in the dissemination of results in astronomy, cosmology, and nuclear and particle physics to the general public, in collaboration with the Copernicus Center for Interdisciplinary Studies, Kraków, Poland, among others.

Ana Alonso Serrano is a junior researcher at the Max Planck Institute for Gravitational Physics (Albert Einstein Institute) in Germany. She is a theoretical physicist working on the interplay between general relativity and quantum mechanics. She developed her thesis research at The Spanish National Research Council (CSIC) in Spain, obtaining her Ph.D. at the Complutense University of Madrid in 2015. Since then, she had several postdoctoral positions at Victoria University of Wellington (New Zealand), Charles University of Prague (Czech Republic) and Max Planck Insitute for Gravitational Physics (Germany). Her work has addressed several aspects of the multiverse, wormholes, quantum cosmology, and the study of singularities, black holes information paradox, quantum field theory in curved spaces and phenomenology of quantum gravity among others. She is also very involved in scientific outreach activities by giving talks, publishing articles, organizing workshops, and being part of the Berlin Soapbox Science team.

Thomas Naumann Prof. Naumann finished his Diploma of physics at the Technical University of Dresden in 1975. He then started to study strongly interacting elementary particles at the Institute of High Energy Physics in Zeuthen near Berlin and finished his Ph.D. at the Humboldt University of Berlin in 1980. In 1987, Prof Naumann joined the H1 experiment at the electro-proto collider HERA at the Deutsches Elektronen-Synchroton DESY where he became a senior scientist in 1992 and worked on the precision measurement of the proton structure function and the coupling constant of the strong interaction. In 2005, he was appointed Honorary Professor at the University of Leipzig. Prof. Naumann joined the ATLAS experiment at the Large Hadron Collider LHC at CERN in 2006. Since 2012 he has led the Particle Physics Group of the DESY site in Zeuthen and represents Germany in the International Particle Physics Outreach Group IPPOG.

Editorial

Post-Editorial of "The Multiverse" Special Volume

Ana Alonso-Serrano [1], Mariusz P. Dąbrowski [2,3,4,*] and Thomas Naumann [5]

1. Max Planck Institute for Gravitational Physics, Albert Einstein Institute, Am Mühlenberg 1, D-14476 Golm, Germany; ana.alonso.serrano@aei.mpg.de
2. Institute of Physics, University of Szczecin, Wielkopolska 15, 70-451 Szczecin, Poland
3. National Centre for Nuclear Research, Andrzeja Sołtana 7, 05-400 Otwock, Poland
4. Copernicus Center for Interdisciplinary Studies, Szczepańska 1/5, 31-011 Kraków, Poland
5. Deutsches Elektronen-Synchrotron DESY, Platanenallee 6, 15738 Zeuthen, Germany; Thomas.Naumann@desy.de
* Correspondence: Mariusz.Dabrowski@usz.edu.pl

Received: 6 January 2020; Accepted: 14 January 2020; Published: 20 January 2020

Abstract: A succesful series of papers devoted to various aspects of an idea of the Multiverse have been gathered together and presented to the readers. In this post-editorial we briefly challenge the content referring to the main issues dealt with by the Authors. We hope that this will inspire other investigators for designing future tests which could make this very notion of the Multiverse falsifiable.

Keywords: philosophy of multiverse; categories of multiverses; different physics universes; superstring multiverse; dark multiverse; multiverse entanglement; universe-antiuniverse pair creation; multiverse habitability: stars, planets, life, consciousness; falsifiability of multiverses

---

Although the idea of the Multiverse as a collection of possible universes has entered the area of physics long time ago, it is right now when it is taking viability and providing alternatives to confront the current cosmological conundrums.

While one may consider the studies related to the concept of the Multiverse as a new revolution that can change the current paradigm in cosmology, in fact, it can rather be understood as the next step in the Copernican transit, where our habitat has gradually lost relevance as unique, special, and also tiny as compared to early science ages thought. Nowadays, the notion of the multiverse emerges naturally from some developments in cosmology and particle physics as a consequence of the same physical theories which we experience on the Earth [1]. Since the multiverse is not a theory by itself, then there is no closed scenario or definition of it. Firstly, it depends on the definition of what we mean under the notion of the universe. Is it the Solar System, as it was thought at Copernicus age? Is it a galaxy, as it was thought till the beginning of the 20th century? Next, is it the Hubble size universe with its outer horizon bulk or something much larger and perhaps hardly achievable by current observations? Depending on the range of it, one then allows for a great diversity of multiverse theories and asks for a deeper debate about the nature of this entity even on the level of a single universe [2].

Whatever the interpretation, it seems that operationally the consideration of an idea of the multiverse could provide solutions to several open problems in physics. This is why we are interested in different approaches and proposals regarding the multiverse in this issue. However, the biggest challenge of the multiverse hypothesis is the possibility to falsify it by some observational or experimental data. Without this most important point, we cannot make it a physical theory in the sense of contemporary definition of the scientific paradigm.

Hoping it will serve as a basic and updated reference, this Special Issue covers all current research avenues on the exciting track to the Multiverse starting from philosophical aspects, throughout the theory, to its possible observational verification.

The area of philosophy and history of physics is where the debate about how to define the category of the multiverse and the need (or not) to endow it with physical meaning exists [3]. In several papers we cover the ideas that philosophy of science provide to falsify multiverse theories and describe the scientific progress.

The diversity of possible physical shapes of a universe within the multiverse can be interpreted in terms of diversity of possible ways to choose physical parameters and can be related to the issue of varying physical constants and varying physical laws [4]. Another idea related to the Multiverse we cover in the issue is the Anthropic Principle which, despite being in some sense tautological, it has been argued how it could give some insight and possible constraints onto the nature of the physics we experience here in "Our Universe", whatever it is.

One strongly studied approach is given by superstring theory which led physicists to an idea of superstring landscape and the swampland, through many ways of choosing the physical vacua due to the symmetry breaking mechanism [5]. This provides a theoretical framework for the multiverse and may as well be related to the eternal inflation theory that constitutes one of the possible mechanisms for the inflation of our universe. In the issue some of these ideas are analyzed, including the discussion about their results and criticisms.

A more recent idea of the multiverse is constituted of the quantum multiverse, in which different individual universes are classically causally separated, but quantum mechanically entangled [4]. This approach has entered strongly in the scenario, because it gives possible predictions and an opportunity to falsify the concept. Also, the problems related to the creation of the multiverse are tackled in the similar sense as the debate of the imposed boundary conditions for the creation of a single universe [6].

Last but not least, the universe and surely the multiverse by its name are everything so it is no wonder that some interesting convergence of topics in an interdisciplinary fashion should appear which can provide some wiser view on the nature and a broader view of the effects under study. For that reason, the issue also contains the investigation of the multiverse from the point of view of astrobiology. This is in tight relation to the Anthropic selection of the three fundamental constants: the fine structure constant α, the electron-to-proton mass ratio β, and the strength of gravity as expressed by the ratio of proton mass to the Planck mass γ. The selection means various levels of habitability criteria in the multiverse, beginning with the number of stars and their properties to host life [7], through the number of habitable planets [8] and the fraction of planets having a chance to develop life [9], and the fraction of that chance admitting the intelligent life [10]. All this is considered in terms of the so-called typicality or the probability of observation of the values of the fundamental constants which do not deviate from the values we know in our Universe. The habitability criterion is then expected to be the useful observational test of the multiverse concept which is the main objective of the whole story about it.

We then finally encourage the reader to dive into the ocean of all these problems since they might be quite fascinating developments of the prospective 21st century physics.

Author Contributions: A.A.-S. and M.P.D. contributed equally to this work. T.N. gave a general feedback and the consent to publish in an agreed form. All authors have read and agreed to the published version of the manuscript.

Funding: This research received no external funding.

Acknowledgments: The guest editors would like to thank all the Authors for their contributions and the Reviewers for the constructive reports. Their work helped the editors to collect this special issue.

Conflicts of Interest: The authors declare no conflict of interest.

References

1. Rebhan, E. Possible Origins and Properties of an Expanding, Dark Energy Providing Dark Multiverse. *Universe* **2019**, *5*, 178. [CrossRef]
2. Alonso-Serrano, A.; Jannes, G. Conceptual Challenges on the Road to the Multiverse. *Universe* **2019**, *5*, 212. [CrossRef]

3. Heller, M. Multiverse–Too Much or Not Enough? *Universe* **2019**, *5*, 113. [CrossRef]
4. Dąbrowski, M.P. Anthropic Selection of Physical Constants, Quantum Entanglement, and the Multiverse Falsifiability. *Universe* **2019**, *5*, 172. [CrossRef]
5. Douglas, M.R. The String Theory Landscape. *Universe* **2019**, *5*, 176. [CrossRef]
6. Robles-Perez, S. Time Reversal Symmetry in Cosmology and the Creation of a Universe?Antiuniverse Pair. *Universe* **2019**, *5*, 150. [CrossRef]
7. Sandora, M. Multiverse Predictions for Habitability: The Number of Stars and Their Properties. *Universe* **2019**, *5*, 149. [CrossRef]
8. Sandora, M. Multiverse Predictions for Habitability: Number of Potentially Habitable Planets. *Universe* **2019**, *5*, 157. [CrossRef]
9. Sandora, M. Multiverse Predictions for Habitability: Fraction of Planets that Develop Life. *Universe* **2019**, *5*, 171. [CrossRef]
10. Sandora, M. Multiverse Predictions for Habitability: Fraction of Life That Develops Intelligence. *Universe* **2019**, *5*, 175. [CrossRef]

© 2020 by the authors. Licensee MDPI, Basel, Switzerland. This article is an open access article distributed under the terms and conditions of the Creative Commons Attribution (CC BY) license (http://creativecommons.org/licenses/by/4.0/).

Article

Multiverse—Too Much or Not Enough?

Michael Heller

Copernicus Center for Interdisciplinary Studies ul. Szczepańska 1/5, 31-011 Cracow, Poland; mheller@wsd.tarnow.pl

Received: 15 April 2019; Accepted: 9 May 2019; Published: 11 May 2019

Abstract: The aim of this essay is to look at the idea of the multiverse—not so much from the standpoint of physics or cosmology, but rather from a philosophical perspective. The modern story of the multiverse began with Leibniz. Although he treated "other worlds" as mere possibilities, they played an important role in his logic. In a somewhat similar manner, the practice of cosmology presupposes a consideration of an infinite number of universes, each being represented by a solution to Einstein's equations. This approach prepared the way to the consideration of "other universes" which actually exist, first as an auxiliary concept in discussing the so-called anthropic principle, and then as real universes, the existence of which were supposed to solve some cosmological conundrums. From the point of view of the philosophy of science, the question is: Could the explanatory power of a multiverse ideology compensate for the relaxation of empirical control over so many directly unobservable entities? It is no surprise that appealing to a possibly infinite number of "other universes" in order to explain some regularities in our world would seem "too much" for a self-disciplined philosopher. With no strict empirical control at our disposal, it is logic that must be our guide. Also, what if logic changes from one world to another in the multiverse? Such a possibility is suggested by the category theory. From this point of view, our present concepts of the multiverse are certainly "not enough". Should this be read as a warning that the learned imagination can lead us too far into the realms of mere possibilities?

Keywords: multiverse; Leibniz; other worlds; multiverse levels

---

Or, had I not been such a commonsensical chap, I might be defending not only a plurality of possible worlds, but also a plurality of impossible worlds, whereof you speak truly by contradicting yourself.
David Lewis

1. Introduction

The editor of the volume *Universe or Multiverse?* makes a funny, albeit fully justified, remark: "The word 'multiverse' is always spelt [in this volume] with a small 'm', since ... there could be more than one of them". How so? Because "the idea arises in different ways" ([1] p. XV). Indeed, ideas arise in many ways. The origin of some of them almost goes back to prehistory, where some evolve slowly through the ages, and some are brought to life by the sudden illumination of a genius. None of these ways gave birth to the idea of a multiverse.[1] It emerged long before it was named—first in poetry and fairy tales, then in philosophy, and has only recently appeared on the fringes of science. This was originally to dramatise the "extreme improbability" of some events, where it then emerged soon after as a real possibility, founding itself at the heart of a heated discussion. Some see in this idea the beginning of a new era in the philosophy of science, whereas others regard it as a serious danger to well-established methods that have, so far, guaranteed the greatest success.

---

[1] In this essay, all "multiverses" and "universes" will be democratically written with small letters.

In this essay (just an essay, not a systematic study), I want to look at the recent adventures (to use Whitehead's phrase) of the multiverse idea—though not so much from the standpoint of contemporary physics or cosmology, but rather from a philosophical perspective. If there are problems with the empirical control of the multiverse models, we should examine their explanatory power. In all these questions, the philosophy of science has something to say. However, we should not forget that the standard philosophy of science has never previously encountered postulated entities in physics which are so distant from any empirical control.

Suppose we have decided to leave the secure region of scientific methodology; then why not go even further into the domains suggested by the developments of mathematics? After all, if the empirical criteria are relaxed, it is only mathematics that remains as our guide, and the mathematical category theory says that even logic can change from one category to another. Should we not take into account the variability of logic in creating "other universes"? From this point of view, our present thinking about multiverses is certainly inadequate.

My way of thinking about these matters will be organised along the following lines. The modern story of the multiverse began with Leibniz. Although he treated "other worlds" as mere possibilities, they played an important role in both his theodicy and his logic (Section 2). In a similar manner, doing cosmology presupposes the consideration of an infinite number of universes, each being represented by a solution to Einstein's equations. This is a non-controversial version of a multiverse, also called an "ensemble of universes" (Section 3). This approach prepared the way to considering "other universes" which actually exist—first as an auxiliary concept in discussing the so-called anthropic principle, and then as real universes, the existence of which were supposed to solve some important cosmological conundrums (Section 4). From the point of view of the philosophy of science, one should ask the following question: Could the explanatory power of a multiverse ideology compensate for the relaxation of the empirical control over so many directly unobservable entities? It is unsurprising that an appeal to a great, perhaps infinite number of "other universes" in order to explain some regularities in our world, must seem "too much" for the orderly philosopher (Section 5). One should agree that without strict empirical control, it is logic and mathematics that must be our guide. And what if logic changes from one world to another in the multiverse? Such a possibility is suggested by the category theory, which is, today, unavoidable in any consideration of the foundations of mathematics, and consequently of physics as well. From this point of view, our present concepts of the multiverse are certainly "not enough" (Section 6). This is a challenge, but could also be read as a warning that the imagination—even the learned imagination—can lead us too far into the realms of mere possibilities.

2. A Philosopher's Paradise

The great story of the multiverse began with Leibniz. When he ordered his God to create the best of possible worlds, he automatically urged him to make some calculations on the set of all possible universes. However, the "all possible worlds" ideology is interesting, not only as far as Leibniz's own philosophical views are concerned. Only some minor intellectual gymnastics need be exercised to show that his doctrine of possible worlds foreshadows the modal semantics of the twentieth century [2].

Modal logic is a chapter of logic in which, besides the usual connectives (implication, negation, etc.) of classical propositional calculus, the functors "it is necessary that" and "it is possible that" are considered. The point is that whereas propositional calculus is extensional, modal logic is intensional. An extension of a term is the set of things to which the term applies; intension of a term involves its meaning. There is a serious problem on how to define semantics for modal logic—that is, "a theory that provides rigorous definitions of truth, validity, and logical consequence for the language" [3]. A solution was suggested by Carnap [4], and developed by Lewis [5] and others. The solution is as follows:

- A sentence is *necessarily* true if, and only if, it is true in every possible world.
- A sentence is *possibly* true if, and only if, it is true in some possible worlds.

In this way, modal logic did not become extensional, but it receives extensional semantics. Lewis did not hesitate to write: "As the realm of sets is for mathematicians, so logical space is a paradise for philosophers. We have only to believe in the vast realm of possibilia, and there we find what we need to advance our endeavours" ([3] p. 4).

3. A Context for Cosmology

Something similar happened in a field of research very distant from logical semantics—that is, cosmology. It turned out that the study of possibilities is essential for the study of the actual universe. To see this, let us start with something that seems to indicate precisely the opposite.

It is commonplace to declare that the peculiarity of cosmology as a science stems from the fact that the object of its study "is given to us in a single copy". This was most eloquently expounded by Bondi in his classic textbook ([6] pp. 9–10): "In physics we are accustomed to distinguish between the accidental and the essential aspects of a phenomenon by comparing it with similar phenomena. ... The uniqueness of the actual universe makes it impossible to distinguish, on purely observational grounds, between its general and its peculiar features, even if such a distinction were logically tenable. ... In either case, we select the important (as opposed to the 'accidental') features of the actual universe by their relation to the theory chosen rather than by any independent criterion". How do we do that? By artificially multiplying the object of our study. Let us take a closer look at this strategy.

For the empirical method, it is essential to have many instances of objects under investigation. The laws of physics are usually formulated in the form of differential equations, which describe general dependencies between the properties of a given class of phenomena, whereas individual characteristics of phenomena are accounted for by selecting initial or boundary conditions, and identifying a particular solution. Also, the very idea of measurement presupposes many copies of the measured object. Because of unavoidable measurement errors, we never refer to a single measurement result, but rather to their class within the box of errors. In this way, we are not dealing with a single object, but rather with a class of objects, only slightly differing from each other. This strategy in cosmology assumes the following form. We (tacitly) assume that the actual universe is described (up to a reasonably good approximation) by a solution to Einstein's field equations, and that any other of its solutions describes a possible universe. Our measurements never single out a unique solution, but rather a class of "nearby" solutions. In this way, the actual universe is placed "in the context" of other universes.

Moreover, in cosmology, this strategy has been exploited in a systematic manner and expanded to the form of a specialized field of research. The space of all solutions to Einstein's equation has even merited a special name: the *ensemble of universes* (or the *ensemble*, for short). This space is extremely rich—so rich that usually only some of its rather restricted subspaces are subject to investigation. For instance, the space of solutions with no symmetries to vacuum Einstein's equations is a smooth, infinite-dimensional manifold, but in the neighbourhood of a solution with symmetries, the smoothness breaks down and a conical singularity intervenes [7].[2]

As we have mentioned above, the very problem of measurement compelled us to consider "nearby solutions". This is also true as far as some not directly observable properties of solutions are considered—for instance, initial conditions leading to inflation, the existence of singularities or horizons. Claiming that the solution has such a property presupposes its *structural stability*. Roughly speaking, a property is said to be structurally stable if it is shared by all "nearby" solutions.[3] Structurally unstable properties are not considered to be realistic. The problem at stake is by no means

---

[2] In general, the space of solutions to the vacuum Einstein equations is a smooth infinite-dimensional manifold. However, in a neighbourhood of a solution with symmetries (it is enough for a solution to have a single Killing vector field) the solutions cannot be parametrised in a smooth way by elements of a linear space. In other words, at such a solution there is no tangent space to the space of solutions; just as there is no tangent space at the vertex of a cone.

[3] Structural stability is, in principle, a property of a dynamical system. It says that small perturbations do not affect the qualitative behaviour of nearby trajectories. In many cases, solutions to Einstein equations can be represented as trajectories of a dynamical system.

trivial. To meaningfully define structural stability (with respect to a given property), one should know which solutions are to be regarded as equivalent ("identical"), and when the distance between two non-equivalent solutions is to be regarded as "small". To this end, the space of solutions must be equipped with a suitable topology, and this is a tricky task which has been elaborated only with respect to some properties (e.g., for space-times that are stably causal [8,9][4]).

We can conclude that the ensemble of universes is a natural environment or a context for cosmology; one could even claim that theoretical cosmology is but a theory of the ensemble ([10] pp. 70–76).

4. Multiverses (In Plural)

From what is natural to what exists is but a small step. Over recent decades, it has become fashionable to look for answers to some deep questions, both from the fields of cosmology and metaphysics, by claiming that a large, possibly even an infinite, number of universes exists. The name "multiverse" was coined for such a huge collection of worlds. One of the first signals that such a fashion was approaching was Brandon Carter's idea of an "ensemble of universes"[5], which he referred to in the context of the so-called anthropic principles [11]. In Carter's intention, it served only as a means to dramatise the situation he was talking about. The drama consisted of the fact that the subset of universes in which observers can exist is extremely small: "Any organism describable as an observer will only be possible within a certain restricted combinations of parameters", and consequently, extremely "improbable" to be established by chance. What was a heuristic picture for Carter, soon became a postulated reality—if any combination of cosmological parameters is implemented somewhere within the ensemble of universes, then the observer permitted by such a combination must also be implemented somewhere, and in this way, "small probability" and "chance" are smoothly eliminated.

The contemporary defenders of the multiverse idea emphasize that this line of reasoning is now, at most, an auxiliary argument on behalf of this idea[6], and that the authentic arguments stem from the fact that the concept of a multiverse emerges, as a kind of side effect, from several physical theories or models. In this sense, it is almost unavoidable. As explained by George Ellis: "It has been claimed that a multi-domain universe is the inevitable outcome of physical processes that generated our own expanding region from a primordial quantum configuration; they would therefore have generated many other such regions" ([12] pp. 387–388). Although he directly refers only to multi-domain universes,[7] the same remains valid, more or less, as far as other types of multiverses are concerned.

Brian Green organizes his best-selling book *The Hidden Reality* [13] around nine types of multiverses (not limiting himself only to those that are implied by some physical theories or models). They are:

1. The multi-domain multiverse: Alluded to by Ellis.[8]
2. The inflationary multiverse: Eternal cosmological inflation leading to innumerable generations of bubble universes.
3. The brane multiverse: This is an outcome of the M-theory.
4. The cyclic multiverse: Collisions between branes producing big bangs separating subsequent universes.
5. The landscape multiverse: A combination of inflationary cosmology with string theory.

---

[4] Strictly speaking, in this context it is not the space of solutions of Einstein's equations that is taken into account, but rather the space of all Lorentz metrics, only some of them could be solutions to Einstein's equations.
[5] Not to be confused with the "ensemble of universes" as the space of solutions to Einstein's equations.
[6] This does not mean, however, that it disappears from cosmological texts; on the contrary, references and longer debates abound.
[7] By a multi-domain universe, Ellis understands an infinite universe in which very far-away space-time domains are regarded as separate universes.
[8] Green claims that in an infinite universe the same "local universe" must replicate itself across space.

6. The quantum multiverse: This is implied by the realistically understood many-worlds interpretation of quantum mechanics.
7. The holographic multiverse: Stems from the hypothesis that the entire universe can be viewed as information on the two-dimensional cosmological horizon.
8. The simulated multiverse: That is, simulated by a supercomputer.
9. The ultimate multiverse: All mathematical structures are instantiated as real universes.

This catalogue can hardly be regarded as complete; for instance, it does not include Penrose's cyclic universe [14]. Such a proliferation of multiverses poses the question of the observational verification of theories that claim their existence. The situation appears dramatic: "Cosmology will only ever get one horizon-full of data. Our telescopes will see so far, and no further. At any particular time, particle accelerators reach to the finite energy scale, and no higher" [15]. The problem is, therefore, about the "inference beyond data" [16]. In their heroic attempts to solve this problem, people use various strategies, frequently turning for help to Bayesian Probabilistic methods.

Although multiverse models are resistant with respect to verification, some of them could offer some hope as far as falsification is concerned (at least in principle). For instance, suppose that a multiverse model predicts that every member of the ensemble of possible universes should share a certain feature (for example, a negative cosmological constant). If this was not observed, this particular model would be falsified. However, in standard physics, falsification of theories is always made, at least implicitly, in the hope that this could help in selecting the right theory, whereas in the case of the multiverse, such a hope is practically non-existent.

5. It Is Too Much

The problem of the unobservable entities that are presupposed by physical theories (their linguistic counterparts are called theoretical terms) is an old problem in the philosophy of science. The works of Carnap [17] and Ramsey [18] (for a review, see [19]) are classic in this respect. However, the situation with multiverses is somewhat different. In classic discussions on the topic, theoretical terms referred to such entities like electrons and neutrons at that time were unobservable but had observable consequences, whereas in the present debate, we have in mind parallel universes that are causally totally disconnected from us. The number of such universes is usually assumed to be infinite, or at least extremely large[9]. In this context, Tavakol and Gironi [20] speak about the infinite turn in contemporary cosmology.

Today, nobody denies scientific theories the right to bring to life theoretical (not directly observable) terms, although their ontological status is an object of highly divergent opinions. Those who adhere to some kind of scientific realism are prone to agree that theoretical terms might correspond to some really existing objects, "provided that theory under consideration is *currently complete*, i.e., capable of unambiguously accounting for the current observations in its domain of applicability, as well as making novel testable predictions" ([20] p. 786). No single theory invoked to justify the existence of a multiverse is complete in this sense. Therefore, if we use the multiverse idea to explain some puzzling data in cosmology, we are, in fact, using an "explanation by unexplained" strategy. To quote the same authors: "The aim of a cosmological explanation should be that of providing a deeper understanding of the Universe, but the Multiverse type scenario rather shifts the target of the explanatory task from the finite observed Universe to a postulated relative or real infinite 'Landscape'" ([20] p. 793)[10]. This consequently opens up a new question about the explanatory power of the multiverse idea.

---

[9] Although a multiverse consisting of a finite number of universes is in principle possible, it seems rather artificial.
[10] Capitalized by the quoted authors.

Once again, the problem of explanation is a frequently discussed problem in the philosophy of science (see, e.g., [21,22]). Very roughly, philosophers distinguish two main types (or models) of explanations: deductive-nomological explanations and causal explanations.

According to the first type, also called Hempel's model (or the Hempel-Oppenheim model, or the Popper-Hempel model) of explanation, a phenomenon is explained if a sentence describing it (*explanandum*) is deduced from a set of true sentences (*explanans*), among which there is at least one law of nature. The first (deductive) part of this model could somehow be adapted to the multiverse situation: some properties of our universe could be deduced from some properties of a multiverse (not necessarily in the logico-linguistic manner, preferred by Hempel, but rather in the form of probabilistic inferences). Whether the second (nomological) part of the explanation model refers to a given multiverse depends on how it is conceived.

As far as the causal model of explanation is concerned, there exists tremendously rich literature (see, for instance, the often-quoted [23–25]), and such a variety of approaches, that even a short review of them would exceed the limits of this study. However, I feel free from this responsibility, since causal explanations hardly apply to our case. I will mention only one proposal which could be of some relevance at least to some versions of multiverse. I have in mind Wesley Salmon's proposal, which he calls the causal mechanical model of explanation [25]. He considers physical processes and distinguishes "pseudo-processes from genuine causal processes". The motion of a material object or the propagation of a sound wave through space are genuine causal processes, whereas a moving shadow is a pseudo-process, since it cannot transmit causal interactions. In his opinion, "a process is causal if and only if it has the capacity to transmit a mark". A mark or signature is a local modification or perturbation in the process that can be transmitted further by this process [26]. In some versions of multiverses, such a signature is indeed claimed to be transmitted to our universe. For instance, in the model of chaotic inflation, one considers the possibility that "baby universes", which nucleated from our universe and then collided with it, could have impressed a trace in the microwave background radiation in our universe (see, e.g., [27]). With respect to all other multiverse models, causal explanations simply do not work.

The standard answer to the above criticism usually given by defenders of the multiverse philosophy is to say that all the above objections are based on the old-fashioned philosophy of science, whereas the multiverse approach opens a new paradigm. If so, this new paradigm would apply only to some unsolved problems of cosmology and fundamental physics, since any relaxation of strict empirical constraints in all other branches of physics would be disastrous.

6. It is Not Enough

As we have remarked above, all theories giving rise to multiverses are incomplete. Therefore, it cannot be excluded that the future complete (or at least, "more complete") theory of fundamental physics will compel us to introduce some changes into the very concept of the multiverse. It is worthwhile asking to what extent the present concepts of multiverse are "stable" with respect to future developments in physics. Of course, the full answer to this question can only be provided by what will happen in the real future. Let us, however, permit ourselves the following thought experiment.

John Baez, a physicist of considerable renown, confesses that, motivated by the diversity of approaches to the search for a "final theory", he has decided, for now, to suspend his work on loop quantum gravity and to "re-examine basic assumptions and seek fundamentally new ideas". This has led him to the program of the categorification of physics [28]. It follows the program of the categorification of mathematics, which has been intensively pursued for some time. It seems reasonable to ask how the categorification of physics could modify our views on the multiverse idea. In trying to tackle this question, I shall not go into the details of the categorification program itself, but rather focus on one, possibly unavoidable, consequence. To see the extent of this consequence, we must first take a look at the connection between the category theory and logic.

The crucial point is that any sufficiently rich category[11] has its internal logic, in the sense that any such category can be used to construct a model for a certain logical system and, *vice versa*, one can extract a logical system out of any sufficiently rich category. The strongest such logical system constitutes the internal logic of this category. These intuitions can be made precise (see [29] or for a short review [30]). For instance, all of the axioms of intuitionistic logic are satisfied[12] in any topos[13]; however, the internal logic of a topos can be stronger than (first-order) intuitionistic logic (depending on the internal architecture of a given topos). Analogously, any complementary topos (also called cotopos) satisfies all of the axioms of paraconsistent logic[14], and the axioms of classical logic are satisfied in the category denoted by SET (which has sets as objects, maps between sets as morphisms, and all set theoretic axioms are expressed in the language of morphisms).

This strict connection between category theory and logic gives a new insight into the role of category theory as a unifying framework for the whole of mathematics. The fact that dependencies between different mathematical theories can be presented with the help of functors[15], and functors between functors, is now common knowledge, but the consequences of the fact that logic is not a sovereign ruling the entire affair from above, but rather a factor strongly involved in the game itself, is something that only laboriously, but irrevocably, emerges. Moreover, the same mathematical theory can be viewed from different "categorical frames". This motivated John Bell to propose a "local interpretation of mathematics". In his words: "With the relinquishment of the absolute universe of sets, mathematical concepts will in general no longer possess absolute meaning, nor mathematical assertions absolute truth values, but will instead possess such meanings or truth values only *locally*, i.e., *relative* to local frameworks"[16] [33].

Can these developments have repercussions in physics? Obviously, the standard macroscopic physics is conducted in the "framework" of SET, but the fact that quantum mechanics presupposes a not-quite-classical logic (the so-called quantum logic) signals that logic could be a "physical variable", meaning it could change depending on a physical theory (in the spirit of Bell's interpretation). The categorical approach to physics, although it is, at present, on the introductory level, reveals an intricate role that logic seems to play in physical theories. It is classical logic that has been abstracted from our everyday experience, but there is no guarantee that the same logic is adequate for all levels of reality. In this respect, the category theory offers entirely new possibilities. Logical tools of analysis are not to be employed from outside, but rather, to work within the physical theory as an aspect of its own structure.[17] This possibility should be taken seriously, especially as far as the candidates for a "final" physical theory are concerned.

As far as I know, all concepts of the multiverse now in circulation are (tacitly) assumed to function within the SET category (perhaps with the exception of Max Tegmark's idea that all mathematical structures are somewhere implemented as "other universes", but to say this is not to say much) and, in light of the above, this is an abject limitation. If we admit "all possibilities", we should not

---

[11] A quick reminder: A category consists of a collection of objects and a collection of morphisms (also called arrows) between objects. Morphisms can be composed (provided the head of one arrow coincides with the tail of the other arrow), and the composition of morphisms is associative. There exist identity morphisms satisfying the usual identity axioms. Richness of categories is enormous: from simple ones (like the category consisting of a single object and a single identity morphism) to categories covering large areas of mathematics.

[12] Intuitionistic logic is the one in which the excluded middle law (p or not p) is not valid and the axiom of choice cannot be used; see [31].

[13] Topoi (or toposes) form a class of categories especially related to the theory of sets. Although a topos can contain objects that are much richer and considerably different from sets, the abstract categorical properties of topoi are essentially the same as those known from the theory of sets; see [31].

[14] Paraconsistent logic is, in a sense, dual with respect to intuitionistic logic; in it the noncontradiction law (it is not true that p and not p) is not valid; see [32].

[15] A functor transforms one category into another category (objects into objects, morphisms into morphisms) in such a way that all defining axioms are preserved.

[16] By the term "local framework" Bell understands a topos with an object of natural numbers.

[17] For the role of category theory in physics, see [34].

impose arbitrary limits. Imagine a multiverse based on all of the consequences of the category theory, with logic changing from one world to another! What a paradise for a gifted mathematician!

In the above remarks, I have taken into account only ordinary categories (1-categories), and what about n-categories and their internal logic? And why exclude the category of all categories? Who said that the logic—our poor classical logic—inscribed by evolution into our brains, is able of comprehending everything that exists? Indeed, multiverses, as we consider them now, are not enough.

Would this idea somehow help us to practice physics, which is wretchedly restricted to the visible horizon of our local universe? This is another story.

Funding: This publication was made possible through the support of a grant from the John Templeton Foundation (Grant No. 60671).

Conflicts of Interest: The authors declare no conflict of interest.

References

1. Carr, B. (Ed.) *Universe or Multiverse*; Cambridge University Press: Cambridge, UK, 2007.
2. Look, B.C. Leibniz's Modal Metaphysics. In *Stanford Encyclopedia of Philosophy*. 2013. Available online: https://plato.stanford.edu/entries/leibniz-modal (accessed on 23 June 2018).
3. Manzel, C. Possible Worlds. In *Stanford Encyclopedia od Philosophy*. 2017. Available online: https://plato.stanford.edu/entries/possible-worlds (accessed on 23 June 2018).
4. Carnap, R. *Meaning and Necessity*; The University of Chicago Press: Chicago, IL, USA, 1947.
5. Lewis, D. *On the Plurality of Worlds*; Blackwell: Oxford, UK, 1986.
6. Bondi, H. *Cosmology*; Cambridge University Press: Cambridge, UK, 1960.
7. Fisher, A.E.; Marsden, J.E.; Moncrief, V. Symmetry Breaking in General Relativity. In *Essays in General Relativity. A Festschrift for Abraham Taub*; Tipler, F.J., Ed.; Academic Press: New York, NY, USA; London, UK, 1980; pp. 79–96.
8. Hawking, S.W. Stable and Generic Properties in General Relativity. *Gen. Relativ. Gravit.* **1971**, *1*, 393–400. [CrossRef]
9. Hawking, S.W.; Ellis, G.F.R. *The Large Scale Structure of Space-Time*; Cambridge University Press: Cambridge, UK, 1973.
10. Heller, M. *Theoretical Foundations of Cosmology*; World Scientific: Singapore; London, UK, 1992.
11. Carter, B. Large Number Coincidences and the Anhropic Principle in Cosmology. In *Confrontation of Cosmological Theories and Observational Data (IAU Symposium)*; Longair, M., Ed.; Reidel: Dordrecht, The Netherlands, 1974; pp. 281–289.
12. Ellis, G.F.R. Multiverses: Description, Uniqueness and Testing. In *Universe or Multiverse?* Carr, B., Ed.; Cambridge University Press: Cambridge, UK, 2007; pp. 387–409.
13. Greene, B. *The Hidden Reality*; Vintage Books: New York, NY, USA, 2011.
14. Penrose, R. *Cycles of Time*; The Bodley Head: London, UK, 2010.
15. Barnes, L.A. Testing the Multiverse; Bayes, Fine-Tuning and Typicality. In *The Philosophy of Cosmology*; Chamcham, K., Silk, J., Barrow, J.D., Saunders, S., Eds.; Cambridge University Press: Cambridge, UK, 2017; pp. 447–466.
16. Sahlén, M. On Probability and Cosmology: Inference Beyond Data? In *The Philosophy of Cosmology*; Chamcham, K., Silk, J., Barrow, J.D., Saunders, S., Eds.; Cambridge University Press: Cambridge, UK, 2017; pp. 429–446.
17. Carnap, R. *Philosophical Foundations of Physics: An Introduction to the Philosophy of Science*; Basic Books: New York, NY, USA, 1966.
18. Ramsey, F.P. Theories. In *Foundations. Essays Philosophy, Logic, Mathematics and Economics*; Mellor, H.D., Ed.; original edition in 1929; Routledge and Kegan Paul: London, UK, 1978; pp. 101–125.
19. Holger, A. Theoretical Terms in Science. In *The Stanford Encyclopedia of Philosophy*; Zalta, E.N., Ed.; 2017. Available online: https://plato.stanford.edu/archives/fall2017/entries/theoretical-terms-science (accessed on 23 June 2018).
20. Tavakol, R.; Gironi, F. The Infinite Turn and Speculative Explanations in Cosmology. *Found. Sci.* **2017**, *22*, 785–798. [CrossRef]

21. Skow, B. Scientific Explanation. In *The Oxford Handbook of Philosophy of Science*; Humphreys, P., Ed.; Oxford University Press: Oxford, UK, 2016; pp. 525–543.
22. Woodward, J. Scientific Explanation. In *The Stanford Encyclopedia of Philosophy*; Zalta, E.N., Ed.; 2017. Available online: https://plato.stanford.edu/archives/fall2017/entries/scientific-explanation (accessed on 23 June 2018).
23. Woodward, J. *Making Things Happen: A Theory of Causal Explanation*; Oxford University Press: Oxford, UK, 2003.
24. Lewis, D. Causal Explanation. In *Philosophical Papers*; Oxford University Press: New York, NY, USA, 1986; Volume II, pp. 214–240.
25. Salmon, W. *Scientific Explanation and the Causal Structure of the World*; Princeton University Press: Princeton, NJ, USA, 1984.
26. Salmon, W. The Causal Structure of the World. *Metatheoria* **2010**, *1*, 1–13.
27. Johnson, M.C.; Peiris, H.V.; Lehner, L. Determining the Outcome of Cosmic Bubble Collisions in Full General Relativity. *Phys. Rev. D* **2012**, *85*, 083516. [CrossRef]
28. Baez, J. Categoryfying Fundamental Physics. Available online: http://math.ucr.edu/home/baez/diary/fqxi_narrative.pdf (accessed on 20 June 2018).
29. Lambek, J.; Scott, P.J. *Introduction to Higher Order Categorical Logic*; Cambridge University Press: Cambridge, UK, 1994.
30. Fu, Y. Category Theory, Topos and Logic: A Quick Glance. Available online: http://charlesfu.me/repository/topos.pdf (accessed on 29 September 2015).
31. Goldblatt, R. *Topoi. The Categorical Analysis of Logic*; Dover Publications: Mineola, NY, USA, 2006.
32. Estrada-González, L. Complement-Topoi and Dual Intuitionistic Logic. *Aust. J. Log.* **2010**, *9*, 26–44. [CrossRef]
33. Bell, J.L. From Absolute to Local Mathematics. *Synthese* **1986**, *69*, 409–426. [CrossRef]
34. nLab. Higher Category Theory and Physics. Available online: https://ncatlab.org/nlab/show/higher+category+theory+and+physics (accessed on 23 June 2018).

© 2019 by the author. Licensee MDPI, Basel, Switzerland. This article is an open access article distributed under the terms and conditions of the Creative Commons Attribution (CC BY) license (http://creativecommons.org/licenses/by/4.0/).

Article

Conceptual Challenges on the Road to the Multiverse

Ana Alonso-Serrano [1,*,†] **and Gil Jannes** [2,†]

1 Max Planck Institute for Gravitational Physics, Albert Einstein Institute, Am Mühlenberg 1, D-14476 Golm, Germany
2 Department of Financial and Actuarial Economics & Statistics, Universidad Complutense de Madrid, Campus Somosaguas s/n, 28223 Pozuelo de Alarcón (Madrid), Spain; gil.jannes@ucm.es
* Correspondence: ana.alonso.serrano@aei.mpg.de
† The authors contributed equally to this work.

Received: 31 July 2019; Accepted: 8 October 2019; Published: 10 October 2019

Abstract: The current debate about a possible change of paradigm from a single universe to a multiverse scenario could have deep implications on our view of cosmology and of science in general. These implications therefore deserve to be analyzed from a fundamental conceptual level. We briefly review the different multiverse ideas, both historically and within contemporary physics. We then discuss several positions within philosophy of science with regard to scientific progress, and apply these to the multiverse debate. Finally, we construct some key concepts for a physical multiverse scenario and discuss the challenges this scenario has to deal with in order to provide a solid, testable theory.

Keywords: multiverse; physical multiverse; philosophy of science; empirical testability; falsifiability; Bayesian analysis; fine-tuning

1. Introduction

Ideas related to what we presently call "the multiverse" have historically always attracted both supporters and detractors. The multiverse is not really a theory, but a scenario that arises in several theories and can be defined in different ways depending on the underlying theory. This apparently vague remark in fact has deep consequences on the discussion about the possible existence of other universes and the associated change of the very definition of "the universe", but also of what a scientific theory is or should be, and how it should be assessed. As we will discuss later in detail, there is no standard or commonly agreed upon definition of universe, and therefore not for the multiverse either. What seems clear along history is that the universe has been defined as the (connected) region of space accessible to us. However, this meaning has evolved depending upon the theory and observations at our disposal.

In the past few years, the amount of papers and ideas about the multiverse have increased tremendously. The sometimes heated discussion about the viability of the multiverse and related ideas [1–3] show that we are far from reaching an agreement in the scientific community. At this stage there are, on the one hand, strong claims about what "could be one of the most important revolutions in the history of cosmogonies" [4], which "changes the way we think about our place in the world" [5] and "if true - would force a profound change of our deep understanding of physics" [4]. On the other hand, there is also a strong opposition to the very mention of the possibility of a multiverse scenario and of the scientific significance of such mentions [6,7].

We therefore believe that it is relevant to examine the fundamental questions that emerge in multiverse ideas, and to emphasize some criteria that the multiverse should meet before being more widely acceptable as a viable scientific proposal. We therefore undertake an analysis based on concepts from the philosophy of science related to the definition of the scientific method, an epistemological theory, and a scientific revolution implying a change of paradigm. This reflection will help us postulate some constraints that should be imposed on multiverse theories and the conceptual challenges they will need to confront.

2. The Multiverse: Nothing New under the Suns?

The idea that there might be other worlds or universes beyond our own has been a recurrent concept throughout history [8,9]. It followed naturally from the desire of knowing whether we are unique observers, and of the possibility of discovering worlds similar to ours. At the same time, it has always been accompanied by skepticism about the possibility of actually answering these questions.

These issues are part of the most ancient and most essential philosophical and cosmological questions. It should therefore come as no surprise that the first recorded notion of a multiverse in occidental intellectual history dates back to the ancient Greeks. Anaximander, in the 6th Century BC, speculated about a plurality of worlds such as our cosmos, appearing and disappearing in an eternal movement of generation and destruction. A few centuries later, Epicurus described how an unlimited number of worlds fills the infinite vacuum [10].

In medieval times, Robert Grosseteste described the condensation of different universes from an initial big-bang-like explosion [11]. Giordano Bruno proposed an infinite "cosmic pluralism" filled with many inhabitable worlds as an alternative to the Copernican heliocentric model [12]. In the 18th century, Emanuel Swedenborg conjectured a model of the evolution of our solar system and the firmament we observe, based on theological and philosophical arguments. He postulated the possible existence of other celestial spheres in the firmament and argued that every of those world-systems would follow the same principles. This argument can be interpreted as the first idea of a nebular hypothesis [13]. Thomas Wright was the first to interpret the astronomical observations of distant faint nebulous structures as other galaxies, suggesting that they could have their own "external creation" [14]. This idea was later elaborated by Immanuel Kant, who popularized the idea of the possible existence of habitable worlds around stars other than the Sun [15]. This theory was baptized "island universes" by von Humboldt a century later [16]. By 1920, as more observations became available, the question led to the "Great Debate" between Shapley and Curtis about the scale of the Universe. Shapley defended that our Milky Way constituted the entire universe and that the other observed nebulae were small entities on its outskirts, while in Curtis' opinion, at least some of them were in fact separate, distant galaxies [17]. It was not until Hubble's decisive observational evidence a few years later that this battle of paradigms, originally started from purely philosophical principles, was finally settled.

From a more metaphysical point of view, Leibniz argued that our universe was the best among an infinity of possible universes [18]. According to Leibniz, logical constraints meant that (even) God could not have made our universe any better. This argument was later turned around by Schopenhauer, who argued that our world must be the worst of all possible worlds, because if it were even slightly worse in any respect, life could not continue to exist [19]. This argument is curiously reminiscent of recent fine-tuning arguments: a slight change in any of several basic constants would have made complexity (and therefore life) impossible [20].

Before turning to contemporary theories of the multiverse, it might be interesting to mention that there also exist (non-occidental) religious and mythological ideas related to the multiverse. Such ideas are implicit in Buddhism, with its cyclical view on the continuous destruction and recreation of the universe. They appear explicitly in Hinduism: "The golden egg that is this universe is wrapped in seven sheaths: the earth and the other elements, each casing being ten times as great as the one it encases. There are millions upon millions of these in each universe. There are millions of such universes. Lord, all these together are like a single atom upon your head! So, we call you Ananta, Infinite One." [21].

In Christianity, the multiverse idea is controversial, as it seems opposite to the uniqueness idea common to most monotheistic religions. However, Page has defended that the multiverse is not in conflict with Christianity [22]. In this context, Page mentions a serious challenge to the multiverse concept, namely the question of whether sinning civilizations in other universes have also been redeemed by the death of Christ. To our great relief, Page himself answers this question: "we could just interpret the Bible to mean that Christ's death here on earth is unique for our human civilization", and so with peace in mind we will focus here on cosmological approaches of the multiverse and its conceptual challenges.

3. Definition and Classification of the Multiverse

The epistemological extension from the universe to a multiverse is often compared to the Copernican revolution, as a further step in the gradual loss of importance of our own habitat (although there seems to be some disagreement about whether this would be the fourth [4,23] or fifth [24] Copernican revolution). However, from a physical point of view, the contemporary cosmological concept of multiverse arises not so much as a direct theory in itself, but as an indirect consequence of problems mainly related to the current cosmological paradigm of an acceleratedly expanding universe governed by the laws of General Relativity [1]. The multiverse is argued to be a natural extension of developments within string theory or early-universe cosmology (in particular, chaotic eternal inflation), and is invoked to solve a series of open problems in theoretical physics, such as the problem of the beginning of the universe, the cosmic coincidence problem, or the smallness of the cosmological constant, as well as the more general fine-tuning of physical constants [1]. In some of those cases, the multiverse in itself only partially solves the problem, but mainly establishes a reformulation of the question. For example, related to the fine-tuning problem, the multiverse suggests a distribution of values of certain fundamental constants among the different possible universes. The question why we live precisely in a universe with the observed values can then be answered by some form of the anthropic principle.[2] The new question which arises then is how likely the values of the physical constants of our universe are across the probability distribution within the multiverse. In this context, Vilenkin introduced the mediocrity principle [27], which defends that we should be "typical" observers, and therefore a priori we are expected to live in one of the most probable universes among all those which allow for the existence of life. Ideally, this principle will be testable by comparison with the probability distribution. We will come back to this issue later in Section 4.2.5. Let us first look at possible definitions of the multiverse.

In general terms, a first attempt to define the concept of multiverse could be that the multiverse encompasses all the multiple possible universes predicted by an underlying theory insofar as they are actually realized, i.e.: everything that physically exists, the totality of space and time and its material-energetic content. However, this (intentionally vague) definition leads to an obvious question from a semantic point of view. If we understand the Universe etymologically as the whole possible entity, all of spacetime and its content, then there is no place for the multiverse. Perhaps one should then redefine the universe itself, depending on concepts such as causal connection or variations of physical laws. The situation is reminiscent of the atom, originally meaning "indivisible". Eventually the physical meaning was overcome even though the name stuck. In the case of the multiverse, this apparently lexicological question bears a direct impact on physical issues. As is well-known, in the

---

[1] Even though Everett's many-worlds interpretation of quantum mechanics is presently considered a multiverse scenario (see Tegmark's classification below), it really stands a bit apart for a variety of reasons, the first one being its historical origin purely within quantum mechanics. We will briefly mention this scenario again in Section 3.1 but otherwise focus mainly on cosmological multiverse scenarios.

[2] For the relationship between the anthropic principle and the multiverse, see e.g., [25]. We will not discuss the anthropic principle here because, first, as paraphrased in [25], "many commentators have already thrown much darkness on this subject, and it is probable that, if they continue, we shall soon know nothing at all about it"; and second because, although the anthropic principle has undoubtedly contributed much to conceptual thinking about the multiverse, it is not clear whether it can also make any real contribution when it comes to empirical predictions, let alone—despite common claims to the contrary—whether it has done this so far [26].

case of a single universe, the boundary conditions are crucial to determine the mathematically and physically acceptable solutions (see the Hartle–Hawking no-boundary proposal [28] or Vilenkin's tunneling proposal [29]). Then in the case of a multiverse, the issue of the boundary conditions between the various universes within the multiverse is probably equally important [30]. In fact, the issue is even more general, since the overall nature of the multiverse depends on the particular definition one uses of the constituent universes. It is possible to consider as universe, for instance, the habitable region we live in (delimited by the Hubble sphere); a causal spacetime region; one of the quantum branches in the Everett interpretation of quantum mechanics; or simply one of the particular solutions to the cosmological equations that appear in string theory.

The best-known classification for the different multiverse hypotheses is due to Tegmark [31]. Tegmark establishes a hierarchical classification, where each higher level includes the lower ones. The level I multiverse consists of a variety of Hubble volumes, causally disconnected but all with the same physical laws and constants. Level II allows for a variation of the physical constants, for example due to bubble-breaking during the inflationary phase. Level III corresponds to quantum many-world branching. Finally, level IV is constituted by all the different possible mathematical structures, all of which are assumed to represent physically real universes.

A point which might be worth mentioning is that different physical multiverse models are not always straightforward to classify in Tegmark's (or some other) scheme. More generally, both when defending or criticizing the multiverse, or when trying to elaborate on "the multiverse", it is not always clearly stipulated which kind of multiverse one is dealing with, and this can seriously complicate the assessment of the arguments.

In the following, we focus on the origin of the most common contemporary multiverse models.

3.1. Physically Motivated Multiverse Scenarios

The earliest multiverse model in modern physics comes from Everett's many-worlds interpretation of quantum mechanics [32]. This interpretation considers the multiverse as all the possible histories from a quantum superposition, with one particular branch corresponding to our universe. This configuration of the multiverse as a host of bifurcating quantum branches leads to a continuous multiplication of parallel universes. Unsurprisingly, this interpretation has historically been quite controversial.

Inflation theory has led to a different multiverse notion. As a consequence of the quantum effects in the early universe, it could be possible to create new universes by a mechanism of eternal chaotic inflation [33–35], in which the different regions of space can transform into macroscopic bubble universes, which split off from a preceding universe and create a new one. In these multiverse models, the fundamental laws of physics are the same in each universe, due to the fact that they were originated in a common inflationary universe. The constants of nature, however, are allowed to have different values, depending on the specific inflationary process of each particular universe.

In the context of string theory, the idea of a multiverse stems from the pocket universes associated with the colossal number of false vacua predicted by the theory, which conform the so-called landscape [35,36]. Each of the universes could have different dimensions, elementary particles or fundamental constants of nature. In this scenario, our universe emerges by a selection procedure, following an anthropic reasoning [36], or arguments from quantum cosmology [37]. This approach can be related with the idea of bubble universes in the sense of the possibility of tunneling among different vacua, giving rise to an eternal inflation that populates the landscape.

Recently, the idea that the landscape should be constrained by consistency conditions has gained momentum. The argument is that the vast range of solutions coming from the landscape are physically restricted to those that give rise to effective field theories, surrounded by a swampland of inconsistent solutions [38]. In this scenario, the huge amount of possible vacua from the string landscape is strongly restricted, possibly leading to a unique vacuum state and hence without room for a string multiverse, except perhaps in the form of a cyclic universe [39]. This idea is closely related with other proposals about cyclic universes, where each end of a universe poses the initial conditions for a next one, thus leading to a conformal cyclic cosmology [40].

There exists a larger variety of multiverse scenarios originating from other physical ideas and proposing different concrete schemes. We should stress that the different multiverse scenarios, although conceptually akin, are so far only vaguely related in terms of physical formulation. So an obvious challenge on the road to the multiverse is to clarify the physical and mathematical relation between the different multiverse scenarios. For example, the relation between inflation and string theory is a subject of ongoing research, and is in fact considered one of the major challenges within string theory research, see e.g., [41] and references therein. According to the present status of research, it seems that only certain very specific string theory scenarios might actually allow for inflation, and the concrete details of the mechanism are only starting to be understood quantitatively. To illustrate this point, ref. [41] states that "At present, we are led to inflation in string theory by a web of inference" and that a better understanding of "non-supersymmetric solutions of string theory, particularly de Sitter solutions (...) continues to be a zeroth-order challenge for deriving inflation from string theory". Curiously, the fact that this embryonic understanding of the relation between inflation and string theory poses a major challenge for the multiverse idea and especially for the interpretation that a common view should exist between string and inflationary multiverse scenarios is a question that (to the best of our knowledge) is barely being addressed in current research. With respect to Everett's many-worlds interpretation, the question is perhaps even less clear. In [42], an argument was made for a relation between the many-worlds interpretation and the (inflationary) multiverse. However, it is probably only fair to say that the argument is mainly qualitative and much further work is required in this area to show whether the conjectured connections between the different types of multiverse can actually be described in a concrete and convincing way.

A related point is that all of these multiverse scenarios currently face a series of unsolved issues and require much more detailing before they can be considered mature physical theories. This, in combination with the simple fact that the multiverse can be argued to constitute a change of paradigm, makes the multiverse subject to criticism. The strongest argument against the multiverse probably lies in the fact that the multiverse is considered a speculative idea that cannot be falsified, perhaps not even in principle. Indeed, one could wonder what physical sense it makes to consider other different universes if these have no detectable effect whatsoever on our universe, possibly not even in principle. Some scientists argue that, if this is indeed the case, then the multiverse cannot be considered a scientific theory, but should at most be included in the field of metaphysics [1]. Regardless of the defense that some authors have made of "non-empirical theory assessment" [43] and related concepts (see also our discussion in Section 4), it bears little doubt that multiverse scenarios would gain strength if they would lead to new and preferably concrete predictions about the properties of our universe [44]. Interesting examples are the prediction that some features of the CMB spectrum, such as the cold spot, could be due to entanglement with another universe in string theory [37,45,46], collisions with other bubble universes in the context of eternal inflation [47–49] or recently topological defect nucleation in bubble universes [50]. From a philosophical point of view, as stressed by Popper [51], it is better to have concrete conjectures that can be tested and possibly refuted, rather than a very general scenario which cannot be confirmed nor refuted.

Before discussing this philosophical question in more detail, let us already mention a lesser-known multiverse scenario, which could alleviate some of the mentioned criticisms. In the proposal of a quantum multiverse [52–55], a quantum cosmology program is developed for the multiverse as a set of entangled universes. Two relevant characteristics of this scenario are the following. First, it does not pre-suppose a specific model for the different universes to pop up. Second, the entanglement between universes could give rise to dynamical and observable effects on each universe due to its interaction with other universes [56–58]. We will come back to this idea in Section 6.

As we have just argued, a question that naturally emerges when dealing with such physical theories that explore the limits of our knowledge concerns their scientific viability. In this context, let us take a look at the descriptions of scientific method, change of paradigm and related issues in the philosophy of science.

4. Philosophical Aspects

There are various reasons why it is relevant to look at philosophical aspects of the multiverse question. Let us just mention two. First, as we already mentioned in the introduction, the generic idea of a multiverse is not really new, and so it might be interesting to look at what philosophers and historians of science have to say about this.

Second, theories about the multiverse almost automatically lead to questions about how to assess these theories. Notions such as theory confirmation, verification and viability then become important. These notions are rarely defined explicitly, but have acquired relatively well-developed meanings in the philosophy of science. The following (basic) definitions might be useful to set the vocabulary. "Theory assessment" consists of submitting a given hypothesis to empirical data. As possible outcomes, "theory confirmation" consists of empirical evidence which supports a given hypothesis, in particular by being in agreement with a prediction from the hypothesis. "Falsification" obviously is the opposite case: empirical evidence which contradicts a given hypothesis. "Verification" is the ideal case in which supporting evidence is so strong that the hypothesis can be considered to be conclusively confirmed. Finally, the "viability" of a theory could be paraphrased as its compatibility with already existing data, irrespectively of whether it has produced any new predictions which could be submitted to additional testing. We will not further discuss these concepts explicitly, but the remainder of this section will make clear that their understanding has evolved over time, and that they are much more involved than the naive definitions just given might suggest, see e.g., [59].

Our third, and most important argument, is that the idea of a multiverse clearly challenges the epistemological boundaries of science, and so enters into the grey zone where physics meets philosophy. In fact, the multiverse is often presented as a change of paradigm which completely alters our understanding of cosmology, and perhaps even more: some physicists claim that the multiverse revolutionizes the way in which we should look at science itself, and the way in which we assess scientific theories. However, these terms, "paradigm" and "scientific revolution" stem from the philosophy of science, where they were studied in great detail. Claims about the character of science and its methodology transcend science itself, and would thus benefit from a broader philosophical context.

The philosophical dispute about the multiverse has so far taken place almost exclusively between physicists, with professional philosophers largely staying safely out of the ring. The argument has centered on (naive interpretations of) Popper's falsificationism criterion, with some recent attempts to reframe the question in terms of Bayesianism. In the light of the philosophy of science of the past century, this is a bit curious. To explain why, it might be useful to go through a crash course which will lead us from Popper to Bayesianism, and then see how these ideas can be interpreted in the context of the multiverse.

4.1. Philosophy of Science and the Description of Scientific Progress

We will limit ourselves to the key elements of Popper's, Kuhn's, Lakatos' and Feyerabend's ideas with relation to the demarcation problem and the formal description of scientific progress, and how Bayesianism can be situated in this context. The interested reader is referred to, e.g., [60] or [61] for good introductions.

Let us start by sketching the historical context in which Popper came to prominence. In the early 20th Century, there was a generalized expectation that all of mathematics could be framed on solid logical foundations. This expectation was epitomized by Russell and Whitehead's "Principia Mathematica" [62–64] and Hilbert's programme [65,66]. A related feeling existed in the philosophy of science: science should obey firm laws of logic, and scientific progress should be formally expressible in terms of logical laws related to deduction and induction. The main discussion in the philosophy of science of the epoch was between proponents and opponents of logical positivism: the idea that (both in science and in philosophy) only verifiable claims about the empirically observable reality are meaningful. However, even the opponents (including Popper) of logical positivism opposed only its positivist part but had no doubt that scientific progress could be described as a cumulative logical process. The Russell–Whitehead–Hilbert ambition was blown to pieces by Gödel's incompleteness theorems [67]. The demise of logic as the foundation of scientific knowledge took a bit longer, starting with Kuhn. However, we are running ahead of our argument.

Karl Popper's landmark "Logik der Forschung" [68] was an attempt to circumvent the problem of induction while retaining the logical mould into which the description of scientific process should be cast. The problem of induction, in its simplest version, is the fact that inference from (no matter how many) concrete observations or experiments can never lead to certain knowledge about a general theory. Popper's replacement of induction as a scientific criterion by falsification, and by a process of conjectures and refutations, is logically water-tight: in principle, a single counter-example suffices to demonstrate the falsity of a general conjecture. However, Popper's programme failed in two senses. First, it failed as a demarcation criterion to distinguish science (which makes falsifiable conjectures) from non-science (which does not). Second, it failed as a description of how scientific progress really works in practice. Enter Thomas Kuhn.

Kuhn held a historical view of science, as opposed to Popper's normative view. In [69], Kuhn defended that the advance of scientific knowledge is not linear and continuous, but proceeds by alternation between normal science and paradigm shifts or scientific revolutions. The paradigm is the basic set of concepts, beliefs and practices shared by a community of scientists. During the normal science phase, scientists attempt to articulate this paradigm in more detail, make empirical interpretations and predictions, but do not question the paradigm itself. The observational interpretations are always framed within the paradigm, and in particular: the further they are distant from direct sensory experiences, the more they depend on theoretical concepts characteristic of the paradigm, an issue called the "theory-ladenness of observations".

Through accumulation of anomalies, i.e.: conceptual or observational challenges that resist easy solution within the paradigm, this paradigm can enter into a state of crisis, thus opening the possibility for a paradigm shift or scientific revolution. Some scientists will swap allegiance to a new paradigm, others will stick to the old paradigm until they retire (or die). However, generally speaking, scientists usually adhere quite strongly to the fundamental assumptions of their own paradigm, which are often not even formulated very explicitly, and use these to judge other paradigms. This lack of an explicit formulation of criteria within each paradigm makes it very hard for scientists belonging to different paradigms to converse rationally about the pros and cons of each approach, a problem called "incommensurability". This is related to the well-known problem of underdetermination of theory by evidence: the idea that the experimental and observational evidence available within a particular branch of science is (even in principle) insufficient to pick out a single "true" theory in a uniquely determined way.[3] Kuhn's response to the problem of underdetermination is that "scientific truth" is not solely determined by objective facts but also by consensus within the scientific community. In this sense, Kuhn was probably the first to emphasize the social aspect of science, an aspect which was later worked out in detail by authors such as Pickering [73]. This, in combination with the lack of a clear criterion for paradigm change, has led to criticisms on Kuhn as proposing a relativist, even irrational view on the progress of science. Kuhn defended himself by pointing out that rational criteria for paradigm choice can indeed be identified, such as empirical accuracy, consistency, broad scope, simplicity, and fruitfulness. However, these criteria are not sufficient, in Kuhn's view: two scientists might agree on the criteria but nevertheless make different paradigm choices.

---

[3] This issue becomes ever more acute in high-energy physics and cosmology, and is related to the question of non-empirical theory assessment that we have mentioned earlier [43]. See also [70–72] for contemporary views on this problem.

So who was right between Popper and Kuhn? There have been arguments in both directions. The two most influential reactions to the Popper versus Kuhn debate were embodied by Lakatos and Feyerabend, which we will briefly discuss in turn.

Imre Lakatos [74] tried to reconcile Popper's and Kuhn's views. He replaced Kuhn's "paradigm" by the concept of "research programme", which consists of a hard core of fundamental assumptions, and a series of auxiliary hypotheses. These auxiliary hypotheses can serve to increase the predictability of the research programme, or to save it from threats. If most auxiliary hypotheses belong to the first category, and most of the novel predictions are confirmed, then the research programme is in a progressive state. If most predictions of the theory are refuted, and auxiliary hypotheses are invoked to save the research programme, then the latter is degenerative. Research programmes are thus not falsified in Popper's naive sense, but they should be abandoned if they have entered a degenerative state, and a progressive alternative is available which has a stronger empirical content. In this way, Lakatos tried to reframe Kuhn's revolutions on a rational basis, by giving concrete criteria for switching allegiance between research programmes, while updating the essence of Popper's falsification idea.

Feyerabend, on the other hand, dismissed both Popper and Kuhn's views on science (and therefore also Lakatos' attempt at a compromise) [75]. According to Feyerabend's "epistemological anarchism," any standard of rationality or universal methodological rule would be too restrictive and, if really applied, in fact hinder science. Feyerabend concludes, based on a historical analysis, that all commonly accepted rules of science are frequently violated, and that new theories are accepted not because of accord with some universal "scientific method", but because its supporters made use of any "trickery"—apart from rational argumentation, Feyerabend mentions propaganda, psychological tricks, and rhetoric, including jokes and *non sequiturs*—to advance their cause. Illustratively, Feyerabend disagreed with the commonly held negative attitude towards ad hoc hypotheses. In Feyerabend's opinion, ad hoc hypotheses are often required to temporarily make things work until a better understanding is achieved. Furthermore, Feyerabend rejected consistency as a criterion for theory-building, since new theories cannot be expected to be as consistent as the old theory they purport to replace. Feyerabend also made controversial claims about the ideological totalitarianism of science, and its negative impact on (western) society. Few scientists and philosophers of science would probably agree with Feyerabend's most radically relativist claims. Nevertheless, the essence of Feyerabend's analysis has withstood criticisms. As a consequence of Feyerabend's work, together with a more general shift of focus within the philosophy of science, the goal of formulating a universal logical-methodological framework for science, or a single and absolute demarcation criterion between science and non-science, has been mostly abandoned.

However, one further attempt at formalizing the progress of science should be mentioned, namely Bayesian epistemology, or simply Bayesianism. Bayesianism relies on Bayesian inference, and especially its so-called "subjective" interpretation. This is historically prior to the Popper–Kuhn–Lakatos–Feyerabend discussion, but has become widely popular as an approach to scientific progress more recently, and in particular is often mentioned in the context of theoretical physics. Bayes' famous formula which relates conditional probabilities can be written as

$$P(H\,|\,E) = \frac{P(E\,|\,H) \cdot P(H)}{P(E)} \qquad (1)$$

where, in this context, H represents a hypothesis and E the evidence, and probabilities are taken as expressing a priori ($P(H)$) and a posteriori ($P(H\,|\,E)$) degrees of belief, and $P(E) = \sum P(E\,|\,H_i)P(H_i)$ the overall probability for the evidence E to actually occur. In practice, only a limited range of hypotheses are summed over, and $P(E)$ becomes a somewhat subjective assessment. The Bayesianists' claim is that Bayes' formula provides a useful model for scientific progress in general [76]. This is certainly true within Kuhnian phases of "normal science". Bayesian inference is commonly and successfully used to assess, for instance, the significance level of the outcomes of particle detector experiments or of cosmological observations. When it comes to paradigm shifts, there is a certain debate about Bayesianism [77]. The right-hand side in (1) contains the a priori or subjective belief $P(H)$ in the hypothesis H. Then, no matter how rigorously one defines $P(E|H)$, and thus no matter how rationally the belief in H is updated, the left-hand side will still be a subjective belief, not a proof for the validity or degree of probability for the truth of hypothesis H. However, two key points stressed by Bayesianists are the following. First, the subjectivity of the prior beliefs can be avoided by working with Bayesian likelihood factors $B_{12} = \frac{P(H_1|E)}{P(H_1)} \left(\frac{P(H_2|E)}{P(H_2)} \right)^{-1}$, representing the relative support of evidence E for H_1 with respect to H_2. These do indeed not depend on the prior beliefs $P(H_i)$, although they still depend on the $P(E|H_i)$, for which an agreement among defenders of different paradigms might be equally hard to achieve. Second, when there is a sufficient accumulation of evidence in favour of a particular hypothesis H, all rational observers will eventually agree on a high probability for H, regardless of their original degree of belief.

4.2. Application to the Multiverse

Let us now see how all these general philosophical arguments relate to the multiverse.

4.2.1. Popper

From a Popperian point of view, "the multiverse" as a generic theory cannot be falsified, and this probably forms the most frequently heard criticism on the multiverse. However, two nuances are immediately in order.

First, as indicated before, Popper's falsificationist programme is no longer seriously upheld within the philosophy of science, certainly not in its naive form: falsification by itself is not the motor of scientific progress. This indicates that, within theoretical physics, we should overcome the very popular discussions about the multiverse and the anthropic principles centred on falsificationism, with one side defending it [78] and the other side claiming that it is about time to throw it overboard [79]. However, the question of falsifiability is not just about "scientific methodology". Popper's insistence on the testing of theories is still as relevant as it was a hundred years ago. You can formulate theories about reality, but if reality disagrees, the only way for nature to "kick back" and tell us whether our theories are tentatively right or wrong is through empirical confrontation with experiment and observation. Said in other words, a scientific theory should be able to make predictions which are testable: it must be possible to formulate what should be the case empirically (at least in principle) if the theory is true, and in which empirical case the theory should be considered as being in trouble. The idea that this combination of verification and falsification is an essential element in the "scientific method" is still largely uncontroversial among the immense majority of scientists and philosophers of science alike. Even [43], which advocates strongly for "non-empirical theory assessment" based on the alleged success of string theory, admits that such non-empirical assessment should ideally be temporary and that "empirical testing must be the ultimate goal of natural science".

A second nuance is that, although the multiverse in general cannot be falsified, this does not mean that concrete multiverse scenarios cannot be falsified. In fact, in our opinion this is perhaps the most crucial challenge for the multiverse in the near future: to work out concrete scenarios with concrete predictions that could be tested, at least in principle. We will discuss this in more detail in Section 6.

4.2.2. Kuhn

Does the multiverse really represent a "paradigm change" or a "scientific revolution", in Kuhn's vocabulary? This question is hard to answer for several reasons. One is that, historically speaking, such paradigm changes are usually not identified as and when they occur, but only a posteriori. Also, despite Kuhn's insistence on the revolutionary character of such paradigm changes, the moment when this revolution has taken place is very hard to pinpoint exactly. The revolutionary process depends on a confluence of several factors. That a single event and/or a single scientist is afterwards highlighted has often more to do with good story-telling than with the real complicated process that has taken place.

Special Relativity is a good example. While often presented as Einstein's first stroke of genius out of the blue, Einstein's treatment was in fact the culmination of a long process with crucial contributions from Lorentz and Poincaré.

These observations are in stark contrast with some messages in the multiverse literature which prophesy a revolution in our understanding of reality (see the examples [4,5,23,24] given earlier). In our opinion, one should be careful with this kind of claims. Announcing a scientific revolution while it is supposedly taking place runs a serious risk of sounding hollow. We are not aware of any research on the frequency of such claims in physics, but it might be interesting to note what is happening in other areas of research. In medical research, for instance, it was found that the frequency of announcements of "unprecedently innovative groundbreaking" ideas has increased up to 15,000% over the past four decades [80]. The authors dryly remark that "whether this perception fits reality should be questioned". The editorial [81] concludes that "it is time to acknowledge that the misrepresentation of research findings through exaggeration or hype is a grave matter for scientific integrity". Despite Kuhn's insistence on the social character of scientific truth-building, for a scientific revolution to take place, it is not sufficient that a particular scientific community claims that it is taking place. Similar feelings have often existed, even in the relatively recent past, and were proven to be wrong much more often than they were right. The development of quantum mechanics is an interesting exception; Chew's bootstrap model, early versions of supergravity and geometrodynamics, and Euclidean quantum gravity are just a few confirming examples.

A related question is: which paradigm is the multiverse supposed to be replacing? The paradigm of "the universe"? Or merely the standard ΛCDM model of cosmology? In the first case, it certainly seems a bit early to argue that "the universe" is in a state of crisis. In the second case: it is true that there are serious puzzles in cosmology, from the nature of dark matter and dark energy to the connection between primordial fluctuations and large-scale structure formation, to name but a few. However, these are challenges for any cosmological model rather than clear-cut problems with the current ΛCDM paradigm. ΛCDM can be interpreted as the simplest cosmological model based on General Relativity which is in agreement with the firmly established bulk of current observations, i.e.: a concordance model with much room for further modifications and extensions. There is at present not a single observation that points towards the assumption of a single universe as the crucial cause of these puzzles. Also, let us not forget that observational cosmology has grown in only a few decades from a phenomenon almost on the margin of science to a blooming area of research, but is still in its infancy. Depending on how one wishes to look at it, one might therefore argue that the ΛCDM model is suffering from serious anomalies, or that it has so far been of an unprecedented success in the history of cosmology. Either way, ΛCDM will undoubtedly require corrections, perhaps even major revisions [82,83]. However, from a purely observational point of view, the case for giving up trying to explain cosmology within a single universe is currently rather thin.

4.2.3. Lakatos

According to Lakatos' criterion, a research programme should be abandoned when it enters into a degenerative state, and at the same time a progressive alternative is available. Recall that a research programme consists of a hard core, which is maintained unaltered until the research programme is completely abandoned, and a series of auxiliary hypotheses. A degenerative research programme is characterized by the formulation of auxiliary hypotheses in order to save it from failures to predict or explain empirical observations, while a progressive one uses auxiliary hypotheses to strengthen its empirical content.

From this point of view, most approaches to the multiverse should probably not (yet) really be classified as a research programme. Rather, in Lakatosian terms, the multiverse is in fact an auxiliary hypothesis which has arisen within various existing research programmes (such as string theory and inflation theory) to justify their lack of empirical success, and more in general: the lack of empirical success of any approach to quantum gravity and/or Planckian physics. We do not wish here to jump to the conclusion that these are all degenerative research programmes. However, it might be relevant to realize that "the multiverse" in its current state does not fit the Lakatosian description of a (mature) research programme.

Moreover, as we already indicated in the Kuhnian discussion above, there is no degenerative research programme in need of replacement (yet). It is certainly true that there are many challenges for the standard ΛCDM model of cosmology, in particular the cosmological constant problem. However, concluding that ΛCDM is in a degenerative state would be a bit overhasty, and at present there is not a progressive alternative with a stronger and more successful empirical record available. With regard to the cosmological constant problem (see also Section 5), the multiverse offers one type of solution, but there also exist several other categories of interesting ideas that do not require positing a multiverse [84], and it is probably fair to say that none of these proposals, neither universe nor multiverse-related, are currently generally accepted as satisfactory.

4.2.4. Feyerabend

Within Feyerabend's vision, it is tempting to highlight that some scientists make use not only of rational argumentation, but also of the various types of "trickery" mentioned by Feyerabend to reinforce the impact of their model. There is a certain truth to this: the multiverse cause is omnipresent, especially in the popular scientific press, but also in the academic literature (with a high publication rate of scientific articles). On the positive side, this illustrates that theoretical physicists are no longer isolated in their academic ivory towers, but make an effort to reach out to the general public and present current ideas about fundamental issues to a wider audience. On the negative side, in the absence of empirical testing, despite the robustness of the underlying theories, criticists might argue that the truth-claims of the multiverse rely mainly on a social consensus.

There is a related point for which we will return to Lakatos' terminology: research programmes define which paths to pursue (positive heuristic) but also which paths to avoid (negative heuristic). Is there a real risk that the increasing influence of multiverse ideas might lead to a gradual decline in explorations of alternative approaches in cosmology and high-energy physics as was argued in related contexts [85–87]? The current situation in cosmology does not seem so alarming. In addition, in order to raise a new issue it is necessary to explore its possibilities. We will therefore not further examine this question here. It should be clear that we agree on the importance of empirical testing, and will therefore insist that this should be crucial also within multiverse approaches.

4.2.5. Bayesianism

The main weakness, in our opinion, of a Bayesian defense of the multiverse, is the following. Bayesianism is very well-suited to formulate logically how the probability for the true occurrence of a certain event should be updated in the light of observational evidence, but is more questionable when it comes to formalizing paradigm changes.

We pointed out earlier—see Equation (1)—that a sufficient accumulation of evidence in favour of a particular hypothesis H will "force" all rational observers to assign a high degree of probability for H. However, it is equally true that anybody who is strongly unconvinced a priori of the hypothesis H will (and, rationally speaking: *should*) refuse to admit a strong probability for the truth of H until there really is overwhelming evidence. In addition, such overwhelming evidence, in the opinion of the large majority of scientists, should still come from the confrontation of the hypothesis with observation. Once such "traditional" scientific proof in the form of empirical verification becomes available, then it is largely irrelevant whether one abides by Bayesian principles or not. It is therefore hard to see how Bayesian inference could formalize scientific progress across the kind of paradigm shift which is currently being defended by some proponents of the multiverse, namely one based largely on theoretical arguments. More generally, despite the unquestionable value of Bayesian inference, the Bayesian view on scientific progress in general (including theoretical paradigm shifts) is a bit curious, because it represents a return to an attempt at a logical formulation of the progress of science, disregarding the historical evolution from Popper to Feyerabend that we have tried to outline earlier.

We will here briefly sketch three further problems related to Bayesianism.

(1) The first problem is well known as the measure problem. Our universe provides us with a sample of fixed size $n = 1$. This means that almost all statistical properties of the alleged multiverse population are ill-defined, unless one somehow defines a concept of measure across the multiverse population. This can be done essentially in two (interrelated) ways. The first way is to assume some simple distribution, such as a uniform distribution of the possible cosmological constants [88] (or of a small set of variables, for example the cosmological and gravitational constants). However, apart from the fact that it is hard to justify a priori why precisely these variables should characterize the distribution, one should also realize that, with a uniform distribution, many statistical characteristics are essentially determined by the limiting values. Just like in the famous German tank problem, estimating these limiting values based on a single observation entails a very high degree of uncertainty. A second method to define a measure is to assume some fundamental theory, typically string theory [36], and use the theoretical knowledge obtained from this theory to derive a measure. However, this has various associated risks. As pointed out by Ellis [87], "the statistical argument only applies if a multiverse exists; it is simply inapplicable if there is no multiverse: we cannot apply a probability argument if there is no multiverse to apply the concept of probability to." Even if the multiverse really does exist, there is still a risk of circularity: to construct a measure from an empirically unverified theory based on an $n = 1$ sample needs some auxiliary hypothesis, the most obvious possibility being related to what has become known as Vilenkin's mediocrity principle [27], namely that the sample lies in the densest part of the probability distribution. If such a measure can then be constructed, it should obviously show that the $n = 1$ sample indeed lies in the densest part of the probability distribution. However, apart from the almost tautological character of this construction, there is no way of empirically contrasting the obtained measure, not even in principle, since we are by definition limited to the $n = 1$ sample size. This does not deny the adequacy of a Bayesian treatment based on relative likelihoods, which can be useful to compare different multiverse scenarios, for example to determine whether certain parameters or observations favour one multiverse scenario over the other, see e.g., [89]. However, there is no well-defined mechanism of correcting the original theoretical assumptions themselves. So apart from the question of which measure is most adequate, there also exists a challenge of understanding how such a construction based on an $n = 1$ sample could help us in deciding whether a multiverse scenario is really needed, rather than a single universe.

(2) A second problem has to do not so much with Bayesianism in itself, but with naive applications of it. Polchinski famously arrived at a 94% probability for the multiverse to exist [90,91]. Polchinski's estimate is based on four yes–no questions (e.g., is there a satisfactory understanding for the cosmological constant value?). Since conventional (non-multiverse) physics answers four times "no", Polchinski arrives at a probability of $1 - (1/2)^4 = 0.9375$ in favour of the multiverse.[4] Let us play the devil's advocate and, for the mere sake of the argument, defend the creationist view of intelligent design in biology. State any four gaps in the evolutionary picture of life and humanity. The Polchinski–Bayesian conclusion would be that there is a 93.75% probability that the universe was literally created by God in seven days. If this argument sounds too far-fetched, let us insist on the key point: the current lack of explanation for any scientific challenge within conventional single-universe relativistic cosmology in itself is not a sufficient support for an anthropic or multiverse argument. We will come back to this question in Section 4.3.

(3) The third problem is closely related to the previous one, namely the risk of accepting Bayesianism in combination with purely theoretical arguments as a substitute for empirical testing.

---

[4] In reality Polchinski's four questions are not independent, so the numerical estimate is incorrect even from a purely probabilistic point of view. However, since Polchinski himself states that the number itself is not important, we will not further dissect this issue.

This is best illustrated by an example. According to a Bayesian reasoning with purely theoretical arguments, there should have been almost 100% certainty in favour of the Georgi–Glashow SU(5) model [92]: this was closely based on some of the best physical theories that mankind has ever produced, it was mathematically elegant and favoured by a large proportion of theoretical physicists, and no alternatives even closely as appealing were available at the time. Yet, proton decay was not observed and so the Georgi–Glashow SU(5) unification model turned out to be wrong. This again illustrates our continuous insistence on empirical assessment.

4.3. Consistency and Uniqueness Claims

In the previous section, we have insisted on empirical theory assessment. Within the multiverse context, some authors propose to diminish the importance thereof, and to replace it (partially) by purely theoretical criteria. This is another line of thought where the philosophy and history of science can be relevant.

The idea that theoretical arguments, for example criteria of mathematical consistency and elegance, can illuminate the path towards a correct "fundamental" theory, is not new. On the contrary, this is closely related to Platonism, one of the oldest branches of western philosophy. It has resurged in theoretical physics repeatedly, especially in the past century or so [93]. The common pattern is striking: a scientist or group of scientists believes in the fundamentality and finality of the theory they are working on, based on the past success of the building blocks of this theory and the elegance of their construction. Empirical predictivity is either looked down upon, or the lack of empirical success is simply disregarded. Eventually, so far at least, the theory turns out to be either completely wrong or, in the best of cases, simply void of empirical content. The best-known example is perhaps Descartes' vortex theory, abandoned in favour of Newton's empirically much more successful laws of motion. However, more recent examples also abound. A ring-vortex theory, highly popular in late 19th century Britain, was developed in quite some mathematical detail by such famous contributors as William "Lord Kelvin" Thomson and FitzGerald. Although it never managed any level of empirical success, it continued to be defended by many scientists for several decades because of its elegance. As another example, Eddington developed a fundamental "Relativity Theory of Protons and Electrons" based on the construction of a series of fundamental constants which were supposed to relate microphysics and cosmology. In Eddington's view, the truth of his theory followed from purely epistemological considerations. Empirical confirmation was completely secondary, and even though he did in fact make quite a few observational predictions, he would simply disregard any disagreement with actual observations rather than let them ruin his beautiful theory. Many more historic and contemporary examples are discussed in detail in the excellent [93].

While [93] cautiously avoids extracting any explicit conclusions with regard to the current situation, authors such as [94,95] have argued that a blind quest for mathematical beauty has indeed led contemporary fundamental physics astray. So does history simply repeat itself? Let us examine some reasons to believe that the case of the multiverse might be different. From a social point of view, the current state with respect to various approaches in quantum gravity represents the first time that a large and international scientific community defend a common idea based on such theoretical criteria. The previous occurrences were mainly of single scientists (including such prestigious ones as Eddington and his "Fundamental Theory" or even Einstein and his "Unified Field Theory"), or had at most "national" success (the late 19th-century vortex theory, which was called a "Victorian theory of everything" by Kragh [96], was very popular in the UK but had limited resonance in the rest of the world). With respect to scientific content, the key argument in string theory and some multiverse-related approaches is that the theoretical "gap" to be bridged is shallow, in other words: that the multiverse is a natural continuation of our best theories, general relativity and quantum field theory; that we are indeed close to finding such a "final theory", and that consistency, elegance and uniqueness should therefore be sufficient arguments to solve the remaining problems (until the solution is eventually confirmed empirically). In this context, there is a statement by Popper that comes to mind: "Whenever a theory appears to you as the only possible one, take this as a sign that you have neither understood the theory nor the problem which it was intended to solve." [97].

Let us try to make a more precise counter-argument.

First of all, it is true of course that unification has been an important motor in the history of science. However, unification and uniqueness are two different concepts. Their apparent relation finds its origin in the reductionist idea that gradual unification will lead to a unique theory of everything, at the top of a pyramid of theories. This idea was strongly criticized by some physicists [98] and philosophers [99] alike, see also [100], who argue that the major advances in fundamental physics in the recent past have relied on a combination of unification and emergence.

Second, it is an interesting question whether the current state of physics, and in particular general relativity and quantum field theory (the precursors of the multiverse), could have been achieved through arguments of consistency, elegance and uniqueness. For general relativity, such arguments have certainly been crucial in Einstein's reasoning, and so one might be tempted to answer "yes". However, for quantum field theory, and its application to particle physics, although we cannot repeat history to answer the hypothetical question, the historical answer is a definite "no", as described in detail for the case of quarks in [73].

Third, the final unification is believed to take place at the Planck scale, and so the "dreams of a final theory" [101] are related to the idea that we are close to uncovering Planckian physics. This third point deserves a more detailed analysis, which we will undertake in the next section.

5. Fine-Tuning and the Multiverse... or Is It Really a Tale of Scales?

One of the strongest arguments in favour of the multiverse is the cosmological constant problem. Since the observed value of Λ is some 120 orders of magnitude smaller than the straightforward theoretical estimation[5], and any value of Λ very different from the actually observed one would probably make life in the universe impossible, it is argued that "the only known way to address [this problem] without invoking incredible fine-tuning [is] related to the anthropic principle, and, therefore, to the theory of the multiverse" [5].

Let us jump back to the Planck-scale unification argument of Section 4.3 for a moment. Some scientists working in quantum gravity believe that we are close to uncovering Planck-scale physics, and that consistency and perhaps uniqueness arguments should therefore be sufficient to bridge the remaining gap towards a final theory [43,90,101], possibly a multiverse theory.

The highest-energy physics that we actively control is the energy produced at the LHC. This is currently on the order of 10 TeV, i.e., 10^4 GeV. Compare this to the Planck scale, 10^{19} GeV, the scale at which quantum gravity supposedly take place. There is a difference of 15 orders of magnitude. Even high-energy cosmic ray detection rarely exceeds 10^4 TeV, still 13 orders of magnitude below the Planck scale. This problem is of course well-known among high-energy physicists, but there seems nevertheless to exist an optimistic view on bridging this gap [90]. However, two simple comparisons might serve as a cold shower. The extrapolation from the highest-energy physics that we control empirically to the physical theories which justify the idea of a multiverse is (literally) still several orders of magnitude stronger than the extrapolation from a grain of salt (size 10^{-4} m) to the size of the moon (diameter 10^6 m). To put another example: imagine that a biologist would claim that,

---

[5] Just in case some reader might benefit from a reminder, the theoretical estimate comes essentially from assuming that the cosmological constant represents the vacuum energy E_{vac}, imposing a cut-off k_c to the theory and calculating E_{vac} by integrating over all degrees of freedom up to k_c, which gives $E_{vac} = \hbar k_c^4$ (a result which is consistent with a straightforward dimensional analysis [102]). Assuming $k_c = E_{Planck}$ immediately leads to the undesired result, while even $k_c = E_{EW}$, with E_{EW} the electroweak scale, still leads to a discrepancy of some 50 orders of magnitude. It might be worth insisting that it is essential to insert a cut-off in the calculation in order to avoid an even more unpleasant prediction for the vacuum energy, namely infinity. Note that the observational energy scale associated with dark energy is in fact small, and might therefore be due to quantum field effects potentially accessible to near-future observations. However, this would still leave the cosmological coincidence problem unexplained, namely why the matter energy density and the dark energy density have the same order of magnitude in the present epoch.

by studying the macroscopic properties of the largest living beings on earth, blue whales, he could infer the biological structure of the smallest known bacterial cells, with sizes 0.1 µm. The mere scale difference of 10^8 is peanuts in comparison with the jump from the LHC to the Planck scale.

There are strong theoretical reasons to believe that the cosmological constant is related to Planck-scale physics. In fact, the argument in favour of the landscape of string theory rests precisely on Planck-scale arguments. Therefore, because of the energy gap just described, perhaps we should simply admit that the "worst theoretical prediction in the history of physics" is due to our (unsurprising) ignorance of physics at the Planck scale, that we are currently exploring many ideas, but that all of these (including the multiverse) are so far still in an embryonic state.

It is tempting to put the blame on the lack of empirical data [43]. It is of course true that there is no empirical data available for physics at the Planck scale. However, two interpretations are possible. One could say that experimentalists have not been able to keep up with theorists. Perhaps a fairer interpretation is that, after the enormous success of quantum field theory and the standard model of particle physics, theorists have run ahead, jumping several scales and constructing theories well above current experimental possibilities. As we have defended above, scientific progress typically rests on a complex interplay between theory and observation, and this might be even more important as we move further and further away from direct sensory experience. Bottom-up and top-down approaches in fundamental physics should be complementary [103]. Presently the equilibrium in the search for quantum gravity is a bit distorted.[6] Near-future observational surveys with respect to the "dark sector" of the universe such as DESI and EUCLID are promising. However, because of the scale problem that we have just stressed, the key message should probably be one of patience and of anticipating slow and indirect progress, rather than immediate spectacular advances.

6. Physical Multiverse and Testability

The previous discussion leads to the following general issue: How could the overall conceptual challenge be met of converting the multiverse from a speculative (or even metaphysical) consideration into a physical theory? The multiverse currently provides an interesting framework to understand reality, but it should also be followed by testable predictions. To come back to the relation between the multiverse idea as an epistemological extension of the Copernican revolution that we mentioned at the beginning of Section 3: Humanity has gradually realized its loss of importance as the center of existence when science has been able to look further and realize that those observed objects were in fact other structures similar to ours: other planets, other stars, other galaxies. The possible extension to the multiverse is not accompanied by any such direct observation, and is therefore of a different speculative order. Ultimately, the multiverse question comes down to determining whether, in order to confront the observational challenges and anomalies of cosmology, it is sufficient to consider a single universe, or whether we need a multiverse scenario. The main element in the effort to answer this question consists of setting up multiverse scenarios and looking for empirical predictions which can be tested. Depending on the type of multiverse scenario, this can be very hard, perhaps even impossible. For instance, in a multiverse scenario where the different universes possess different physical laws or mathematical structures, it is hard to see how to look for interactions with our universe. Alternative ways of assessment might then still lead to a certain degree of confidence, but always provided that other parts of the theory can be tested empirically.

In this section, we want to sketch a possible way of approaching the empirical multiverse question, i.e.: to establish empirical predictions in the traditional physical sense, limited to our universe but nevertheless allowing us to find some hint of interactions with other universes. In order to describe

---

[6] The only well-developed bottom-up approach to "quantum gravity phenomenology" is the ongoing search for Lorentz Invariance Violations [104,105]. However, it might be useful to stress that neither string theory nor loop quantum gravity make clear and unambiguous predictions about Lorentz Invariance, not even at a qualitative level.

such "physical multiverse" scenarios, no specific multiverse model is required. It is sufficient to impose certain minimal requirements on the multiverse scenario.

The first of these requirements is the classical independence of the spacetimes of each composing universe, at the level of the standard phenomena in General Relativity. This degree of independence is essential in order to consider each universe as a separate and differentiable entity. In the opposite case, it would be possible to define the different components as different regions of a single universe. If we understand as standard spacetime connections the ones allowed by General Relativity (without introducing exotic issues such as closed timelike curves) we assume that such causally completely determined relations do not exist between the spacetimes of the different universes that compose the multiverse. Note that the existence of non-standard (classical or quantum) connections is not prohibited by this definition. On the contrary, these are essential in order to have some possibility of interaction among universes and therefore some empirical imprint to look for.

The second requirement is that each of the universes must be potentially observable, by direct or indirect measures, from some other universe. In this way, physical predictions in our universe can be established as a consequence of the physics of the whole multiverse. This shows that the independence among universes works only at the (classical) level of the spacetimes per se, not of all its components. There must exist some degree of interaction among them.

Finally, we also impose that, if the constants of nature are allowed to vary from one universe to another, then the values of these constants must be linked through the physical laws governing the overall multiverse. In other words, these physical constants cannot emerge independently but must be correlated, for instance, through quantum entanglement effects between universes.

One could paraphrase these three conditions by saying that the different universes in the multiverse should have causally independent spacetimes but with correlations among them. These requirements do not impose any specific physical scenario, but are sufficient to define testable consequences of any multiverse scenario which obeys them.

In order to clarify this concept one can consider a classification of these correlations in terms of their classical or quantum nature. The classical correlations could be given, for instance, by considering the multiverse as a multiply connected spacetime, where each universe is connected with other by means of Lorentzian tunnels [106,107]. It is important to note that, from the first condition given above, there cannot exist causal relations among the different universes. The existence of causal relations would entail the existence of a common time between both universes which could then not be considered independent. The connections could therefore be formed by wormholes converted into time machines, providing closed timelike curves in the interior of the tunnel [108]. Potential observable effects of wormholes were studied in several papers [109–114]. The current challenge in the multiverse context is the search of an unequivocal observable effect of such a wormhole connection with another universe [115].

Quantum correlations could come from the quantum entanglement between universes. According to the first physical multiverse requirement, the spacetimes of each component universe must be classically differentiable. However, in this scenario they would not be quantum separable, giving rise to an entangled multiverse [52]. In this context, one could determine the effects on our universe that show up as a consequence of these inter-universe quantum correlations. In the absence of a classical channel, these correlations cannot be directly detected. However, the influence of such correlations can be examined, for example on the value of the cosmological constant. It might be hard to imagine such an indirect effect which could not equally be explained within a single-universe scenario. However, the study of the different schemes of interaction among universes and the development of a toy-model catalogue of observable effects and predictions allows an important progress towards multiverse phenomenology and of the types of effects that could be expected in more detailed scenarios [52,53,56–58]. The investigation of these quantum correlations is complicated by the lack of a quantum theory describing our universe, and most current models therefore focus on qualitative approximations to the collective phenomena that can arise in the consideration of a quantum multiverse. Alternatively, an exhaustive description of the spacetimes in the framework of quantum mechanics could be attempted. This allows a more rigorous description of the interactions, but at the cost of a great technical complexity which limits the conception of different universes [116].

We are still very far from having a complete theory that would allow us to settle the present discussion, or a fully systematic way of deriving empirical consequences from concrete multiverse scenarios. Nonetheless, the physical multiverse ideas just described show that, at least for certain classes of multiverse models, it should be possible to extract empirical predictions based on relatively general considerations. So it is interesting to keep them in mind when dealing with these issues. Also, if such inter-universe correlations as just described really exist, then this would indicate the necessity of considering the multiverse as an indivisible framework. This in itself should be sufficient motivation to construct such generic physical multiverse models and study their possible empirical effects.

7. Conclusions

In our view, it is not so important to determine whether speculations about "the multiverse" are part of science or not. Only time will tell. However, multiverse scenarios should certainly be recognized for what they (still) are: an embryonic framework which can be useful to understand and formulate certain problems related to cosmology, but which is still far away from being testable in any general physical sense. Care should perhaps be taken with claims that "multiverse theories are utterly conventionally scientific" [117], or that Bayesian arguments show that, by a "conservative" estimate, "the likelihood that the multiverse exists [is] 94%" and that "those who find this calculation amusing (...) should be a bit more humble" [90]. Such claims might be more counter-productive than anything else. Indeed, they do not fairly reflect the current scientific status of the field, nor do they agree with historical and philosophical analyses and in particular the importance of empirical content and of the complex interplay between theory-building and observation required for scientific progress.

Vice versa, despite all the warnings that we have formulated, it is certainly not our intention to dismiss the general idea of a multiverse as a developing physical theory. This would imply closing the door to a whole range of ideas and techniques that are currently being developed, and some of which could indeed turn out to be fundamental in our understanding of the nature of spacetime. However, empirical testing should always remain the central aim of science, even (or perhaps: especially) in the multiverse epoch. In that sense, the general framework for a physical multiverse that we have discussed could be useful as a guide for the development of empirical multiverse scenarios, and more generally: to discriminate emergent ideas and to look for the possible testability of different cosmological scenarios involving either a single universe or a multiverse in any of the various multiverse definitions.

Author Contributions: Both authors contributed equally to this work.

Funding: The authors acknowledge Project No. FIS2017-86497-C2-2-P (A.A.-S.) and FIS2017-86497-C2-1 (G.J.) from the Spanish Mineco.

Acknowledgments: A.A.-S. wants to acknowledge the memory of Pedro Félix González-Díaz as a source of inspiration for her research and for originating the discussion about the physical multiverse.

Conflicts of Interest: The authors declare no conflict of interest.

References

1. Carr, B. (Ed.) *Universe or Multiverse?*; Cambridge University Press: Cambridge, UK, 2007.
2. Chamcham, K.; Silk, J.; Barrow, J.D.; Saunders, S. (Eds.) *The Philosophy of Cosmology*; Cambridge University Press: Cambridge, UK, 2017.
3. Dardashti, R.; Dawid, R.; Thébault, K. (Eds.) *Epistemology of Fundamental Physics: Why Trust a Theory?*; Cambridge University Press: Cambridge, UK, 2019.
4. Barrau, A. Physics in the multiverse: An introductory review. *CERN Cour.* **2007**, *47*, 13–17.
5. Linde, A. A brief history of the multiverse. *Rep. Prog. Phys.* **2017**, *80*, 022001. [CrossRef] [PubMed]

6. Ellis, G.; Silk, J. Scientific method: Defend the integrity of physics. *Nature* **2014**, *516*, 321–323. [CrossRef] [PubMed]
7. Ellis, G. Does the Multiverse Really Exist? *Sci. Am.* **2011**, 38–43. [CrossRef] [PubMed]
8. Kragh, H. Contemporary History of Cosmology and the Controversy over the Multiverse. *Ann. Sci.* **2009**, *66*, 529–551. [CrossRef]
9. Bettini, S. A Cosmic Archipelago: Multiverse Scenarios in the History of Modern Cosmology. *arXiv* **2005**, arXiv:physics/0510111.
10. Rioja, A.; Ordoñez, J. *Teorías del Universo*; Síntesis: Madrid, Spain, 2006
11. Bower, R.G.; McLeish, T.C.B.; Tanner, B.K.; Smithson, H.E.; Panti, C.; Lewis, N.; Gasper, G.E.M. A medieval multiverse?: Mathematical modelling of the thirteenth century universe of Robert Grosseteste. *Proc. R. Soc. Lond. A* **2014**, *470*, 20140025. [CrossRef]
12. Bruno, G. *De la causa, principio et Uno 1584. For a Modern Translation, see Bruno, G. Cause, Principle and Unity*; Translated and Edited by De Lucca, R.; Cambridge University Press: Cambridge, UK, 1998.
13. Swedenborg, E. *Principia Rerum Naturalium 1734*; Translated by J. R. Rendell and I. Tansley as: The Principia Or The First Principles Of Natural Things; The Swedenborg Society: London, UK, 1912.
14. Wright, T. *An Original Theory or New Hypothesis of the Universe, Founded upon the Laws of Nature*; Modern Edition; Cambridge University Press: Cambridge, UK, 2014.
15. Kant, I. *Universal Natural History and Theory of the Heavens, 1755*; Cambridge University Press: Cambridge, UK, 2012.
16. Von Humboldt, A. *Cosmos: A Sketch of a Physical Description of the Universe, 1845–1862*; Cambridge University Press: Cambridge, UK, 2011.
17. Shapley, H.; Curtis, H.D. The scale of the universe. *Bull. Natl. Res. Council* **1921**, *2*, 171–217.
18. Leibniz, G.W. *Essais de Théodicée sur la Bonté de Dieu, la Liberté de L'homme et L'origine du mal. Amsterdam 1710*; For a Modern Translation, see e.g., Leibniz, G.W. Theodicy: Essays on the Goodness of God, the Freedom of Man, and the Origin of Evil; Huggard, E.M., Translated; Open Court: Lasalle, IL, USA, 1985.
19. Schopenhauer, A. *Von der Nichtigkeit und dem Leiden des Lebens. Chapter 46 in: Die Welt als Wille und Vorstellung, First Included in the 2nd Expanded Edition 1844*; For a Modern Translation, see e.g., Schopenhauer, A. On the Vanity and Suffering of Life. Chapter 46 in: The World as Will and Representation; Payne, E.F.J., Translated; Dover: New York, NY, USA, 1969.
20. Rees, M. *Just Six Numbers: The Deep Forces that Shape the Universe*; Weidenfeld & Nicolson: London, UK, 1999.
21. *Śrīmad-Bhāgavatam (Bhāgavata Purāṇa) 6.16.37*; The Translation Is Taken from Ramesh Menon, Bhagavata Purana, The Holy Book of Vishnu (2 vols.); Vedic Books: Delhi, India, 2007.
22. Page, D. Does God So Love the Multiverse? In *Science and Religion in Dialogue*; Stewart, M.Y., Ed.; Blackwell Publishing: Hoboken, NJ, USA, 2010.
23. Rees, M. *On the Future—Prospects for Humanity*; Princeton University Press: Princeton, NJ, USA, 2018.
24. Livio, M.; Rees, M. Fine-Tuning, Complexity, and Life in the Multiverse. *arXiv* arXiv:1801.06944. To appear in: Consolidation of Fine Tuning (forthcoming)
25. Vilenkin, A. *Many Worlds in One: The Search for Other Universes*; Hill and Wang: New York, NY, USA, 2007.
26. Kragh, H. An anthropic myth: Fred Hoyle's carbon-12 resonance level. *Arch. Hist. Exact Sci.* **2010**, *64*, 721. [CrossRef]
27. Vilenkin, A. The principle of mediocrity. *Astron. Geophys.* **2011**, *52*, 5–33. [CrossRef]
28. Hartle, J.B.; Hawking, S.W. Wave function of the Universe. *Phys. Rev. D* **1983**, *28*, 2960. [CrossRef]
29. Vilenkin, A. Boundary conditions in quantum cosmology. *Phys. Rev. D* 1986, 33 , 3560. [CrossRef] [PubMed]
30. Gott, J.R., III; Li, L.X. Can the universe create itself? *Phys. Rev. D* **1998**, *58*, 023501 [CrossRef]
31. Tegmark, M. The multiverse hierarchy. In *Universe or Multiverse?*; Carr, B., Ed.; Cambridge University Press: Cambridge, UK, 2007.
32. Everett, H. Relative state formulation of quantum mechanics. *Rev. Mod. Phys.* **1957**, *29*, 454. [CrossRef]
33. Linde, A.D. Eternal chaotic inflation. *Mod. Phys. Lett. A* **1986**, *1*, 81. [CrossRef]
34. Linde, A.D. Eternally existing selfreproducing chaotic inflationary Universe. *Phys. Lett. B* **1986**, *175*, 395. [CrossRef]
35. Linde, A.; Vanchurin, V. How many universes are in the multiverse?. *Phys. Rev. D* **2010**, *81*, 083525. [CrossRef]

36. Susskind, L. The anthropic landscape of string theory. In *Universe or Multiverse?*; Carr, B., Ed.; Cambridge University Press: Cambridge, UK, 2007.
37. Holman, R.; Mersini-Houghton, L.; Takahashi, T. Cosmological avatars of the landscape. II. CMB and LSS signatures. *Phys. Rev. D* **2008**, *77*, 063511. [CrossRef]
38. Vafa, C. The String landscape and the swampland. *arXiv* **2005**, arXiv:hep-th/0509212v2.
39. Johnson, M.C.; Lehners, J.L. Cycles in the Multiverse. *Phys. Rev. D* **2012**, *85*, 103509. [CrossRef]
40. Penrose, R. Before the Big Bang: An outrageous new perspective and its implications for particle physics. *Conf. Proc. C060626* **2006**, 2759.
41. Baumann, D.; McAllister, L. *Inflation and String Theory. Cambridge Monographs on Mathematical Physics*; Cambridge University Press: Cambridge, UK, 2015.
42. Bousso, R.; Susskind, L. The Multiverse Interpretation of Quantum Mechanics. *Phys. Rev. D* **2012**, *85*, 045007. [CrossRef]
43. Dawid, R. *String Theory and the Scientific Method*; Cambridge University Press: Cambridge, UK, 2013.
44. Rees, M.J. Cosmology and the multiverse. In *Universe or Multiverse?*; Carr, B., Ed.; Cambridge University Press: Cambridge, UK, 2007.
45. Di Valentino, E.; Mersini-Houghton, L. Testing Predictions of the Quantum Landscape Multiverse 1: The Starobinsky Inflationary Potential. *J. Cosmol. Astropart. Phys.* **2017**, *1703*, 002. [CrossRef]
46. Kinney, W.H. Limits on Entanglement Effects in the String Landscape from Planck and BICEP/Keck Data. *J. Cosmol. Astropart. Phys.* **2016**, *1611*, 013. [CrossRef]
47. Aguirre, A.; Johnson, M.C.; Shomer, A. Towards observable signatures of other bubble universes. *Phys. Rev. D* **2007**, *76*, 063509. [CrossRef]
48. Wainwright, C.L.; Johnson, M.C.; Aguirre, A.; Peiris, H.V. Simulating the universe(s) II: phenomenology of cosmic bubble collisions in full General Relativity. *J. Cosmol. Astropart. Phys.* **2014**, *1410*, 024. [CrossRef]
49. Zhang, P.; Johnson, M.C. Testing eternal inflation with the kinetic Sunyaev Zel'dovich effect. *J. Cosmol. Astropart. Phys.* **2015**, *1506*, 046. [CrossRef]
50. Zhang, J.; Blanco-Pillado, J.J.; Garriga, J.; Vilenkin, A. Topological Defects from the Multiverse. *J. Cosmol. Astropart. Phys.* **2015**, *1505*, 059. [CrossRef]
51. Popper, K.R. *Conjectures and Refutations: The Growth of Scientific Knowledge*; Routledge: Abingdon, UK, 1963.
52. Robles-Perez, S.; Gonzalez-Diaz, P.F. Quantum state of the multiverse. *Phys. Rev. D* **2010**, *81*, 083529. [CrossRef]
53. Robles-Perez, S.; Gonzalez-Diaz, P.F. Quantum entanglement in the multiverse. *J. Exp. Theor. Phys.* **2014**, *118*, 34. [CrossRef]
54. Kanno, S.; Shock, J.P.; Soda, J. Entanglement negativity in the multiverse. *J. Cosmol. Astropart. Phys.* **2015**, *1503*, 015. [CrossRef]
55. Kanno, S. Quantum Entanglement in the Multiverse. *Universe* **2017**, *3*, 28. [CrossRef]
56. Robles-Perez, S.; Alonso-Serrano, A.; Gonzalez-Diaz, P.F. Decoherence in an accelerated universe. *Phys. Rev. D* **2012**, *85*, 063511. [CrossRef]
57. Alonso-Serrano, A.; Bastos, C.; Bertolami, O.; Robles-Perez, S. Interacting universes and the cosmological constant. *Phys. Lett. B* **2013**, *719*, 200. [CrossRef]
58. Robles-Pérez, S.; Alonso-Serrano, A.; Bastos, C.; Bertolami, O. Vacuum decay in an interacting multiverse. *Phys. Lett. B* **2016**, *759*, 328. [CrossRef]
59. Hacking, I. *Representing and Intervening: Introductory Topics in the Philosophy of Natural Science*; Cambridge University Press: Cambridge, UK, 1983.
60. Okasha, S. *Philosophy of Science: A Very Short Introduction*; Oxford University Press: Oxford, UK, 2002.
61. Chalmers, A. *What Is This Thing Called Science?*, 4th ed.; Queensland University Press: Brisbane, Australia, 2013.
62. Whitehead, A.N.; Russell, B. *Principia Mathematica*; Cambridge University Press: Cambridge, UK, 1910; Volume 1.
63. Whitehead, A.N.; Russell, B. *Principia Mathematica*; Cambridge University Press: Cambridge, UK, 1912; Volume 2.
64. Whitehead, A.N.; Russell, B. *Principia Mathematica*; Cambridge University Press: Cambridge, UK, 1913; Volume 3.
65. Hilbert, D. Mathematische Probleme. Nachrichten von der Königlichen Gesellschaft der Wissenschaften zu Göttingen, Math. *Phys. Klasse* **1900**, 253–297

66. Hilbert, D. *Grundlagen der Mathematik. Vorlesung, Winter-Semester 1921/22. Lecture notes by Paul Bernays*; Universität Göttingen: Göttingen, Germany, 1922.
67. Gödel, K. Über formal unentscheidbare Sätze der Principia Mathematica und verwandter Systeme I. *Monatshefte für Mathematik* **1931**, *38*, 173–198. [CrossRef]
68. Popper, K.R. *Logik der Forschung. Zur Erkenntnistheorie der modernen Naturwissenschaft*; Extended English version: K.R. Popper, The Logic of Scientific Discovery: Hutchison 1959; Springer: Berlin, Germany, 1934.
69. Kuhn, T. *The Structure of Scientific Revolutions*; University of Chicago Press: Chicago, IL, USA, 1962..
70. Dardashti, R. Physics without Experiments? In *Epistemology of Fundamental Physics: Why Trust a Theory?*; Dardashti, R., Dawid, R., Thébault, K., Eds.; Cambridge University Press: Cambridge, UK, 2019.
71. Oriti, D. No alternative to proliferation. In *Epistemology of Fundamental Physics: Why Trust a Theory?*; Dardashti, R., Dawid, R., Thébault, K., Eds.; Cambridge University Press: Cambridge, UK, 2019.
72. Sahlén, M. On Probability and Cosmology: Inference Beyond Data? In *The Philosophy of Cosmology*; Chamcham, K., Silk, J., Barrow, J.D., Saunders, S., Eds.; Cambridge University Press: Cambridge, UK, 2017.
73. Pickering, A. *Constructing Quarks: A Sociological History of Particle Physics*; Edinburgh University Press: Edinburgh, UK, 1984.
74. Lakatos, I. Falsification and the Methodology of Scientific research programmes. In *Criticism and the Growth of Knowledge*; Lakatos, I., Musgrave, A., Eds.; Cambridge University Press: Cambridge, UK, 1970.
75. Feyerabend, P. *Against Method: Outline of an Anarchistic Theory of Knowledge*; New Left Books: London, UK, 1975.
76. Howson, C.; Urbach, P. *Scientific Reasoning: The Bayesian Approach*; Open Court: La Salle, IL, USA, 1989.
77. Ortovela, P. Modeling the Change of Paradigm: Non-Bayesian Reactions to Unexpected News. *Am. Econ. Rev.* **2012**, *102*, 2410–2436.
78. Smolin, L. Scientific alternatives to the anthropic principle. In *Universe or Multiverse?*; Carr, B., Ed.; Cambridge University Press: Cambridge, UK, 2007.
79. Susskind, L. *The Cosmic Landscape: String Theory and the Illusion of Intelligent Design*; Little Brown & Co.:Boston, MA, USA, 2005.
80. Vinkers, C.; Tijdink, J.; Otte W. Use of positive and negative words in scientific PubMed abstracts between 1974 and 2014: retrospective analysis. *BMJ* **2015**, *351*, h6467. [CrossRef]
81. Scott, S.L.; Jones, C.W. Superlative Scientific Writing. *ACS Catal.* **2017**, *7*, 2218–2219.. [CrossRef]
82. Silk, J. Towards the limits of cosmology. *Found. Phys.* **2018**, *48*, 1305–1332. [CrossRef]
83. Turner, M.S. ΛCDM: Much More Than We Expected, but Now Less Than What We Want. *Found. Phys.* **2018**, *48*, 1261–1278. [CrossRef]
84. Nobbenhuis, S. Categorizing different approaches to the cosmological constant problem. *Found. Phys.* **2006**, *36*, 613. [CrossRef]
85. Woit, P. *Not Even Wrong: The Failure of String Theory and the Search for Unity in Physical Law*; Basic Books: New York, NY, USA, 2006
86. Smolin, L. The Trouble with Physics: The Rise of String Theory, The Fall of a Science, and What Comes Next. *Houghton Mifflin Harcourt* **2006**, *30*, 66–69.
87. Ellis, G. Multiverses, Science, and Ultimate Causation. In *Georges Lemaître: Life, Science and Legacy*; Holder, R.D., Mitton, S., Eds.; Springer: Berlin, Germany, 2012.
88. Weinberg, S. Anthropic Bound on the Cosmological Constant. *Phys. Rev. Lett.* **1987**, *59*, 2607. [CrossRef]
89. Barnes, L.A.; Elahi, P.J.; Salcido, J.; Bower, R.G.; Lewis, G.F.; Theuns, T.; Schaller, M.; Crain, R.A.; Schaye, J. Galaxy formation efficiency and the multiverse explanation of the cosmological constant with EAGLE simulations. *Month. Not. R. Astron. Soc.* **2018**, *477*, 3727. [CrossRef]
90. Polchinski, J. String Theory to the Rescue. In *Epistemology of Fundamental Physics: Why Trust a Theory?*; Dardashti, R., Dawid, R., Thébault, K., Eds.; Cambridge University Press: Cambridge, UK, 2019.
91. Polchinski, J. Why trust a theory? Some further remarks (part 1). In *Epistemology of Fundamental Physics: Why Trust a Theory?*; Dardashti, R., Dawid, R., Thébault, K., Eds.; Cambridge University Press: Cambridge, UK, 2019.
92. Georgi, H.; Glashow, S. Unity of All Elementary-Particle Forces. *Phys. Rev. Lett.* **1974**, *32*, 438. [CrossRef]
93. Kragh, H. *Higher Speculations: Grand Theories and Failed Revolutions in Physics and Cosmology*; Oxford University Press: Oxford, UK, 2011

94. Baggott, J. *Farewell to Reality: How Modern Physics Has Betrayed the Search for Scientific Truth*; Pegasus Books: New York, NY, USA, 2014.
95. Hossenfelder, S. *Lost in Math: How Beauty Leads Physics Astray*; Basic Books: New York, NY, USA, 2018.
96. Kragh, H. The Vortex Atom: A Victorian Theory of Everything. *Centaurus* **2003**, *44*, 32–114. [CrossRef]
97. Popper, K.R. *Objective Knowledge: An Evolutionary Approach*; Oxford University Press: Oxford, UK, 1972.
98. Anderson, P.W. More Is Different. *Science* **1972**, *177*, 393–396. [CrossRef]
99. Battermnan, R.W. *The Devil in the Details: Asymptotic Reasoning in Explanation, Reduction, and Emergence*; Oxford University Press: Oxford, UK, 2001.
100. Jannes, G. Some comments on "The Mathematical Universe". *Found. Phys.* **2009**, *39*, 397–406. [CrossRef]
101. Weinberg, S. *Dreams of a Final Theory: The Scientist's Search for the Ultimate Laws of Nature*; Vintage Book: New York, NY, USA, 1992.
102. Carroll, S.M. The Cosmological Constant. *Living Rev. Relativ.* **2001**, *4*, 1. [CrossRef] [PubMed]
103. Dieks, D. Bottom-Up versus Top-Down: The Plurality of Explanation and Understanding in Physics. In *Scientific Understanding: Philosophical Perspectives*; de Regt, H.W., Leonelli, S., Eigner, K., Eds.; University of Pittsburgh Press: Pittsburgh, PA, USA, 2009.
104. Mattingly, D. Modern Tests of Lorentz Invariance. *Living Rev. Relat.* **2005**, *8*, 5. [CrossRef] [PubMed]
105. Liberati, S. Tests of Lorentz invariance: A 2013 update. *Class. Quantum Grav.* **2013**, *30*, 133001. [CrossRef]
106. Visser, M. *Lorentzian Wormholes: From Einstein to Hawking*; American Institute of Physics Press: Woodbury, NY, USA, 1995.
107. Morris, M.; Thorne, K. Wormholes in space-time and their use for interstellar travel: A tool for teaching general relativity. *Am. J. Phys.* **1988**, *56*, 395. [CrossRef]
108. Morris, M.; Thorne; K.; Yurtsever, U. Wormholes, time energy condition. *Phys. Rev. Lett.* **1988**, *61*, 1446. [CrossRef] [PubMed]
109. González-Díaz, P.F. Observable effects from space-time tunneling. *Phys. Rev. D* **1997**, *56*, 6293. [CrossRef]
110. Torres, D.F.; Romero, G.E.; Anchordoqui, L.A. Might some gamma-ray bursts be an observable signature of natural wormholes?. *Phys. Rev. D* **1998**, *58*, 123001. [CrossRef]
111. Safonova, M.; Torres, D.F.; Romero, G.E. Macrolensing signatures of large scale violations of the weak energy condition. *Mod. Phys. Lett. A* **2001**, *16*, 153. [CrossRef]
112. Cramer, J.G.; Forward, R.L.; Morris, M.S.; Visser, M.; Benford, G.; Landis, G.A. Natural wormholes as gravitational lenses. *Phys. Rev. D* **1995**, *51*, 3117. [CrossRef]
113. Eiroa, E.; Romero, G.E.; Torres, D.F. Chromaticity effects in microlensing by wormholes. *Mod. Phys. Lett. A* **2001**, *16*, 973. [CrossRef]
114. Shatskiy, A. Passage of photons through wormholes and the influence of rotation on the amount of phantom matter around them. *Astron. Rep.* **2007**, *51*, 81. [CrossRef]
115. González-Díaz, P.F.; Alonso-Serrano, A. Observing other universe through ringholes and Klein-bottle holes. *Phys. Rev. D* **2011**, *84* 023008. [CrossRef]
116. Alonso-Serrano, A.; Garay, L.J.; Marugán, G.A.M. Correlations across horizons in quantum cosmology. *Phys. Rev. D* **2014**, *90*, 124074. [CrossRef]
117. Carroll, S. Beyond Falsifiability: Normal Science in a Multiverse. In *Epistemology of Fundamental Physics: Why Trust a Theory?*; Dardashti, R., Dawid, R., Thébault, K., Eds.; Cambridge University Press: Cambridge, UK, 2019.

© 2019 by the authors. Licensee MDPI, Basel, Switzerland. This article is an open access article distributed under the terms and conditions of the Creative Commons Attribution (CC BY) license (http://creativecommons.org/licenses/by/4.0/).

Review

Anthropic Selection of Physical Constants, Quantum Entanglement, and the Multiverse Falsifiability

Mariusz P. Dąbrowski [1,2,3]

1. Institute of Physics, University of Szczecin, Wielkopolska 15, 70-451 Szczecin, Poland; Mariusz.Dabrowski@usz.edu.pl or mpdabfz@wmf.univ.szczecin.pl
2. National Centre for Nuclear Research, Andrzeja Sołtana 7, 05-400 Otwock, Poland
3. Copernicus Center for Interdisciplinary Studies, Szczepańska 1/5, 31-011 Kraków, Poland

Received: 6 June 2019; Accepted: 11 July 2019; Published: 14 July 2019

Abstract: This paper evaluates some important aspects of the multiverse concept. Firstly, the most realistic opportunity for it which is the spacetime variability of the physical constants and may deliver worlds with different physics, hopefully fulfilling the conditions of the anthropic principles. Then, more esoteric versions of the multiverse being the realisation of some abstract mathematics or even logic (cf. paper by M. Heller in this volume). Finally, it evaluates the big challenge of getting any signal from "other universes" using recent achievements of the quantum theory.

Keywords: varying constants; anthropic principle; multiverse levels; multiverse entanglement; multiverse tests

1. Introduction

1.1. Scientific Method

According to the philosopher Karl Popper [1] "the scientific method assumes the existence of a theory which is described by mathematical notions which, in order to be the scientific theory, must fulfil the criterium of falsifiability i.e., it should contain a predictive result of an experiment or an explanation of the phenomenon allowing to conclude if such a theory is wrong". Remarks related to such a definition are as follows: (1) a scientific theory does not necessarily have to be in agreement with the experiment; (2) alternative theories can also be falsifiable though they either do not apply in our reality or, for example, it is not possible to perform any proposed experiment to falsify them on the current level of the human development. An experimental method is a scientific method which allows a quantitative investigation of scientific theories by continuously repeating a certain process or a phenomenon (i.e., one can actively modify this phenomenon). This method is applied commonly in fundamental sciences such as physics, chemistry, and biology. In astronomy and cosmology (and apparently also in economics) one applies the observational method which does not allow any possibility of changing such a phenomenon (for example, a supernova explosion). In fact, within the scientific community, the observational method is treated on the same footing as the experimental method. It is worth saying that nowadays one commonly accepts Einstein's view that "the only criterion of validity of a theory is an experiment".

The reason for mentioning the above is the fact that the rest of the article will be devoted to cosmology, which is often not considered to be strongly supported by local experiments and further to the multiverse, which is considered even worse in that respect.

The question about the multiverse also touches the question about the boundaries of our knowledge of the universe and about the extrapolation of our known physics into the distant and the unexplored regions of space and time.

1.2. Cosmology as an Experimental Science

The scientific method relies on the theories which are described by some specific laws expressed in terms of mathematics—the physical laws. These laws, however, are verified in *local experiments* i.e., the Earth experiments or Earth's neighbourhood experiments. In fact, we *do not know* if these laws also apply in the distant parts of the universe, but usually assume that they do so. In other words, we *extrapolate* the local laws into the whole observable universe. Such an approach legitimises validity of the most observational facts related to cosmology such as the universe expansion [2], its hot and dense phase in the past [3,4], its current acceleration [5,6], etc. In that sense cosmology is a science and so all its aspects which are based on the scientific method—including the multiverse concept—should seriously be taken into account.

One interesting issue is that cosmology deals with a unique object—the universe—and it is considered as all which surrounds us. There is a question as to whether cosmology deals with all the possible mathematical structures and whether these structures are physical reality somewhere in the universe, which is often called *the multiverse*. Another point is whether our physical theories and the views given by physics as the fundamental science can easily be extended onto such fragile phenomena as life (biology) and consciousness (psychology).

A more recent view of a theory to be scientific was presented by a cosmologist George Ellis [7,8] who strongly differentiates between cosmology as the theory which is based on contemporary achievements of physics and mathematics and verified by observations and "cosmologia" which adds more aspects to the investigations which are related to philosophy, social sciences, biology, and even metaphysics. One of his criterion of a scientific theory is the observational and experimental support which composes firstly of ability to make a quantitative prediction to be tested and secondly of its confirmation. His worry is if some theories which predict the multiverse are really scientific in the observational and experimental sense. It seems that this concern is not so much a problem in view of the Popper's criterion.

1.3. Physical Laws and Constants

Physical laws are verified by measurements of the physical quantities entering these laws including the physical constants which basically seem to be "constants" in the pure sense of their merit. *However, one asks a fundamental question: why are the laws of physics of the form they are, and why are physical constants of the values they are?* One may ask: why is Newton's force of gravity inversely proportional to the second power of the distance between the masses? Why not to the third power or perhaps to some other fractional power? Surely, it is not forbidden to have any other power in any way as a mathematical law, but not as it relates to the physical, since it does not explain our universe. We may also ask why the interaction between electric charges and masses lowers with the distance, while it grows with distance for quarks endowed with colour charges. Could gravitational force be also growing with the distance? The mathematical answer is simply yes, but not in any kind of (at least local) physical universe.

However, even if the laws have similar mathematical structure, they may still give different quantitative output due to the different values of the physical constants. The Newton's law and the Coulomb's law have the same mathematical structure, but the constants they contain: the gravitational constant G and the electric constant k, are many orders of magnitude different. This very fact leads to important consequences for living organisms since this is the electric force which guarantees their integrity—gravity is too weak to do so. However, we may imagine a different world or a period of evolution of our world in which constants G and k are of similar value. Then, one would perhaps create a "gravitationally bound" life rather than an "electrically bound" one, though not in our current universe.

1.4. What Is the Multiverse?

Assuming that our universe is equipped with some specific set of physical constants, we immediately come into a philosophical problem of potential existence of the whole set of universes (or perhaps pieces of our universe) equipped with different sets of physical constants and/or physical laws—*the multiverse*.

The very term "Multiverse" comes from the works of a philosopher, William James [9], in 1895 in which he defines *Visible nature is all plasticity and indifference, a multiverse, as one may call it, and not a universe*. One of the problems with the above formulation, which we will discuss in more detail later, is whether those universes evolve independently or whether there is some physical relation between them. The latter would be the only option which could allow us to test their existence by our own universe's experiments.

Sticking to the scientific method of Popper [1] we may also ask the question of initial conditions—i.e., ask if there was any freedom of the choice of physical constants and laws initially, and why they have been chosen in a way they are now. In other words, we may investigate the problem of how different our world would have been, if the laws and constants had had different values from what they are now. This further can be extended into the existence and the type of life problems, i.e., asking if there are any universes in which physical laws and physical constants do not allow life or even better if they do not allow *our type* of life allowing or not some *different* type of life (provided we know what this "different" type of life is).

The content of this paper is as follows. In Section 2 we briefly discuss an idea of varying constants then formulate appropriate theories in Section 3. In Section 4 we discuss the relation between varying constants theories and the anthropic principles. In Section 5 we concentrate on the definitions of the multiverse and the multiverse hypothesis falsifiability. In Section 6 we give some afterword.

2. Some History and Remarks on Varying Constants

Physical laws do not exist without physical constants which at the first glance seem to be quite numerous, though after a deeper insight only a few of them seem really fundamental. According to the famous discussion between three eminent physicists, Duff, Okun, and Veneziano [10] based on the famous Bronshtein-Zelmanov-Okun cube [11] (cf. Figure 1), at most three of the constants are necessary: the gravitational constant G, the velocity of light c, and the Planck constant h. Clearly, speed of light is for relativity, gravitational constant for gravity, and Planck constant for quantum mechanics. As it is argued by string theorists, one needs even less: the speed of light c and the fundamental string length λ_s [10].

These three constants can be considered to be of the so-called class C [12,13] because they build bridges between quantities and allow new concepts to emerge: c connects space and time together; h relates the concept of energy and frequency; while G appears in Einstein equations and creates links between matter and geometry.

The gravitational constant, measured by Cavendish in 1798 [14], is historically the oldest known one and occurs in Newton's law of gravitation and its generalisation, the Einstein equations of general relativity.

Already the 19th century physicists started thinking of a basic set of physical units ("natural" units) of which all the other physical units could be derived. Johnstone-Stoney [15] introduced the "natural" units of charge, mass, length, and time as the "electrine" unit $e = 10^{-20}$ ampere-seconds (ε_0—permittivity of space in the Coulomb's law), and the mass, length, time respectively

$$M_J = \sqrt{\frac{e^2}{4\pi\varepsilon_0 G}} = 10^{-7}\,\text{g}, \tag{1}$$

$$L_J = \sqrt{\frac{Ge^2}{4\pi\varepsilon_0 c^4}} = 10^{-37}\,\text{m}, \tag{2}$$

$$t_J = \sqrt{\frac{Ge^2}{4\pi\varepsilon_0 c^6}} = 3 \times 10^{-46}\,\text{s}. \tag{3}$$

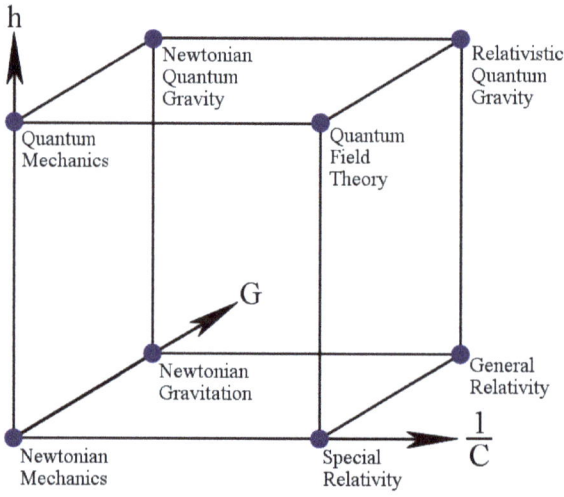

Figure 1. Bronshtein-Zelmanov-Okun cube, or the cube of the physical theories [11]. Three orthonormal axes are marked by $1/c$, h and G. The vertex $(0,0,0)$ corresponds to non-relativistic mechanics; $(c,0,,0)$ to special relativity; $(0,h,0)$ to non relativistic quantum mechanics; $(c,,h,0)$ to quantum field theory; $(c,0,G)$ to general relativity, (c,h,G) to relativistic quantum gravity.

This was later modified by Planck in 1899, following the discovery of the electron charge by Thompson in 1897 as $e = 1.6 \times 10^{-19}$ Coulombs, the introduction of the Planck constant $h = 6.6 \times 10^{-34}$ J·s, and the Boltzmann constant $k = 1.38 \times 10^{-23}$ J/K into the Planck mass, length, time, temperature, respectively:

$$M_{pl} = \sqrt{\frac{hc}{G}} = 5.56 \times 10^{-5}\,\text{g}, \tag{4}$$

$$L_{pl} = \sqrt{\frac{Gh}{c^3}} = 4.13 \times 10^{-35}\,\text{m}, \tag{5}$$

$$t_{pl} = \sqrt{\frac{Gh}{c^5}} = 1.38 \times 10^{-43}\,\text{s}, \tag{6}$$

$$T_{pl} = \sqrt{\frac{hc^5}{k^2 G}} = 3.5 \times 10^{32}\,\text{K}. \tag{7}$$

Nowadays, we use the International System of Units (SI) which contains seven units for the basic physical quantities (meter, kilogram, second, ampere, kelvin, mole, candela) out of which all the other units are supposed to be derived [16].

At the beginning of the 20th century Weyl [17] found that the ratio of the electron radius to its gravitational radius was about 10^{40}. Further Eddington [18] found the proton-to-electron mass ratio to be

$$\frac{1}{\beta} = \frac{m_p}{m_e} \sim 1840 \tag{8}$$

and further took an inverse of the fine structure constant to be

$$\frac{1}{\alpha} = 4\pi\varepsilon_0 \frac{\hbar c}{e^2} \sim 137, \tag{9}$$

then found the ratio of electromagnetic to gravitational force between a proton and an electron as

$$\frac{e^2}{4\pi\epsilon_0 m_e m_p} \sim 10^{40}, \tag{10}$$

which led him to the introduction of the then so-called Eddington number $N_E = 10^{80}$.

Eddington's ideas were put into a deeper physical context in the Dirac's Large Numbers Hypothesis [19], who besides calculating static ratio of electromagnetic and gravitational forces between proton and electron, also took into account the dynamics of cosmology calculating the ratio of the observable universe to the classical radius of the electron

$$\frac{c/H_0}{e^2/4\pi\epsilon_0 m_e c^2} \sim 10^{40}, \tag{11}$$

and the number of protons in the observable universe

$$N = \frac{4}{3} \frac{\pi \rho (c/H_0)^3}{m_p} \sim N_E \sim (10^{40})^2, \tag{12}$$

where the critical density $\rho = (3H_0^2)/(8\pi G) \sim 10^{-29} gcm^{-3}$, H_0 is the Hubble parameter at present. However, due to the evolution of the universe, the Hubble parameter changes with time $H(t) \equiv [da(t)/dt]/a(t)$, where $a(t)$ is the scale factor which describes the expansion rate. So, in order to keep ratios (11) and (12) the same as the ratio of the electromagnetic to gravitational force throughout the whole evolution, some of the physical constants involved e, G, c, m_e, m_p, must vary in time. Since the changes of e, m_e, m_p would need reformulation of the atomic and the nuclear physics which is difficult to observe, the only option is that this is gravitational "constant" G which changes. Dirac's choice was just that

$$G \sim H(t) = \frac{(da/dt)}{a}, \tag{13}$$

so that the scale factor evolved as $a(t) \sim t^{1/3}$, and $G(t) \sim 1/t$. This gave a simple answer to the problem of gravity being so weak compared to electromagnetism because the ratio of the forces was proportional to the age of the universe i.e.,

$$\frac{F_e}{F_p} \sim \frac{e^2}{m_e m_p} t \sim t. \tag{14}$$

3. Theories Incorporating Varying Constants

3.1. Formulations

3.1.1. Varying Gravitational Constant G

The first physical theory which started from the proper Lagrangian was formulated by Jordan [20] in his conformal gravity and further extended in Brans-Dicke scalar-tensor gravity [21]. Brans and

Dicke assumed that the gravitational constant G should be associated with an average gravitational potential represented by a scalar field surrounding a given particle, i.e.,

$$<\phi> = \frac{GM}{c/H_0} \sim 1/G, \qquad (15)$$

which is responsible for the gravitational force. In fact, this was an attempt to incorporate the idea of Ernst Mach, who first suggested that the inertial mass of a given particle in the Universe is a consequence of its interaction with the rest of the mass of the Universe.

In general relativity the gravitational constant is *a constant*, and so the Einstein equations read as ($g_{\mu\nu}$—metric tensor, $R_{\mu\nu}$—Ricci tensor, R—curvature scalar, $T_{\mu\nu}$—energy-momentum tensor, $\mu, \nu = 0, 1, 2, 3$)

$$R_{\mu\nu} - \frac{1}{2}g_{\mu\nu}R = \frac{8\pi G}{c^4}T_{\mu\nu}, \qquad (16)$$

while in Brans-Dicke scalar-tensor gravity one uses the time-varying field ϕ instead of G which results in generalised gravitational (Einstein) equations and the equation of motion for this scalar field:

$$R_{\mu\nu} - \frac{1}{2}g_{\mu\nu}R = \frac{8\pi}{c^4\phi}T_{\mu\nu} + \frac{1}{\phi}\left(\nabla_\mu\nabla_\nu\phi - g_{\mu\nu}\Box\phi\right) + \frac{\omega}{\phi}\left(\partial_\mu\phi\partial_\nu\phi - \frac{1}{2}g_{\mu\nu}\partial_\beta\phi\partial^\beta\phi\right) \qquad (17)$$

$$\Box\phi = \frac{8\pi}{c^4(3+2\omega)}T. \qquad (18)$$

where ω is the Brans-Dicke parameter, ∂_μ - partial derivative, ∇_μ - covariant derivative, and $\Box \equiv \nabla_\mu\nabla^\mu$—the d'Alambert operator. Einstein's general relativity theory is recovered in the limit $\omega \to \infty$. This idea of making a constant to be the field is used in theories of unification of fundamental interactions such as the superstring theory, where the variability of the intensity of the interactions is a rule. There, we have the dimensionless "running" coupling constants of electromagnetic (α), weak (α_w), strong (α_s), and gravitational (α_g) interactions. An example of such a unification is the low-energy-effective superstring theory which is equivalent to Brans-Dicke theory for $\omega = -1$. In such a theory the running coupling constant $g_s = \exp(\varphi/2)$ evolves in time with φ being the dilaton and the Brans-Dicke field $\phi = \exp(-\varphi)$ [22]. Besides, Jordan's conformally invariant theory can be obtained from Brans-Dicke action in the case when $\omega = -3/2$.

In this superstring theory context some interesting examples of multiverses being considered as universes cyclic in time appear. One of them is the pre-big-bang model in which the universe evolution is split into the two phases in time: one "before" big-bang taking place at the moment $t = 0$ and another "after" big-bang [22]. Another is the ekpyrotic and cyclic model based on an extension of the superstring theory onto the M-Theory (brane theory) [23]. In such models, the big-bang is repeated many times as a collision of two lower-dimensional branes (treated as gravitating thin films) in a higher-dimensional spacetime.

3.1.2. Varying Fine Structure Constant α or the Electric Charge e

In recent years, two other theories with varying constants have been intensively studied [24]. First is the varying fine structure constant [25] and the second is the varying speed of light (VSL) [26]. Phenomenologically, these constants are related by the definition of $\alpha \sim 1/c$. However, α is dimensionless while c (as well as the electric charge e and the Planck constant $\hbar$) is dimensionful so that there is an equivalent formulation of varying α theory in terms of varying e theory in which a change in the fine structure constant α was fully identified with a variation of the constant electric charge, e_0 developed as Bekenstein–Sandvik–Barrow–Magueijo (BSBM) model [27].

Let us first examine the varying fine structure constant α models. Such models were first proposed by Teller [28], and later by Gamow [29], following the original path of the Large Number Hypothesis by Dirac [30]. A fully quantitative framework was developed by Bekenstein [31]. The electric charge variability was introduced by defining a dimensionless scalar field, $\epsilon(x^\mu)$, and as a consequence, e_0 was replaced by $e = e_0 \epsilon(x^\mu)$. The electromagnetic tensor was then redefined to the form

$$F_{\mu\nu} = [(\epsilon A_\nu)'_\mu - (\epsilon A_\mu)'_\nu]/\epsilon,$$

where the standard form of it can be restored for the constant ϵ. For simplicity, in [27] an auxiliary gauge potential, $a_\mu = \epsilon A_\mu$, and the electromagnetic field strength tensor, $f_{\mu\nu} = \epsilon F_{\mu\nu}$, were introduced, as well as a variable change: $\epsilon \to \psi \equiv \ln \epsilon$ was performed. The field ψ in this model couples only to the electromagnetic energy, disturbing neither the strong, nor the electroweak charges, nor the particle masses.

The BSBM field equations are

$$R_{\mu\nu} - \frac{1}{2}g_{\mu\nu}R = \frac{8\pi G}{c^4}\left[\Omega \partial_\nu \psi \partial^\nu \psi - \frac{1}{2}\Omega g_{\mu\nu}\left(\partial_\beta \psi \partial^\beta \psi\right)^2 - \left(\frac{1}{4}g_{\mu\nu}f_{\alpha\beta}f^{\alpha\beta} - f_{\sigma\nu}f_\mu^\beta\right)e^{-2\psi}\right], \qquad (19)$$

and the equation of motion for the field ψ is:

$$\Box \psi = \frac{2}{\Omega}e^{2\psi}\mathcal{L}_{em},\qquad (20)$$

where $\Omega = \hbar c/\lambda$, is introduced for the dimensional reason, λ is considered to be the length scale of the electromagnetic part of the theory, and the dimensionless field ψ is given by:

$$\psi = \frac{1}{2}\ln\left|\frac{\alpha}{\alpha_0}\right|. \qquad (21)$$

3.1.3. Varying Speed of Light c

As for varying c the best known (though not derived from the proper action) is the Barrow-Magueijo (BM) [32] model which follows the formulation of Petit [33] and basically makes c the function of time in the standard Einstein general relativistic field equations. A better formulation is by Moffat [34,35] where $\Phi = c^4$ and it is the proper physical model with c being the field with an appropriate kinetic term, i.e.,

$$R_{\mu\nu} - \frac{1}{2}g_{\mu\nu}R = \frac{8\pi G}{\Phi}T_{\mu\nu} + \frac{1}{\Phi}\left(\nabla_\mu \nabla_\nu - g_{\mu\nu}\nabla^\alpha \nabla_\alpha\right)\Phi + \frac{\kappa}{\Phi^2}\left(\partial^\mu \Phi \partial_\mu \Phi - \frac{1}{2}g_{\mu\nu}\partial^\alpha \Phi \partial_\alpha \Phi\right), \qquad (22)$$

and $\kappa = $ const. (it is different from Brans-Dicke parameter ω, though it is introduced analogously into the kinetic term)

$$\Box \Phi = \frac{8\pi G}{3 + 2\kappa}T. \qquad (23)$$

Some other varying c theories are the bimetric theory with different speeds of photons and gravitons [36] and the theories with modified dispersion relation [37]. Other models being considered, though not so strictly based on theoretical formulation, are the varying electron-to-proton mass $\beta = m_e/m_p$ [38].

3.2. Bounds on Variability of Fundamental Constants

3.2.1. Varying G

The strongest limit comes from the lunar laser ranging (LLR), i.e., from the observations of the light reflected by mirrors on the Moon left there by Apollo 11, 14, and 15 missions [39] and reads

$$\left|\frac{\dot{G}}{G}\right| < (4 \pm 9) \times 10^{-13} \text{year}^{-1}. \qquad (24)$$

On the other hand, by using the orbital period of binary pulsars $PSRB1913 + 16$ and $PSRB1855 + 05$ [40], the limits are

$$\frac{\dot{G}}{G} = (4 \pm 5) \times 10^{-12} \text{year}^{-1}, \qquad (25)$$

and

$$\frac{\dot{G}}{G} = (-9 \pm 18) \times 10^{-12} \text{year}^{-1}. \qquad (26)$$

3.2.2. Varying α

The most spectacular bound comes from the Samarium 149 capture process in the natural nuclear reactor Oklo in Gabon which took place about 2 billion years ago [41–43]. The process can be driven by different interactions and the appropriate bounds in the electromagnetic case are

$$\frac{\dot{\alpha}}{\alpha} = 3.85 \pm 5.65 \times 10^{-18} \text{year}^{-1}, \qquad (27)$$

and

$$\frac{\dot{\alpha}}{\alpha} = (-0.65 \pm 1.75) \times 10^{-18} \text{year}^{-1}. \qquad (28)$$

Another method is studying absorption lines of distant quasars. Among the most updated constraint, from a sample of 23 absorption systems along the lines of sight towards 18 quasars, we have [44]

$$\frac{\dot{\alpha}}{\alpha} = (0.22 \pm 0.23) \times 10^{-15} \text{year}^{-1}, \qquad (29)$$

in the redshift range $0.4 \leq z \leq 2.3$. The strongest constraint from a single quasar was given in [45] and reads

$$\frac{\dot{\alpha}}{\alpha} = (-0.07 \pm 0.84) \times 10^{-17} \text{year}^{-1} \qquad (30)$$

at $z = 1.15$.

3.2.3. Varying c

In fact, speed of light has been declared a constant by Bureau International Poids et Mesures (BIPM) and officially has the value $c = 299,792.458$ km/s [46] and its measurement is so far accurate up to 10^{-9} [47]. However, we cannot tell Nature what is the value of c so still the measurement of it makes sense. For example, from the definition of the fine structure constant and based on the measurements of variability of α one finds the bound:

$$\frac{\Delta c}{c} = \frac{\Delta \alpha}{\alpha} \sim 10^{-5}. \qquad (31)$$

Measurement of c in particular is important on the cosmological distance scale since it may also be a sign of an alternative scenario for the cosmological inflation. In fact, recently some new methods of cosmological measurement of c have been introduced [48,49] by using Baryonic Acoustic Oscillations (BAO) to the formula

$$D_A(z_m) H(z_m) = c(z_m), \qquad (32)$$

where D_A is the angular diameter distance of an object in the sky, H—the Hubble parameter, z_m—redshift at maximum of angular diameter distance D_A. The latter has a peculiar property of being small for closer objects and then growing for farther objects (i.e., starts growing with redshift), reaching a maximum and then finally decreasing. When measuring D_A and H (the cosmic "ruler" and the cosmic "clock") for some sample of objects at z_m, one can find the value of c from (32). The method was actually applied to 613 ultra-compact radio quasars and has given the value $c = 3.039 \pm 0.180 \times 10^8$ m/s [50] which is within the figure fixed by BIPM.

3.3. Varying Constants as the Path to the Multiverse

The bounds presented above can be interpreted twilights—as the bounds or as the "evidence" for the variability of constants of nature. Taking the second position, one can say at least that the constants must vary very slowly to fulfil these current observational bounds. This statement is especially suitable for the whole universe evolution, since even a very slow change of a constant, while multiplied by the age of the universe, may give quite a significant result. Finally, the variability of physical constants at least makes us aware of some other possibilities for the physical constants and physical laws being different in some other universes forming the multiverse, which is the main objective of this paper.

Some issues related to the varying constants problem should also be mentioned. Above, we have considered the theories in which just one of the set of the fundamental constants was varying. However, the constants are mutually tightened by various relations so that varying one of them may lead to a change of another. A simple example is the relation (9) which defines the fine structure constant α which involves two (or even three) other fundamental constants and each of them may vary if α is supposed to vary. Furthermore, there are theories which involve more than one constant changing. These are varying both G and c models [51–53] and the justification for such theories is obvious—these constants show up together in the Einstein-Hilbert action for gravity and in the Einstein field equations. There are also models with varying both G and α [54]. The combination of varying both α and c models would exactly refer to the constraint given by the definition of the fine structure constant (9) and so would be hardly distinguishable, though not excluded.

In general, one may think of some "trajectory" of the universe in the "phase space" (in Ref. [55] it is called a state space S) of the physical constants and ask how such a trajectory would be related to the mutual changes of these constants. One may perhaps end up in the situation of some "structural stability" of these trajectories. Of course not all of them would be stable in such a space, probably giving tight constraints on a possibility that any of the fundamental constants can vary.

Another problem is whether the varying constants theories and so the multiverse are in conflict with the principles of general covariance (independence of the physical laws of the coordinate transformations) and with the principle of manifest covariance (tensorial nature of the physical laws) [56]. For our set of constants which were investigated in Section 3: (G, α, c) or (G, e, c), we should consider the problem of their possible invariance (covariance) with respect to the local point transformations (LPT) of the form [57]

$$r = r'(r), \tag{33}$$

where r and r' are arbitrary sets of coordinates. In the standard case of general relativity the constants (G, α, c) or (G, e, c) have the same values *everywhere* on spacetime manifold and so they can be considered constant 4-scalars. However, once they start to vary with respect to the spacetime positions, they obviously are different in different 4-positions and so cannot be considered constant scalars globally on the manifold. However, as it can be seen from the presentations of the varying constants theories of Section 3, now the constants (G, α, c) become (scalar) fields $\phi(x), \psi(x), \Phi(x)$, and as such they of course change their values with spacetime positions. At each point of the spacetime these values should not depend on any particular choice of coordinates, so if x^μ is an old system of coordinates and x'^μ is a new one, then according to a general definition of a scalar field, ϕ' in new coordinates should have the same value as in the old ones, i.e.,

$$\phi'(x'^\mu(x^\nu)) = \phi(x^\mu). \tag{34}$$

A problem appears when one considers some varying-c theories since in these theories c also enters the definition of a time coordinate $x^0 = c(x^\mu)t$ and in such cases some special frame (coordinates)—the light frame—has to be chosen [58]. This is definitely a preferred frame which contradicts both general and manifest covariance which require the physical quantities are represented by invariant under general coordinate transformations objects such as tensors[1]. Besides, these theories lead to the Lorentz symmetry violation and as we know Lorentz symmetry is the pillar of contemporary physics.

However, even for the varying speed of light case, one can construct the theory which still allows general covariance [60], by formulating it in terms of the x^0 coordinate with the dimension of length rather than time (not appealing to its definition as $x^0 = c(x^\mu)t$. This allows the metric, Ricci and other tensors to transform as tensors, as it should be. Besides, the local Lorentz symmetry is also preserved since the theory uses the local value of c which is constant at any particular point of spacetime.

In the Moffat's varying speed of light theory [34], Lorentz symmetry is broken spontaneously and the full theory possesses exact Lorentz invariance while the vacuum fails to exhibit it.

Despite all that, it is commonly believed nowadays that Lorentz symmetry is violated in quantum gravity regime [61]. Then, since we are talking about such an esoteric notion as the multiverse, it does not seem that it may be formulated without any appeal to quantum gravity, so one may not necessarily expect that the covariance and Lorentz invariance will have to be the property of the multiverse. It also might be that the multiverse which would not take any quantum effects into account would perhaps be still manifestly covariant and Lorentz invariant, but the multiverse in its fully quantum picture (see Section 5) would not necessarily be so.

4. Varying Constants and Anthropic Principles

4.1. Coincidences

Let us now ask the fundamental question about the position of a human in the Universe. The question is both of a physical and of a philosophical nature. Starting from the observation of masses and sizes of physical objects in our Universe, one notices that they are not arbitrary—it is rather that the mass is proportional to the size and so both quantities can be placed in linear dependence. In other words, the space of values of masses and sizes is not filled in randomly and only the structures which obey roughly the linear dependence can exist [62]. This linear law shows some kind of coincidence which allows living organisms to evolve because they are subject to the same fundamental interactions (gravitational, electromagnetic, nuclear strong, and nuclear weak with appropriate dimensionless coupling constants α, α_g, α_s, and α_w). Bearing in mind physics, we know the reason—they exist due to stable equilibria between these fundamental interactions. For example, common objects of our every day life (a table, a spoon) exist due to a balance between an attractive force between the protons and electrons and a repulsive force (pressure) of degenerated electrons.

Actually, there are numerous facts both related to every day physics and to the universe's physics which can be called coincidences. Let us enumerate some of them. Practical life shows us the benefit of the fact that water shrinks at 0–4 degrees Celsius which allows fish to survive the winter. The Earth is accompanied by a comparable mass natural satellite—the Moon—which prevents the Earth from wobbling chaotically. This does not happen for Mars, which has only tiny moons (presumably captured asteroids) and its chaotic change of obliquity can be as large as 45 degrees [63], which can dramatically influence the climate and so prevents a possibility for the life to survive. The next example is the enormous variety of chemicals being the result of the fact that the electrons are light enough compared to the nuclei and so atoms do not form any kind of "binary" systems, allowing chemical bonds to develop. In fact, the ratio of the size of the nucleus and the size of an atom is about

$$\alpha\beta = (1/137)(1/1836) \ll 1. \tag{35}$$

---

[1] A famous example of a quantity which is frame-dependent in general relativity is the gravitational energy represented by a pseudotensor [59].

Another issue is the influence of varying constants on chemical bonds. There are basically three types of inter-atomic bonds: ionic, covalent, and metallic [62]. The ionic bonds are characterised by the strong electric interaction between the positive and negative ions of two different atoms which either take or donate electrons. Classic examples are the sodium chloride $NaCl$, the sodium fluoride NaF, and the magnesium oxide MgO. In fact, the Coulomb electric force which makes the bond is different for each of these molecules because of difference in the charge and also in the distance between the ions. This results in different melting temperatures of these molecules in particular or in general in different physical characteristics of them. In our case of varying constants the strength of Coulomb force would also be varying with respect to a possible change of electric charge e or alternatively the fine structure constant α and so would effect the chemical bonds resulting in different physical characteristics of the molecules destroying some coincidences such as the anomalous properties of water at 0–4 degrees. If one appealed to the relation (9) defining the fine structure constant α one could perhaps also find a possibility that the varying speed of light c would equally be influencing the ionic Coulomb force, so changing the chemical structures and physical properties of various molecules.

The covalent bonds are also due to the electric force though they are characteristic for the same atoms exchanging the electrons. The metallic bonds are due to positive ions of metal interacting with the free electron gas moving between these ions. Since they both are of the electric nature, they would be sensitive to a change of the values of e, α, and c in the similar way as the ionic bonds.

The influence of varying G into the chemical bonds seems to be negligible since they are not gravitational in nature. However, once considering the bonds in gravitational field, their strength may be important in view of the macroscopic gravitational influence on larger structures in order to prevent them from being fractured. One should also mention possible influence of the quantum interactions on the molecular bonds which may also be the result of the change of the classical interactions.

Fine-tuned is also the nucleosynthesis (formation of atomic nuclei) in the early Universe which only takes place in a fixed period of time after big-bang ($0.04s < t < 500s$) and is governed by the fine structure constant α and the ratio β leading to the condition

$$\alpha > \beta, \tag{36}$$

which is really the case in our universe. It is interesting to note that the nucleosynthesis would not be possible, if an electron was replaced by its heavier version—the muon.

On a more general level, one realises a very narrow range of physical parameters admissible in our universe—something we can acknowledge as the fine-tuning to fit the conditions for life to exist. Fine-tuned are the values of the fundamental constants α and α_s (cf. Figure 2) [64].

In fact, a slight change of α would prevent the possibility of existence of life in the Universe. We can see that the "inhabitable" zone for life is set respectively to $\alpha \approx 1/137$ and $\alpha_s \approx 0.1$. These values of α and α_s are fine-tuned for our existence, which is indicated by the white cross in the red striped region in the graph. In the orange region, deuteron is unstable and the main nuclear reaction in the star cannot proceed. For $\alpha_s \lesssim 0.3\alpha^{1/2}$ (dark blue region) carbon and higher elements are unstable. On the other hand, unless $\alpha \ll 1$ the electrons in atoms and molecules are unstable to pair creation (top-left part of green region). The requirement that the typical energy of chemical reactions is much smaller than the typical energy of nuclear reactions excludes bottom-right part of the green region. Besides, the light-blue region is excluded because there proton and diproton are not stable, affecting stellar burning and big bang nucleosynthesis. Finally, the electromagnetism is weaker than gravity in the region to the very left.

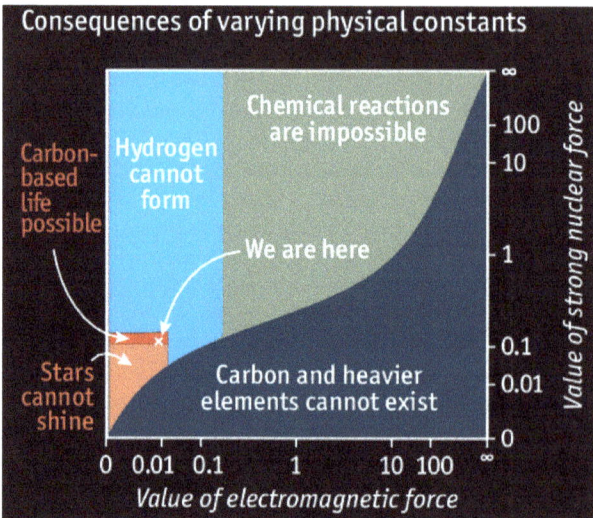

Figure 2. Fine-tuning of the parameters α and α_s. Life-permitting parameter zone of these constants is marked in red [65].

Last but not least, let us mention that the values of the fundamental interactions coupling constants α, α_g determine various "physical" conditions for life [62]. For example, if gravitational energy on the surface of a planet is smaller than the energy required for fracture—any animal, including human, can exist—the condition involves the values of both electromagnetic and gravitational coupling constants α and α_g.

Such kind of argumentation reflected by the above mentioned coincidences leads physicists to create the notion of the anthropic principles, which expressed the fact that possibly the Universe is extremely "fine-tuned" to host humans!

4.2. Anthropic Principles (AP)

What are the anthropic principles? Opponents say that they are just tautologies or some self-explanatory statements with no practical meaning. Some physicists take them into account seriously. Others consider trivial statements with no meaning for the contemporary scientific method of physics. We take a position that they should be explored as kind of "boundary" options for the evolution of the physical universe. In fact, these principles were first presented by Brandon Carter [66,67] on the occasion of 500th anniversary of the birth of Copernicus during the IAU meeting in Kraków, Poland in 1974 and further developed by Barrow and Tipler in their book about the topic [62].

A fundamental problem raised by AP refers to the question about the reason that out of many possible ways of the evolution of the Universe just one specific way was chosen—that one which led to formation of galaxies, stars, planetary systems, and finally both the unconscious and the conscious life. Of course, while using the term "many possible ways" we immediately touch the problem of "other" worlds (at least hypothetically) potentially and in that sense we refer to the notion of the multiverse. Let us then discuss some formulations of the anthropic principles in order to insight their relation to the multiverse concept.

4.2.1. Weak Anthropic Principle (WAP)

The statement is as follows [62]: "the observed values of all physical and cosmological quantities are not equally probable, but that they take on values restricted by the requirement that there exist sites where carbon-based life can evolve and by the requirement that the Universe be old enough for it to have already done so". In other words, life may have evolved in the Universe. In terms of statistics, WAP is usually expressed by the application of the famous Bayes theorem [68].

The Bayes theorem is based on the notion of the conditional probability $P(B \mid A)$, i.e., the probability of proposition B assuming that proposition A is true. The conditional probability fulfils the multiplication rule $P(A \wedge B) = P(A)P(B \mid A)$ which due to symmetry leads to the formulation of the theorem

$$P(A \mid B) = \frac{P(B \mid A)P(A)}{P(B)}, \tag{37}$$

where $P(A \mid B)$ is the probability of proposition A assuming that proposition B is true. In Bayesian inference one considers set of models $\{M_i\}$, $i = 1, 2, \ldots, n$ and using (37) one constructs the so-called posterior probability in view of some data set D [69]:

$$P(M_i \mid D) = \frac{P(D \mid M_i)P(M_i)}{P(D)}. \tag{38}$$

The best model based on set D is the one with the largest value of posterior probability. Before applying the data set D the so-called prior probability for each model M_i is usually assumed to be equal, i.e., that $P(M_i) = 1/n$. The probability $P(D \mid M_i)$ is called the evidence of model likelihood E_i.

The relative plausibility of any model M_i compared to some base model M_0, also called the posterior odds is

$$O_{i0} = \frac{P(M_i \mid D)}{P(M_0 \mid D)} = \frac{P(D \mid M_i)P(M_i)}{P(D \mid M_0)P(M_0)}. \tag{39}$$

In view of the equality of prior probabilities $P(M_i) = P(M_0)$, one gets the so-called Bayes factor

$$B_{i0} = \frac{P(D \mid M_i)}{P(D \mid M_0)} = \frac{E_i}{E_0}, \tag{40}$$

which for $B_{i0} = 1$ says that both models are equally good in view of the data, while for $B_{i0} < 1$ model M_0 is preferred.

An example of the application of the Bayes theorem is when one considers the large size of the Universe as related to the origin of life on Earth [70]. If M_1 is the model which says that the large size of the Universe is superfluous for life, then $P(M_1) = P(M_1 \mid D) \ll 1$, while for M_0 saying that the large size is necessary for life to appear, then $P(M_0) \ll 1$, while $P(M_0 \mid D) \approx 1$, so $B_{i0} < 1$ and it prefers model M_0.

4.2.2. Strong Anthropic Principle (SAP)

It says that "the Universe must have the properties which allow life to develop within it at some stage of its history" [62]. In other words (or more physically) it says that the constants of Nature (e.g., gravitational constant) and laws of Nature (e.g., Newton's law of gravity) must be such that life has to appear. It is interesting to look into various interpretations of the SAP because some of them are quite extreme.

- *Interpretation A* says that "there exists only one possible Universe designed with the goal of generating and sustaining observers" [62]. In fact, it is very teleological, and this is why it sometimes is also called "An Intelligent Project" interpretation.
- *Interpretation B* is very radical and says that "observers are necessary to bring the Universe into being" [62]. Such a statement is based on the philosophical ideas of Berkeley and developed by John Archibald Wheeler as the *Participatory Anthropic Principle* [62].
- *Interpretation C* says that "the whole ensemble of other and different universes is necessary for the existence of our Universe" [62]. The best-known, though most controversial version of this

interpretation is the many-worlds theory of Everett [71,72] having recently some strong support from superstring theory [73].

4.2.3. Final Anthropic Principle (FAP)

It says that "the intelligent information-processing must come into existence in the Universe, and, once it comes into existence, it will never die out" with an alternative statement that "no moral values of any sort can exist in a lifeless cosmology" [62]. As it is easily noticed, it has some broader than physical meaning.

4.2.4. Minimalistic Anthropic Principle (MAP)

This is the most reserved of anthropic principles saying that "ignoring selection effects while testing fundamental theories of physics using observational data may lead to incorrect conclusions" [67].

4.2.5. Going Beyond

The most challenging for the Author is the interpretation C, since it has some strong support from the most advanced and avant-garde theories of contemporary physics such as the many-world interpretation [71] and the superstring theory [73]. The former says that each time one makes a measurement of a physical observable in the quantum world in one of the universes, one also has infinitely many other universes which are equally real and instantaneous in which the result of the measurement is different. In other words, we have a completely different world history in all the other universes (e.g., in one of them, one is married, and in another one, one is single)—called parallel universes.

As for the latter concept, there is the so-called superstring landscape [74] (nowadays even extended into the so-called swampland [75]) which allows to generate

$$(10^{100})^5 = (1 googol)^5 = 10^{500} \tag{41}$$

different vacua (being the properties of individual universes) which determine different sets of physical laws governing the evolution. It is worth noticing that this number is much larger then even the Eddington number $N_E = 10^{80}$.

The number 10^{500} comes because in superstring theory there are many ways of the symmetry breaking and the choices of quantum mechanical vacua. Since the basic space-time of the superstring theory is 10-dimensional, then a 9-dimensional space (plus time) is compactified into a 3-dimensional world in a couple of hundreds (500) ways (called topological cycles). There are about 10 fluxes which can wrap on these topological cycles giving 10^{500} options [74,76] (see also the discussion of M. Douglas in this volume [77]).

It is worth emphasizing that the Interpretation C somehow moves back SAP to WAP because in some rough interpretation both suggest that our Universe is one of many other options and the other options do not necessarily show any (specific to our Universe) coincidences and fine-tuning.

5. The Multiverse and Its Testability

Once we raised the concept of the multiverse as a collection of the universes with one of them inhabited by us—humans—we need to have a discussion on whether there is any practical way to falsify the existence of the "other universes". Before that let us first say what can be meant by the multiverse as an entity—something we can call the hierarchy of multiverses.

5.1. Multiverse Hierarchy

The many-worlds interpretation of quantum mechanics by Everett [71] is very philosophical and it was contested strongly by founders of quantum mechanics (especially those who believed in statistical interpretation and in Copenhagen interpretation). However, as we have mentioned already, the contemporary superstring theory seems to give firm framework for considering different scenarios of evolution of the fundamental sets of constants of nature and the laws of nature (or "different worlds") fitting perhaps an idea of Everett—the multiverse. Such an idea has, among others, its own practical realisation in cosmology throughout the idea of eternal inflation—each emerging vacuum bubble has its own vacuum with different laws of nature [78].

Max Tegmark [64,65] has differentiated four levels of the multiverse which are characterised as follows:

- *Level I* obeys the pieces of our universe which are outside the cosmic horizon (behind our reach due to finite speed of information transmission by the speed of light), it presumably allows the same laws of physics, though different initial conditions.
- *Level II* are the other bubbles (universes) created during the process of eternal cosmic inflation which allows the same laws of physics but the different values of physical constants and different dimensionality of space. Our universe (set of constants) is likely as one of the many options.
- *Level III* is what is essentially the many-wolds of quantum physics proposed by Everett. It is the same as the level II but with some quantum curiosities such as superpositions of "alive" and "dead" cats, etc. The main issue is that there is no collapse of the total wave function of the Universe so that the decoherence (appearance of a classical world) happens only for a branching piece of the whole multiverse.
- *Level IV* is very extreme since it contains "any mathematical structure which is realised somewhere in the multiverse and it is fully materialisable". Within this multiverse, one can have different physical constants, different laws of physics, dimensionality, etc. Since on this level we can make an equality between "mathematical existence" and "physical existence", then we can answer the famous Wheeler and Hawking question: "why these equations (laws of physics), and not others?

For the Author, the most fascinating is the level IV since the universe which is understood as "all which exists" is quite natural to accept, though here by "all" one understands "all which can be thought of in terms of the mathematics" and not necessarily real in our every day life "common sense reasoning" which refers to something which exists as a physical object. However, the meaning of the term "all", after a deeper though about it, can easily be extended into not only "all" which exists in a material world, but also "something" which can exist as an idea a human can think of, and so something which just exists "hypothetically" (like fairy tail giants much larger than dinosaurs being in contradiction to our physical laws applied on Earth [62]).

An interesting and perhaps even further going idea which perhaps rises the level IV onto the "level V" is due to Heller [79] who considers the universes with different logic (reasoning comes from the category theory) rather than just different mathematics.

Despite level IV seeming esoteric, for physicists still probably the most difficult to accept conceptually is the level III which is the many-world interpretation of Everett. This is because Tegmark classifies level III as a generalisation of the level I, so all the phenomena related to branching of the (never collapsing) wave function of the universe happen at the same place in space, which sounds really obscure.

5.2. Multiverse and Our Vision of Life

We suggest a new anthropic hierarchy of the universes in the multiverse, i.e., the hierarchy which can be defined in the context of having observers (or life) in those universes. We can start with the biological and psychological (related to the consciousness) meaning of life defining first the universes which can be inhabitable (I) in the sense of our life (OUL) or in the sense of other life (OTL) as in Table 1. Then, we define the universes which are uninhabitable (UI) both in the sense of OUL or of OTL.

Table 1. Hierarchy of multiverses in view of our definition of life.

	Inhabitable	Uninhabitable
our life	I, OUL	UI, OUL
other life	I, OTL	UI, OTL

If one sticks to the physical merit, one can differentiate the universes which possess our set of physical laws (OUS) and those which possess other set of physical laws (OTS) as in Table 2.

Table 2. Hierarchy of multiverses in view of the hierarchy of physical laws.

	Inhabitable	Uninhabitable
our set of laws	I, OUS	UI, OUS
other set of laws	I, OTS	UI, OTS

5.3. Falsifying the Multiverse

There is a question as to which level of the multiverse can really be falsifiable in the Popper's sense. If one believes in inflationary picture [80,81] and the generation of the quantum fluctuations during this epoch, one can surely appeal to the problem of quantum entanglement [82] which together with the above mentioned problem of variability of constants seems to allow to test the levels I and II. As for the level III it is perhaps also possible to study the entanglement within the branching of the total wave function of the universe and its decoherence within individual branches. However, the level IV (and perhaps "level V" [79]) does not seem to be testable, though one would attempt to formulate an idea of some "signals" of different mathematics or even logics (what?) in the multiverse, but they seem to be much behind the reach of "observable quantities" in the regular sense of physics though still can perhaps be falsifiable in the sense of Popper.

From the practical point of view of contemporary physics we may define the multiverse as: (1) The set of pieces of our universe each of them having different physical laws; (2) The set of completely independent entities each of them having different physical laws. While the possibility (1) could possibly be testable, the possibility (2) does not seem to be easily testable.

In that context one would think of some interesting research challenges:

Challenge I: Inventing an alternative to ours scenario of the evolution of the universe which would be consistent and would allow for life though not necessarily of OUL.

Challenge II: Constructing a consistent scenario of the evolution of the universe which would not allow for life of OUL, i.e., being OTL. This seems to be much harder because we simply do not know what OTL is.

By constructing an alternative scenario we mean the construction of all important physical processes which presumably took place in the universe such as: big-bang, inflation, nucleosynthesis, galaxy formation, etc. (an interesting example of such universes can be the ones which do not exhibit weak interactions [83]).

5.4. How Many Universes?

There is an issue of how large the sets are (or ensembles [55]) of universes and whether these sets are finite or infinite and if the latter is the case, then whether they are at least countable perhaps in the sense of cardinal numbers or uncountable [84]? The question is also what would be the measure of the set of universes like OUL in the space of all universes in the multiverse? These issues have already been discussed by Ellis and collaborators [55]. First of all, appealing to the views of Hilbert [85], the really existing infinite set of anything—including the universes—is just impossible. This is because infinity is not an actual number one can ever specify or reach and in fact it only replaces our statement that "something continues without end". Even a prove of infinity of Euclidean geometry is an untestable concept and so its infiniteness is likely not to be realised in practice. Based on this argument one can say [55] that bearing in mind the properties of infinity, even if the number of really existing models are infinite (or just finite), they form a set of measure zero in the set of all possible universes, which makes a big problem.

Contrary to this infinitude, Paul Davies [86] considers only four (and each finite) numbers which are unique among all the possible numbers of universes in the multiverse. These are: $0, 1, 2$, or 10^{500}. Zero is distinguished, but not experimentally verified, 1 is obviously our every day experience, and 10^{500} is just the number which comes from superstring landscape considerations. However, the number 2 needs more attention, since this is exactly the number of universes being created if one considers the process of universe creation in analogy to a quantum field theoretical process of pair creation. In such a case, the universes can be created as a single universe—antiuniverse pair or 5×10^{499} pairs within the superstring framework [87].

5.5. Falsifying due to Quantum Entanglement

Quantum entanglement is an intensively studied problem in contemporary physics which is related to quantum information, quantum cryptography, quantum algorithms, atoms and particles quantum physics [82] and can also be applied to cosmology on the same footing as applying classical general relativity to cosmology or quantum mechanics to quantum cosmology. In our cosmological context, for the universes which are not fully independent (i.e., quantum mechanically entangled), the underlying idea is to investigate the signal from quantum entanglement of the classically disconnected (causally) pieces of space.

Quantum entanglement effect of the multiverse can be observed due to an appropriate term of quantum interaction in any universe of the multiverse i.e., also in our universe. Practical realisation is by an extra term in the basic cosmological (and classical) Friedmann equation

$$H^2(t) = \frac{8\pi G}{3c^4}\rho(t) + \text{"quantum entanglement"}, \qquad (42)$$

where the H is the Hubble parameter, ρ is the mass density, and there is an extra term coming from quantum entanglement to this classical equation. If such a term is non-zero, then the entanglement signal is imprinted in the spectrum of the cosmic microwave background (CMB) in the form of an extra dipole which is a cause of dark matter flow [88]. Besides, the entanglement influences the potential of a scalar field which drives cosmological inflation and so it induces a change of the CMB temperature [89,90].

In fact, the entanglement can weaken the power spectrum of density perturbations in the universe in large angular scales [91]. It can also lead to a change of the spectral index of density perturbations and so can influence galaxy formation process. It may also create an extra "entanglement temperature" which is added to the CMB temperature [92]. The whole quantum field theoretical approach relies on the option that there is a creation of the pairs of universes in an analogical way as creating the pairs of particles in the second quantisation description (here it is called third quantisation [92]).

It is worth mentioning that the idea of testing the multiverse due to quantum entanglement above relies on the assumption of validity of the quantum laws throughout the whole multiverse (or "meta-laws" which all universes in the multiverse have in common [55]). Otherwise, we would not be able to say what physical phenomenon we could measure, in this particular case—the inter-universal quantum entanglement in the multiverse—a phenomenon known to hold in our universe, too. However, in general we can imagine the situation of the multiverse which allows some different physical phenomena which may or may not be present in our universe which could allow us to test the concept. In the former case we would have to define them from the phenomena we know from our universe physics, while for the latter case they could probably only be showing up in the cosmological equations like (42) as some extra terms which cannot be identified (and not overlap) with anything else we know. In other words, there exists a series of multiverses which still are falsifiable and scientific, but they must be falsified by different methods than quantum entanglement.

5.6. Redundant or Necessary?

One may ask the question: do we really need the vast amount of universes in the form of the multiverse? In order to answer this question let us notice that the classical cosmology (based on the Einstein field equations) selects only one solution (our universe) out of an infinite number of solutions (aleph-one number of solutions equal to the set of real numbers). On the other hand, quantum cosmology needs all the classical solutions to be present in the quantum solution which is the wave function of the universe in order to get the probability of creating one universe. Then, in classical cosmology, one needs some initial conditions as a physical law to resolve the problem of choosing "this" solution (our universe) and not "that" solution (other universe), while in quantum cosmology all the initial points (classical solutions) are present in the quantum solution, and there is no need for any initial conditions to be introduced as an extra law [62]. In that sense, in quantum approach we enlarge ontology (all universes instead of one), but reduce the number of physical laws (no need for initial conditions). It seems to be quite reasonable to do so, and what is more, this enlargement is an analogue of the spatial enlargement of the universe by the Copernican system, which was once criticised by the opponents of Copernicus based on the Occam's razor (something which has proven to be wrong after a couple of centuries). In the case of the multiverse, we enlarge whatever the size of our universe is by adding extra universes which possess different physical laws and which are real, though they may seem to be virtual once we look at them classically as one of the realisations of our hypothetical opportunities reflected in mathematical formulation of some field equations with a number of different solutions.

6. Afterword

It is very hard to say what "the multiverse" actually is. One can easily imagine the notion of the "multi-world" being for example understood as "multi-Earth" in view of the fact that nowadays we know that there exist extrasolar planets—some of them perhaps habitable. If one sticks to "the world" being a solar system, then of course there are many "multi-world' solar systems. Further, we may define hierarchically "multi-galaxy", "multi-cluster", etc., reaching the Hubble horizon limited observable universe. All this does not seem to be abstract and is easily reachable by the current observational data. However, the real concept of the multiverse is something more than that. It could be the entity still based on our known physics, but equally it could go beyond that possibility reaching the level of our virtual imagination presented in terms of mathematics or even logic. Furthermore, these abstract options are the strongest challenge for our observational reach since they require the tools which allow the falsifiability at least understood as in the definition of the scientific method by Popper. Without such tools most of our discussion about the multiverse will remain quite impractical, though very attractive intellectually and philosophically.

Funding: This research received no external funding.

Acknowledgments: I am indebted to John D. Barrow for providing me with the reference to original works of William James which rooted the term "multiverse". I also acknowledge the discussions with Paul C.W. Davies, Michael Heller, Claus Kiefer, Joao Magueijo, John Moffat, Salvador Robles-Perez, Vincenzo Salzano, and John Webb. I should also give some credit to Hussain Gohar who has elaborated his own versions of the figures I used.

Conflicts of Interest: The author declares no conflict of interest.

References

1. Popper, K. *The Logic of Scientific Discovery*; Routledge-Taylor & Francis: Abingdon, UK, 2002.
2. Hubble, E. A Relation Between Distance and Radial Velocity Among Extra-Galactic Nebulae. *Proc. Natl. Acad. Sci. USA* **1929**, *15*, 168. [CrossRef] [PubMed]
3. Penzias, A.A.; Wilson, R.W. A Measurement of Excess Antenna Temperature at 4080 Mc/s. *Astrophys. J. Lett.* **1965**, *142*, 419–421. [CrossRef]

4. Ade, P.A.; Aghanim, N.; Arnaud, M.; Ashdown, M.; Aumont, J.; Baccigalupi, C. Planck 2015 results-xiii. cosmological parameters. *Astron. Astrophys.* **2016**, *594*, A13.
5. Spergel, D.N.; Bean, R.; Doré; O; Nolta, M.R.; Bennett, C.L.; Dunkley, J. Three-year Wilkinson Microwave Anisotropy Probe (WMAP) observations: Implications for cosmology. *Astrophys. J. Suppl. Ser.* **2007**, *170*, 377. [CrossRef]
6. Ade, P.A.R.; Akiba, Y.; Anthony, A.E.; Arnold, K.; Atlas, M.; Barron, D. Measurement of the cosmic microwave background polarization lensing power spectrum with the POLARBEAR experiment. *Phys. Rev. Lett.* **2014**, *113*, 021301. [CrossRef]
7. Ellis, G.F.R. On the philosophy of cosmology. *Studi. Philos. Mod. Phys.* **2014**, *46*, 5–23. [CrossRef]
8. Ellis, G.F.R. The Domain of Cosmology and the Testing of Cosmological Theories. In *The Philosophy of Cosmology*; Chamcham, K., Silk, J., Barrow, J.D., Saunders, S., Eds.; Cambridge University Press: Cambridge, UK, 2017; pp. 3–39.
9. James, W. Is Life Worth Living? *Int. J. Ethics* **1895**, *6*, 1–24. [CrossRef]
10. Duff, M.J.; Okun, L.B.; Veneziano, G. Trialogue on the number of fundamental constants. *J. High Energy Phys.* **2002**, *2002*, 023. [CrossRef]
11. Okun, L.B. The fundamental constants of physics. *Sov. Phys. Usp.* **1991**, *34*, 818. [CrossRef]
12. Levy-Leblond, J.M. The Importance of Being (a) Constant. In Proceedings of the Problems in the Foundations of Physics, Varenna, Italy, 25 July–6 August 1977.
13. Uzan, J.-P. Varying constants, gravitation and cosmology. *Living Rev. Relat.* **2011**, *14*, 2. [CrossRef]
14. Cavendish, H. XXI. Experiments to determine the density of the earth. *Philos. Trans. R. Soc. Lond.* **1798**, *88*, 469–526.
15. Johnstone-Stoney, G. LII. On the physical units of nature. *Philos. Mag.* **1881**, *5*, 381–390. [CrossRef]
16. Uzan, J.-P. The fundamental constants and their variation: Observational and theoretical status. *Rev. Mod. Phys.* **2003**, *75*, 403. [CrossRef]
17. Weyl, H. Eine neue erweiterung der relativitaetstheorie. *Annalen der Physik* **1919**, *59*, 101–133. [CrossRef]
18. Eddington, A.S. *New Pathways in Science*; Cambridge University Press: Cambridge, UK, 1934.
19. Dirac, P.A.M. A new basis for cosmology. *Proc. R Soc. A* **1938**, *165*, 199–208. [CrossRef]
20. Jordan, P. The present state of Dirac's cosmological hypothesis. *Z. Phys.* **1959**, *157*, 112–121. [CrossRef]
21. Brans, C.; Dicke, R.H. Mach's principle and a relativistic theory of gravitation. *Phys. Rev.* **1961**, *124*, 925.
22. Gasperini, M.; Veneziano, G. The pre-big bang scenario in string cosmology. *Phys. Rep.* **2003**, *373*, 1–212. [CrossRef]
23. Khoury, J.; Steinhardt, P.; Turok, N. Inflation versus cyclic predictions for spectral tilt. *Phys. Rev. Lett.* **2003**, *91*, 161301. [CrossRef]
24. Barrow, J.D. *The Constants of Nature*; Vintage Books: London, UK, 2002.
25. Barrow, J.D. Varying alpha. *Annalen der Physik* **2010**, *19*, 202–210. [CrossRef]
26. Albrecht, A.; Magueijo, J. Time varying speed of light as a solution to cosmological puzzles. *Phys. Rev. D* **1999**, *59*, 043516. [CrossRef]
27. Sandvik, J.H.B.; Barrow, J.D.; Magueijo, J. A Simple Cosmology with a Varying Fine Structure Constant. *Phys. Rev. Lett.* **2002**, *88*, 031302. [CrossRef] [PubMed]
28. Teller, E. On the change of physical constants. *Phys. Rev.* **1948**, *73*, 801. [CrossRef]
29. Gamow, G. Electricity, gravity, and cosmology. *Phys. Rev. Lett.* **1967**, *19*, 759. [CrossRef]
30. Dirac, P.A.M. The cosmological constants. *Nature* **1937**, *139*, 323. [CrossRef]
31. Bekenstein, J.D. Fine-structure constant: Is it really a constant? *Phys. Rev. D* **1982**, *25*, 1527. [CrossRef]
32. Barrow, J.D.; Magueijo, J. Solutions to the quasi-flatness and quasi-lambda Problems. *Phys. Lett. B* **1999**, *447*, 246–250. [CrossRef]
33. Petit, J.-P. Cosmological model with variable light velocity: The interpretation of red shifts. *Mod. Phys. Lett. A* **1988**, *3*, 1733–1744. [CrossRef]
34. Moffat, J.W. Quantum gravity, the origin of time and time's arrow. *Found. Phys.* **1993**, *23*, 411–437. [CrossRef]
35. Moffat, J.W. Variable speed of light cosmology, primordial fluctuations and gravitational waves. *Eur. Phys. J. C* **2016**, *76*, 130. [CrossRef]
36. Clayton, M.A.; Moffat, J.W. A scalar-tensor cosmological model with dynamical light velocity. *Phys. Lett. B* **2001**, *506*, 177–186. [CrossRef]
37. Magueijo, J.; Smolin, L. Gravity's rainbow. *Class. Quantum Gravity* **2004**, *21*, 1725. [CrossRef]

38. Ivanchik, A.; Rodriguez, E.; Petitjean, P.; Yarshalovich, D. Do the fundamental constants vary in the course of cosmological evolution? *Astron. Lett.* **2002**, *28*, 423–427. [CrossRef]
39. Moeller, J.; Biskupek, L. Variations of the gravitational constant from lunar laser ranging data. *Class. Quantum Gravity* **2007**, *24*, 4533. [CrossRef]
40. Kaspi, V.M.; Taylor, J.H.; Riba, M.F. High-precision timing of millisecond pulsars. 3: Long-term monitoring of PSRs B1855+ 09 and B1937+ 21. *Astrophys. J.* **1994**, *428*, 713–728. [CrossRef]
41. Shlyakhter, A.I. Direct test of the constancy of fundamental nuclear constants. *Nature* **1976**, *264*, 340. [CrossRef]
42. Petrov, Y.V.; Nazarov, A.I.; Onegin, M.S.; Petrov, V.Y.; Sakhnovsky, E.G. Natural nuclear reactor at Oklo and variation of fundamental constants: Computation of neutronics of a fresh core. *Phys. Rev. C* **2006**, *74*, 064610. [CrossRef]
43. Fujii, Y.; Iwamoto, A.; Fukahori, T.; Ohnuki, T.; Nakagawa, M.; Hidaka, H.; Oura, Y; Moeller, P. The nuclear interaction at Oklo 2 billion years ago. *Nucl. Phys. B* **2000**, *573*, 377–401. [CrossRef]
44. Wilczynska, M.R.; Webb, J.K.; King, J.A.; Murphy, M.T.; Bainbridge, M.B.; Flambaum, V.V. A new analysis of fine-structure constant measurements and modelling errors from quasar absorption lines. *Mon. Not. R. Astron. Soc.* **2015**, *454*, 3082–3093. [CrossRef]
45. Levshakov, S.A.; Centurion, M.; Molaro, P. Most precise single redshift bound to. *Astron. Astrophys.* **2006**, *449*, 879–889. [CrossRef]
46. Resolution 2 of the 15th CGPM (1975). Available online: http://www.bipm.org/en/CGPM/db/15/2/ (accessed on 13 July 2019).
47. Evenson, K.M.; Wells, J.S.; Petersen, F.R.; Danielson, B.L.; Day, G.W.; Barger, R.L.; Hall, J.L. Speed of light from direct frequency and wavelength measurements of the methane-stabilized laser. *Phys. Rev. Lett.* **1972**, *29*, 1346. [CrossRef]
48. Salzano, V.; Dąbrowski, M.P.; Lazkoz, R. Measuring the speed of light with Baryon Acoustic Oscillations. *Phys. Rev. Lett.* **2015**, *114*, 101304. [CrossRef] [PubMed]
49. Salzano, V.; Dąbrowski, M.P. Statistical hierarchy of varying speed of light cosmologies. *Astrophys. J.* **2017**, *851*, 97. [CrossRef]
50. Cao, S.; Biesiada, M.; Jackson, J.; Zheng, X.; Zhao, Y.; Zhu, Z.-H. A new test of f(R) gravity with the cosmological standard rulers in radio quasars. *J. Cosmol. Astropart. Phys.* **2017**, *2017*, 012. [CrossRef]
51. Barrow, J.D. Cosmologies with varying light speed. *Phys. Rev. D* **1999**, *59*, 043515. [CrossRef]
52. Barrow, J.D.; Magueijo, J. Solving the flatness and quasi-flatness problems in Brans-Dicke cosmologies with a varying light speed. *Class. Quantum Gravity* **1999**, *16*, 1435. [CrossRef]
53. Balcerzak, A. Non-minimally coupled varying constants quantum cosmologies. *J. Cosmol. Astropart. Phys.* **2015**, *4*, 019. [CrossRef]
54. Barrow, J.D.; Magueijo, J.; Sandvik, H.B. A cosmological tale of two varying constants. *Phys. Lett. B* **2002**, *541*, 201–210. [CrossRef]
55. Ellis, G.F.R.; Kirchner, U.; Stoeger, W.R. Multiverses and physical cosmology. *Mon. Not. R. Astron. Soc.* **2004**, *347*, 921–936. [CrossRef]
56. Schutz, B.F. *A First Course in General Relativity*; Cambridge University Press: Cambridge, UK, 2009.
57. Tessarotto, M.; Cremaschini, C. Theory of non-local point transformations-Part 1: Representation of Teleparallel Gravity. *arXiv* **2016**, arXiv:1601.03940.
58. Magueijo, J. New varying speed of light theories. *Rep. Prog. Phys.* **2003**, *66*, 2025. [CrossRef]
59. Landau, L.D.; Lifschitz, E.M. *The Classical Theory of Fields*; Addison-Wesley: Boston, MA, USA, 1957.
60. Magueijo, J. Covariant and locally lorentz-invariant varying speed of light theories. *Phys. Rev. D* **2000**, *62*, 103521. [CrossRef]
61. Mattingly, D. Modern tests of Lorentz invariance. *Living Rev. Relat.* **2005**, *8*, 5. [CrossRef] [PubMed]
62. Barrow, J.D.; Tipler, F.J. *The Anthropic Cosmological Principle*; Oxford University Press: Oxford, UK, 1986.
63. Laskar, J.; Robutel, P. The chaotic obliquity of the planets. *Nature* **1993**, *361*, 608. [CrossRef]
64. Tegmark, M. Parallel universes. *Sci. Am.* **2003**, *288*, 40–51. [CrossRef] [PubMed]
65. Tegmark, M. *Science and Ultimate Reality*; Barrow, J.D., Davies, P.C.W., Harper, C.L., Eds.; Cambridge University Press: Cambridge, UK, 2004.

66. Carter, B. Large Number Coincidences and the Anthropic Principle in Cosmology. In *Confrontation of Cosmological Theories with Observation, (IAU Symposium No 63)*; Longair, M., Ed.; Reidel: Dordrecht, The Netherlands, 1974; pp. 291–298
67. Carter, B. The anthropic Principle and its Implication for Biological Evolution. *Philos. Trans. R. Soc. Lond. A* **1983**, *310*, 347. [CrossRef]
68. Lee, P. *Bayesian Statistics: An Introduction*; Wiley: Hoboken, NJ, USA, 2012.
69. Sahlén, B. On Probability and Cosmology: Inference Beyond Data. In *The Philosophy of Cosmology*; Chamcham, K., Silk, J., Barrow, J.D., Saunders, S., Eds.; Cambridge University Press: Cambridge, UK, 2017; pp. 429–446.
70. Rees, M.J. Cosmological significance of e^2/Gm^2 and related "large numbers". *Comments Astrophys. Space Phys.* **1972**, *4*, 179.
71. Everett, H. Relative state formulation of quantum mechanics. *Rev. Mod. Phys.* **1957**, *29*, 454. [CrossRef]
72. DeWitt, B.; Graham, R.N. (eds.) *The Many Worlds Interpretation of Quantum Mechanics*; Princeton University Press: Princeton, NJ, USA, 1973.
73. Polchinski, J. *String Theory*; Cambridge University Press: Cambridge, UK, 1998.
74. Douglas, M.R. The Statistics of string/M-theory vacua. *J. High Energy Phys.* **2003**, *2003*, 046. [CrossRef]
75. Vafa, C. The string landscape and the swampland. *arXiv* **2005**, arXiv:hep-th/0509212.
76. Susskind, L. The anthropic landscape of string theory. *arXiv* **2003**, arXiv:hep-th/0302219.
77. Douglas, M.R. The string theory landscape. *Universe* **2019**, submitted.
78. Pogosian, L.; Vilenkin, A.; Tegmark, M. Anthropic Predictions for Vacuum Energy and Neutrino Masses. *J. Cosmol. Astropart. Phys.* **2004**, *2004*, 005. [CrossRef]
79. Heller, M. Multiverse – too much or not enough? *Universe* **2019**, *5*, 113. [CrossRef]
80. Guth, A.H. Inflationary universe. A possible solution to the horizon and flatness problem. *Phys. Rev. D* **1981**, *23*, 347. [CrossRef]
81. Linde, A.D. A new inflationary universe scenario: A possible solution of the horizon, flatness, homogeneity, isotropy and primordial monopole problems. *Phys. Lett. B* **1982**, *108*, 389. [CrossRef]
82. Horodecki, R.; Horodecki, P.; Horodecki, M.; Horodecki, K. Quantum entanglement. *Rev. Mod. Phys.* **2009**, *81*, 865. [CrossRef]
83. Harnik, R.; Kribs, G.D.; Perez, G. A universe without weak interactions. *Phys. Rev. D* **2006**, *74*, 035006. [CrossRef]
84. Gauthier, Y. A general no-cloning theorem for an infinite multiverse. *Rep. Math. Phys.* **2013**, *72*, 191–199. [CrossRef]
85. Hilbert, D. *On the Infinite. Philosophy of Mathematics*; Benacerraf, P., Putnam, H., Eds.; Prentice Hall: Englewood Cliffs, NJ, USA, 1964; pp. 134–151.
86. Davies, P.C.W. (Beyond Center, Arizona State University, Tempe, AZ, USA). Personal communication, 2019.
87. Robles-Perez, S. Time reversal symmetry in cosmology and the creation of a universe-antiuniverse pair. *Universe* **2019**, *5*, 150. [CrossRef]
88. Mersini-Houghton, L.; Holman, R. 'Tilting' the Universe with the Landscape Multiverse: The 'Dark' Flow. *J. Cosmol. Astropart. Phys.* **2009**, *2009*, 006. [CrossRef]
89. di Valentino, E.; Mersini-Houghton, L. Testing Predictions of the Quantum Landscape Multiverse 1: The Starobinsky Inflationary Potential. *J. Cosmol. Astropart. Phys.* **2017**, *2017*, 002. [CrossRef]
90. di Valentino, E.; Mersini-Houghton, L. Testing predictions of the quantum landscape multiverse 2: The exponential inflationary potential. *J. Cosmol. Astropart. Phys.* **2017**, *2017*, 020. [CrossRef]
91. Kinney, W.H. Limits on Entanglement Effects in the String Landscape from Planck and BICEP/Keck Data. *J. Cosmol. Astropart. Phys.* **2016**, *2016*, 013. [CrossRef]
92. Robles-Perez, S.; Balcerzak, A.; Dąbrowski, M.P.; Krämer, M. Interuniversal entanglement in a cyclic multiverse. *Phys. Rev. D* **2017**, *95*, 083505. [CrossRef]

© 2019 by the authors. Licensee MDPI, Basel, Switzerland. This article is an open access article distributed under the terms and conditions of the Creative Commons Attribution (CC BY) license (http://creativecommons.org/licenses/by/4.0/).

Review
The String Theory Landscape

Michael R. Douglas

Simons Center for Geometry and Physics, Stony Brook University, Stony Brook, NY 11794, USA; mdouglas@scgp.stonybrook.edu

Received: 11 June 2019; Accepted: 18 July 2019; Published: 20 July 2019

Abstract: String/M theory is formulated in 10 and 11 space-time dimensions; in order to describe our universe, we must postulate that six or seven of the spatial dimensions form a small compact manifold. In 1985, Candelas et al. showed that by taking the extra dimensions to be a Calabi–Yau manifold, one could obtain the grand unified theories which had previously been postulated as extensions of the Standard Model of particle physics. Over the years since, many more such compactifications were found. In the early 2000s, progress in nonperturbative string theory enabled computing the approximate effective potential for many compactifications, and it was found that they have metastable local minima with small cosmological constant. Thus, string/M theory appears to have many vacuum configurations which could describe our universe. By combining results on these vacua with a measure factor derived using the theory of eternal inflation, one gets a theoretical framework which realizes earlier ideas about the multiverse, including the anthropic solution to the cosmological constant problem. We review these arguments and some of the criticisms, with their implications for the prediction of low energy supersymmetry and hidden matter sectors, as well as recent work on a variation on eternal inflation theory motivated by computational complexity considerations.

Keywords: string theory; quantum cosmology; string landscape

---

1. Introduction

Superstring theory and M theory are quantum theories of matter, gravity and gauge forces, in which the fundamental degrees of freedom are not particles but extended objects: one-dimensional strings and higher dimensional branes. This solves the problem of the nonrenormalizability of quantum gravity, at the cost of requiring the dimension of space-time to be ten (for superstrings) or eleven (for M theory). Nevertheless, by following the Kaluza–Klein approach of taking the extra dimensions to be a small compact manifold, one can argue that the resulting four dimensional theory can reproduce the Standard Model at low energies. This was first done by Candelas et al. in 1985 [1] and, ever since, superstring theory has been considered a leading candidate for a fundamental theory describing all physics in our universe.

In the years since, not only were the original arguments developed and sharpened, the most attractive competing candidate theories were shown to be equivalent to other regimes of string theory, obtained by taking the string coupling large or by taking the size of the extra dimensions to be sub-Planckian. In particular, eleven-dimensional supergravity, arguably the most symmetric extension of general relativity, turned out to be the strong coupling limit of type IIa superstring theory [2]. Conversely, 11d supergravity contains a membrane solution, and one can obtain string theory from it by compactifying on a sub-Planckian circle, so that a membrane wound around the circle becomes a string. This larger picture containing both string theory and 11d supergravity is sometimes called M theory; we will call it string/M theory to emphasize that all of the superstring theories and 11d supergravity are contained within this single framework. At present, it is the only theoretical framework that has been convincingly shown to quantize gravity in more than three space-time dimensions.

In this brief review, we explain how string/M theory realizes the concept of a multiverse. The primary argument is to look at the construction of quasi-realistic four-dimensional compactifications (by which we mean those which are similar to the Standard Model but not necessarily agreeing with it in all detail) and enumerate the choices which enter this construction. This includes the choice of topology and geometry of the compactification manifold, the choice of auxiliary physical elements such as branes and generalized magnetic fluxes and how they are placed in the compact dimensions, and the choice of metastable minimum of the resulting effective potential. One can roughly estimate the number of choices at each step, and argue that they combine to produce a combinatorially large number of metastable vacua. These arguments are still in their early days and there is as yet no consensus on the number; estimates range from 10^{500} [3] which at the time it was made seemed large, to the recent $10^{272,000}$ [4].

Any of these compactifications are a priori candidates to describe the observed universe. Having chosen one, the next step in analyzing it is to compute or at least estimate the effective potential. This is a function of the scalar fields or "moduli" which parameterize the Ricci flat metric and other fields in the extra dimensions, including the overall volume of the extra dimensions, the string coupling constant (or "dilaton") and typically hundreds or even thousands of additional fields. As in nongravitational physics, the effective potential has both classical contributions (for example, see Equation (5) below) and quantum contributions (Casimir energies, instanton effects, etc.), and must be computed by approximate methods. One then looks for its metastable minima and analyzes the small fluctuations around it, to get the four-dimensional particle spectrum and interactions. To be clear, the definition of "vacuum" in this review is a metastable minimum of the effective potential.[1] This is to be distinguished from "universe," "pocket universe," or "bubble," terms which denote a causally connected region of the multiverse in which the compactification takes a particular size and shape, and which thus sits in a single vacuum. Many universes in a multiverse could sit in the same vacuum, and this is why cosmology will predict a nontrivial probability distribution over vacua.

The effective potential of a string/M theory compactification, while mathematically just a single real-valued function, is a very complex object that summarizes a vast range of possible physical structures of the vacua and phase transitions between them. The set of effective potentials for all the compactifications is clearly far more complex. While computing them is a tall order, the rich mathematical structure of string/M theory compactification has led to amazing progress in this endeavor. While this rapidly gets very technical, it is here that we see how important it is that solutions of string/M theory are mathematically natural and—yes, beautiful—constructs. Although this beauty is subjective and cannot be regarded as an argument for or against their relevance to nature, it is what allows us to compute their properties and get objective information we can use to judge this point. In addition, this study is in its early days; we can be confident that progress in classifying and computing the ab initio predictions of string/M theory vacua will continue.

There are far too many vacua to study each one individually. In studying the string landscape, the next step is to estimate the distribution of observables among the vacua, using statistical techniques surveyed in [5]. A particularly important example is the distribution of values of the effective potential at the minima, in other words the cosmological constant (or c.c.). This is an important input into the arguments for the multiverse from cosmology and especially for the anthropic solution to the cosmological constant problem, for which we refer to [6,7] and the article [8] in this issue. This argument requires the existence of a large number of vacua such that the a priori probability that we will observe a given vacuum is roughly uniform in the cosmological constant at small values.

The a priori probability for making a specified observation, say that the cosmological constant Λ sits in a given range of values, is a central theoretical concept in quantum cosmology, usually called the "measure factor." At first, one might think to define it as a sum over each of the vacua V_i which

---

[1] We might impose additional physical conditions if convenient, such as positive cosmological constant.

realize the specified physical observable, weighted by the probability $P(i)$ with which we as observers believe we live in a universe described by V_i. However, it is better to break down these probabilities further and write the measure as a product of **three** factors. The first is a sum over vacua, while the second "cosmological" factor, call this $U(i)$, is the expected number of universes in the multiverse which realize the vacuum V_i. The third "anthropic" factor, which we call $A(i)$, is the expected number of observers in a universe of type i. We can then write the expectation value of an observable O as

$$\mathbb{E}[O] = \frac{\sum_i U(i)\, A(i)\, O(i)}{\sum_i U(i)\, A(i)}, \qquad (1)$$

where $O(i)$ is the value of the observable in the vacuum labelled i. Thus, the probabilities are

$$P(i) = \frac{U(i)\, A(i)}{\sum_i U(i)\, A(i)}. \qquad (2)$$

A reason to distinguish these three factors is that they are governed by different laws and the problems of understanding them can to some extent be separated. The set of vacua $\{V_i\}$ and many of the corresponding observations $O(i)$ are "static" questions for the fundamental theory (say string/M theory) which could be answered without knowing how the vacua were created. As an example, in a string compactification realizing the Standard Model, we can compute the fine structure constant by combining ideas from Kaluza–Klein theory with quantum field theory and the renormalization group, without knowing its cosmological origins. Crucially, to compute $O(i)$, we only need information about the vacuum V_i, making such problems relatively accessible to present knowledge.[2]

By contrast, the factors $U(i)$ summarize information about how vacua are created in quantum cosmology. As we will discuss in Section 3, this subject depends on sweeping and difficult to justify theoretical assumptions, and the analysis requires knowledge about all of the vacua. While we may be at the point of beginning to ask meaningful questions here, it will be some time before we can have any confidence in statements whose justification requires knowledge about all vacua.

Finally, the numbers $A(i)$ depend on a vast panoply of scientific theories, ranging from topics in cosmology such as baryogenesis and big bang nucleosynthesis, through the astrophysics of structure formation at many scales, to some of the deepest questions in biology and philosophy (what is an observer, anyways?). One would be tempted to dismiss the factor $A(i)$ as totally intractable, were it not that the anthropic solution to the cosmological constant problem requires us to discuss it. Fortunately, the more physical determinants of $A(i)$, such as the size of the universe, number of galaxies and stars and the like, can be estimated, as discussed in the article [9] in this issue.

Given a particular observable O, we can identify two limiting possibilities. One is that $P(i)$ is highly correlated with the value $O(i)$. In this case, clearly it is crucial to estimate $P(i)$ to compute $\mathbb{E}[O]$. The other extreme is when $P(i)$ is totally uncorrelated with $O(i)$, so that

$$\sum_i P(i)\, O(i) \sim \frac{1}{N_{vacua}} \sum_i P(i) \times \sum_i O(i). \qquad (3)$$

Since $\sum_i P(i) = 1$, this means that

$$\mathbb{E}[O] \sim \frac{1}{N_{vacua}} \sum_i O(i). \qquad (4)$$

---

[2] Some observables, for example $\delta\rho/\rho$, do depend on the early history of the universe. To discuss these we would need to generalize Equation (1), but we can still work in terms of the four dimensional degrees of freedom visible in the vacuum V_i.

In other words, the probability that we observe a given value for O will be proportional to the number of vacua that realize it. This is the vacuum counting measure, whose study was formalized in [10].

Almost all statistical studies of the string landscape to date study the vacuum counting measure. For it to make sense, N_{vacua} must be finite. This was shown in [11] using ideas we review in Section 2.

One of the most important claims about the string vacuum counting measure is that it is roughly uniform in the cosmological constant near zero. This was first argued in a toy model by Bousso and Polchinski [12], and then in successively more realistic models [3,13]. Note, however, that it is not at all obvious that the factor $U(i)$ is uniform near $\Lambda = 0$, as we will discuss in Section 3.

One can also study the counting distribution of other observables. An example of great interest is the distribution of the supersymmetry breaking scale, and the distribution of masses of superpartners. One of the primary motivations for building the Large Hadron Collider was the hope that supersymmetry will be discovered there, which is feasible if there are superpartners with mass up to about 1 TeV. In addition, the discovery of supersymmetry was widely predicted, based on the "naturalness" argument. This is the idea is that there should be an explanation within the fundamental theory for the small number $M_{Higgs}/M_{Planck} \sim 10^{-17}$. One might try to trace this to a small parameter in the fundamental theory, but since renormalization in quantum field theory affects the masses of scalars additively, $M^2 \to M^2 + c\Lambda^2$ in terms of the cutoff energy Λ and some order one c, achieving the small observed Higgs mass requires not a small bare mass but rather a bare mass finely tuned to cancel the effects of renormalization. By the same arguments made for the cosmological constant [6,7], this is implausible, posing the "hierarchy problem." By showing that the bare Higgs mass squared is uniformly distributed in quasi-realistic vacua, one can quantify this implausibility: the fraction of vacua in which fine-tuning is needed to obtain the observed Higgs mass is about 10^{-34}.

During the 1970s and 1980s, a major search was made for mechanisms to produce small M_{Higgs} without fine-tuning, with the leading candidates being technicolor and dynamical supersymmetry breaking. Technicolor made other predictions, such as flavor changing neutral currents, which were falsified during the 1990s by precision experiments. This left low energy supersymmetry as the favored explanation. To realize small M_{Higgs} without fine-tuning in this framework, one postulates an additional matter sector with an asymptotically free gauge group, such that supersymmetry is spontaneously broken at the energy scale where its gauge coupling is strong. One then arranges the total theory to mediate this supersymmetry breaking to the Standard Model, either through additional matter (gauge mediated supersymmetry breaking) or through gravitational couplings (gravity mediated). This mediation will generically give mass to all of the superpartners of the observed Standard Model particles, and can give mass to the Higgs boson(s) as well. In this way, one can get natural classes of models in which a large fraction of vacua have small M_{Higgs}. String compactifications in which all of these ingredients appear were constructed in many works—see, for example, [14–17] and the review [18].

Upon formulating the problem of the distribution of supersymmetry breaking scales systematically in a multiverse theory, one realizes that there are major loopholes in the naturalness argument. The main loophole is that it completely ignores the measure factor. Considering each of the three factors in Equation (1), it is reasonable to expect $A(i)$ to be independent of the supersymmetry breaking scale as long as this is well above the scales (atomic, nuclear, etc.) which enter anthropic arguments. As for the cosmological factor $U(i)$, it is true that some of the ingredients in the eternal inflation discussion of Section 3 depend on the supersymmetry breaking scale, but not in a direct way and as yet there is no clear argument either way about such dependence.

However, let's grant for the sake of argument that $U(i)$ is not correlated with the supersymmetry breaking scale. Even so, if we cannot estimate the vacuum counting measure, we still cannot say whether string theory predicts low energy supersymmetry. To see this, suppose there are two classes of vacua, class A with a mechanism such as supersymmetry which leads to low Higgs mass in a large fraction of vacua, and class B in which this can only come from fine-tuning, so that one expects a fraction

10^{-34} of class B vacua to realize a low Higgs mass. Nevertheless. if the counting measure Equation (4) favored class B by a factor greater than 10^{34}, then string theory would predict we live in a class B vacuum. Thus, one can envisage a scenario in which the discovery of low energy supersymmetry would be evidence against string theory! This was realized around 2004 [19,20] and efforts were then made to estimate the counting part of this measure, reviewed in [5]. Although this discussion was not conclusive, my own opinion at present [21] is that the suggestions made at that time for a large number of vacua with high scale supersymmetry breaking were not confirmed, and that vacuum counting will prefer low energy supersymmetry, though this is by no means proven.

In any case, the scenario in which low energy supersymmetry provides a natural solution to the hierarchy problem meets another fatal difficulty in string theory, the moduli problem [22,23]. This starts with the observation that scalars with gravitational strength couplings to the Standard Model with mass $M \lesssim 30$ TeV lead to an overabundance of dark matter which is incompatible with the theory of big bang nucleosynthesis. Then, one can show on very general grounds [13,24] that string compactifications always contain at least one scalar with $M \lesssim M_{susy}$, placing a lower bound $M_{susy} \gtrsim 30$ TeV. While solutions to the moduli problem have been suggested, they are unnatural in other ways. Since taking $M_{susy} \sim 30$ TeV brings the fine-tuning required to reproduce the observed Higgs mass down to a modest 10^{-3} or so, this relatively straightforward scenario seems plausible. Thus, if we accept all this, a combination of phenomenological and landscape arguments favor low energy supersymmetry around the scale 30–100 TeV. If so, while supersymmetry may be out of reach at LHC, a next generation accelerator could discover it.

After the supersymmetry breaking scale, arguably the next most important "axis" along which string vacua differ in a potentially testable way is the number of hidden sectors, meaning matter sectors whose couplings to the Standard Model are comparable to that of gravity. This could be because the couplings are suppressed by powers of large energy scales M_{Planck} or M_{GUT}, or it could be because the couplings violate an approximate symmetry. An example of the former might be an additional x sector, perhaps similar to the Standard Model. An important example of the latter is the axion, a scalar field whose interactions respect a shift symmetry $\phi \to \phi + a$ to all orders in perturbation theory, but which is violated nonperturbatively. Such axions are generic in string theory [25].

An attractive feature of the heterotic models of [1] was that, not only did they naturally lead to grand unified theories, they also led to a simple hidden sector (the extra E_8). However, this simplicity did not survive the second superstring revolution, and the models which solve the cosmological constant problem have many hidden sectors as well as hundreds or more axions. These are all dark matter candidates, and can lead to many other testable predictions. A very striking example is the superradiance of rotating black holes discussed in [26], which can rule out axions in certain mass windows without any other assumptions. In fact, it seems likely that generic string vacua with hundreds of axions whose masses are uniformly distributed on a log scale will be ruled out this way.

This axis of variation can be referred to as the "simplicity–complexity" axis. Is the typical string vacuum simple, having few matter fields, or complex, having many? One would expect on general grounds that there will be many more complex configurations than simple ones because they have more parameters to vary, so that the complex vacua will dominate the vacuum counting measure. Indeed, this is what has come out so far from the arguments we will discuss in Section 2.

Even if the vacuum counting measure leads to definite predictions, we still need to estimate the cosmological factor $U(i)$, for which we need to study the quantum cosmology of string/M theory. A satisfactory discussion of this would start from a precise definition of string/M theory which includes both the quasi-realistic vacua and the other regimes that are important in quantum cosmology, such as eternally inflating solutions and possible non-geometric solutions. Since at present we have no such definition, this discussion is also in its infancy. What has mostly been done is to make contact with the analysis based on eternal inflation and vacuum tunneling, developed in many works of which [27,28] have particularly influenced the string theory work. It seems clear to most researchers that string/M theory satisfies the general conditions required to realize eternal inflation, and thus, by the arguments of that literature, one expects the measure to be dominated by that part of the multiverse, which is described by eternal inflation. Granting this, one can try to use the master equation developed in that work to determine the measure factor. This was studied in the works [28,29], with general results we outline in Section 3.

2. String Compactification and the Effective Potential

There are many good reviews of string compactification such as [30], and the general picture of the subject has been stable for about a decade now, so we refer readers who want the basics to the reviews and here give a very brief overview with comments on the current situation.

Most work on compactification of string/M theory follows the general approach initiated in the 1970s for higher dimensional supergravity theories. One starts with the low energy limiting effective field theory, a ten- or eleven-dimensional supergravity, and adds various terms to the action which describe important stringy and quantum effects. One then assumes that space-time takes the form $M_4 \times K$, where M_4 is a maximally symmetric four-dimensional space-time and K is a compact six- or seven-dimensional manifold, and looks for solutions of the quantum corrected equations of motion.

To get started on this, almost always one takes a manifold K with a Killing spinor, in other words which preserves some supersymmetry in M_4. This would seem to be a very restrictive assumption which rules out de Sitter M_4 from the start. Since the simplest way to model the observed dark energy is as a positive c.c., and we also want to find solutions which describe inflation, it is important to construct de Sitter vacua; indeed, when we talk about quasi-realistic vacua in this survey, we will include the condition of a small positive cosmological constant. Thus, this starting point should be examined critically.

Could there be other solutions not based on supersymmetry and Killing spinors? A heuristic argument for supersymmetry at and near the compactification scale is that instabilities are very generic—this is one of the main lessons of the literature on string compactification. One wants mechanisms to stabilize the solution, and supersymmetry tends to do this because the potential is a sum of squares (we will talk about the negative term in supergravity below). Nevertheless, constructions not relying on special holonomy and Killing spinors have been suggested—see particularly [31] and references there. How can we evaluate them?

Mathematically, one expects K to fall into a "trichotomy" of positive, zero or negative Ricci curvature [32]. For nonzero curvature, one needs internal stress-energy, and for negative curvature this has to violate positive energy conditions, leading to a restrictive no-go theorem [33]. Of course, there are many positive curvature solutions—these naturally have negative c.c. proportional to the curvature of K, and furthermore the diameter of K is related to its curvature as $R \sim c/d^2$ for some order one c, so realizing small negative c.c. generally makes the diameter large, eventually running into the phenomenological bounds from non-observation of "Kaluza–Klein particles" or even from the validity of the inverse square law of gravity down to the few micron scales. Considerations along these lines also lead to arguments that only a finite number of vacua satisfy the diameter constraint [11]. This leaves the case of Ricci flat K, which is accessible to mathematical analysis. All known Ricci flat manifolds in fact have special holonomy and Killing spinors, and proving this for four-dimensional K is an active research topic. If this is always so, then there is a strong argument for supersymmetry at some energy scale. Then, although the scale might be high, one can estimate the dependence of the vacuum counting measure on this scale, and make a clear argument for whether low or high scale is favored.

Once we assume supersymmetry, there is a short list of constructions. Probably, the most general is F theory compactified on a Calabi–Yau fourfold, and the others are M theory compactified on a manifold of G_2 holonomy, and the five superstrings compactified on Calabi–Yau threefolds. By "most general", we mean the construction which realizes the most choices and makes the physics clearest from a purely geometric analysis, based on supergravity plus corrections. The conjecture is that, once we take all stringy and quantum effects into account, all the constructions lead to the same landscape, but the simplicity of a geometric analysis is a great advantage. Having said this, it should also be said that working with F theory requires going beyond perturbative string theory and many physical computations are still only possible in the superstring/CY_3 constructions.

In the Calabi–Yau constructions, one is relying on the famous theorem of Yau which guarantees a Ricci-flat metric, which is uniquely determined given a finite number of parameters or "moduli" which include the overall volume and the complete list of integrals of two preferred differential forms (the Kähler form and the holomorphic three-form or four-form) over respective bases of homology classes. One then must make many additional discrete and continuous choices—higher dimensional branes can fill M_4 and wrap cycles of K; these can carry holomorphic vector bundles; one can postulate generalized magnetic fields or "flux." While all of these choices are classified topologically, the observed fact that they are finite in number—indeed, one can find Calabi–Yau compactifications with arguably "the largest gauge group" [34] or "the largest number of flux vacua" [4]—does not follow from topology but rather from still somewhat mysterious algebraic geometric arguments.

At first sight, the most striking thing about these compactifications is how many fields they have compared to the Standard Model. While the particle physicists of the 1930s were surprised by the muon, asking "Who ordered that?", string theorists should be even more surprised by the thousands or even hundreds of thousands of fields in the most complex compactifications. Still, this is not in contradiction with observation as these fields will be light only in the most symmetric compactifications; symmetry breaking can give large masses to almost all of these fields. Others could be in hidden sectors, coupling only gravitationally to the Standard Model and to the inflaton, so that they are not produced in reheating. While there are also compactifications with relatively few fields, certainly we should look for evidence of many fields in our universe. One interesting possibility is that compactifications often contain many axions, supersymmetry partners of the moduli, which only get masses nonperturbatively and could realize a log uniform mass spectrum. This leads to many potential signatures, as first discussed at length in [26].

Another approach to making predictions is to argue that no string theory compactification, or perhaps even no quantum theory of gravity, can have a particular low energy property. Some of the first examples were the argument that global symmetries must be broken in quantum gravity [35], and the weak gravity conjecture [36], which requires the existence of light charged particles. A recent survey of these "swampland" arguments is [37].

To more deeply understand the physics of a string compactification, one needs to derive the low energy effective theory. One can push this very far using powerful methods of algebraic geometry, especially the theory of variation of Hodge structure which allows for computing the periods of the holomorphic three and four-form as a function of the moduli. This determines the flux superpotential [38], which is a major part of the complete effective superpotential, and which by string dualities encodes a great deal of nonperturbative four-dimensional physics, in particular the nonperturbative effects which were called upon in work on dynamical supersymmetry breaking in beyond the Standard Model physics.

A central property of the string landscape, as discovered by Bousso and Polchinski [12], is that it contains the large set of vacua with different values of the vacuum energy required by the anthropic solution to the cosmological constant problem. They showed this by looking at vacua with nonzero values of higher form gauge fields, constant on M_4 and consistent with its maximal symmetry, but contributing to the vacuum energy by the usual Maxwell-type magnetic terms, in IIb compactification schematically

$$V = \int_K d^6 y \sqrt{g} F \wedge *F + e^{-2\phi} H \wedge *H, \qquad (5)$$

where F and H are three-forms. This expression depends on the metric on K, the "dilaton" scalar field ϕ, and many other fields which we have suppressed. Now, it is not possible to explicitly solve the equations for these fields and directly evaluate V. However, what matters for their argument is that configuration space breaks up into sectors with fixed values of the quantized "fluxes"

$$N_{RR}^i \equiv \int_{\Sigma_i} F; \qquad N_{NS}^i = \int_{\Sigma_i} H, \tag{6}$$

and that within each sector there is a solution. In this case, one can model the dependence of the vacuum energy on the flux quanta N as a quadratic function

$$V \sim M^4 \sum_{i,j} G_{i,j} N^i N^j - \Lambda_0, \tag{7}$$

where M is an energy scale set by the quantization condition Equation (6), and $G_{i,j}$ is determined by the weak dependence of the other fields on the N's. Neglecting this and assuming that the $G_{i,j}$ take generic fixed values, one finds a "discretuum" of possible values of V which will be uniformly distributed near $V \sim 0$ with average spacings $M^4/\bar{N}^{2b_3}$. Here, $b_3 = \dim H^3(K; \mathbb{Z})$ is the third Betti number of K, which ranges from 4 up to 984. The maximal flux $\bar{N}$ is set by string theoretic details (the "tadpole condition") but typically ranges from 10 to 100, so, even with $M \sim M_{Planck}$ (it is smaller), this construction can easily achieve the required spacing.

While this is a good qualitative argument, to go any further and especially to say anything about the distribution of supersymmetry breaking scales, one needs to put in more about string theory. Thus, let us write out the flux superpotential for IIb string compactified on CY_3; it is

$$W = \sum_{1 \leq i \leq b_3} \left(N_{RR}^i + \tau N_{NS}^i\right) \Pi_i(\vec{z}). \tag{8}$$

The flux numbers N_{RR}^i, N_{NS}^i are integers, τ and $\vec{z}$ are the moduli, and the Π_i are the periods of the holomorphic three-form, which can be written explicitly as generalized hypergeometric functions, using methods of mirror symmetry [39]. Following the rules of 4d, $N = 1$ supergravity, this determines the effective potential, as a sum of terms quadratic in W. This can be shown to be equal to the minimum of the potential terms for higher form gauge fields F and H, schematically with F and H satisfying the Dirac quantization conditions at the solutions of the equations of motion. Thus, we can compute the exact effective potential due to flux, despite our inability to explicitly solve for the Ricci flat metric or the other fields on K.

Besides its computability, the main point we want to make about the flux superpotential here is that it provides a large number of independent contributions to the vacuum energy that preserve supersymmetry. This has the important consequence that, as explained in [5], the distribution of cosmological constants is independent of the supersymmetry breaking scale. One might have thought a priori that this distribution was uniform only for $|\Lambda| < M_{susy}^4$, in which case the need to get small cosmological constant would favor low energy supersymmetry. Because Λ also gets supersymmetry preserving contributions, this is not the case.

By itself, the physics of fluxes stabilizes many but not all of the moduli. In particular, one finds that the overall volume of K is not fixed. While it must eventually be fixed when supersymmetry is broken, having this happen at observable scales is very problematic and it is much easier to get a metastable vacuum with reasonable cosmology if all of the moduli are stabilized before supersymmetry breaking. Such models have been constructed, but they rely on partially understood stringy and quantum corrections to the effective potential. Thus, the further discussion is not yet based on exact results but on expansions around controlled limits and physical intuition.

There are two popular classes of moduli stabilizing potentials, the KKLT[3] construction [40] and the large volume construction [41,42]. Both are most often applied to type IIb and F theory compactifications. In KKLT, one relies on nonperturbative effects in the superpotential. These include the very well understood nonperturbative superpotentials of supersymmetric four-dimensional gauge theory, as well as stringy instanton effects (in some cases, these are two dual descriptions of the same thing). These depend on the Kähler moduli and thus a fairly generic superpotential will have supersymmetric minima at which these are stabilized (which are necessarily anti de Sitter (AdS)). Because the new terms are nonperturbative, they can be very small even if the compactification manifold is string scale, and these constructions can make contact with the original picture of dynamical supersymmetry breaking as envisioned in [1]. The analysis also benefits from the protected nature of the superpotential, which makes computations tractable and holds out the possibility of exact results.

In the large volume construction, one relies on a particular correction to the Kähler potential, which appears in world-sheet perturbation theory [43], to find an effective potential with a nonsupersymmetric AdS minimum. While this may sound a bit arbitrary, this correction is believed to be protected as well. Then, because the terms which are balanced to find a minimum are perturbative, one finds a universal structure of the potential near the minimum, and energy scales which are simple powers of the volume of the internal dimensions. This leads to rather different predictions from the KKLT and dynamical supersymmetry breaking models.

Once one has stabilized the moduli, the next step in the analysis is to argue that the vacuum exhibits supersymmetry breaking and has a positive cosmological constant (we refer to these as de Sitter or dS vacua). From a conceptual point of view, once one has accepted that the main point is to show that the effective potential has these properties, such vacua are expected to be generic as there are many candidate supersymmetry breaking effects which make positive contributions to the vacuum energy, while there is nothing special about the value zero for the sum of all of these effects. However, what is special about the value zero is that the controlled limits of the problem (large volume and classical limits) about which we must expand around to do our computations have zero vacuum energy. Furthermore, these limits can be reached by varying moduli (the overall volume modulus and the dilaton). Thus, one needs not only positive energy for the candidate vacuum, but an even larger energy for all the nearby configurations, so that it will be a local minimum. Even if we have this, there will inevitably be an instability to tunneling back to large volume. One of the main points of the original KKLT work was to argue that this decay is usually highly suppressed.

The original KKLT work also argued that one could start with a supersymmetric AdS compactification, which is an easier theoretical problem, and then obtain a dS compactification by inserting an anti D3-brane into the compactification. Since one has combined branes and antibranes which can annihilate, this will always break supersymmetry, while one can show that this annihilation process is also very slow [44]. This mechanism was also called upon in [42].

The anti D3-brane has the advantage of simplicity, but any additional source of positive energy, such as one would get from sufficiently many distinct fluxes or nonperturbative effects, could play the same role. Indeed, in [13], it was argued that generic flux potentials depending on a single modulus would have metastable supersymmetry breaking critical points. If such were the case in models with large numbers of moduli, one might find a vast number of high scale supersymmetry breaking vacua. This point was considered in [45] who introduced a toy random matrix model of the flux superpotential, and showed in their model that such critical points are almost always unstable unless almost all the moduli have masses well above the supersymmetry breaking scale. This is good news for hopes of observing low energy supersymmetry; however, the assumptions behind the toy model are not beyond criticism [46,47]. Still, since one can get metastable minima even with a potential depending on a single field, and there is no way for local physics to distinguish between small positive and negative

---

[3] Kachru, Kallosh, Linde and Trivedi.

values of the c.c. (this was one of the central points of [6,7]), it would require a vast and mysterious conspiracy to eliminate all metastable de Sitter compactifications.

Problems with the String Compactification Analysis

At present, these analyses go beyond the controlled computations and not all string theorists are convinced, see [48–52] for some of the criticisms and [53–55] for a few of the many rebuttals. I cannot do justice to this controversy here but will only comment that to my mind; the criticisms fall into three classes. One class is to my mind superficial and is largely based on properties of the large volume limit and the difficulty of the extrapolation we must do to get to the regime where metastable de Sitter vacua can exist, which as was known for a long time must involve significant stringy and quantum corrections [56–59]. Arguments such as that of [51] which postulate simple constraints on all possible vacua without a mechanism which enforces them on all vacua are not persuasive.

Another class of criticisms [60–62] tries to construct explicit solutions with quantum corrections and meets specific difficulties. For example, after inserting an antibrane to uplift to positive vacuum energy, one must check that the backreaction of the metric and other fields does not make the total vacuum energy negative. While technical, it is crucial to settle such points as these explicit constructions are the foundation for the entire discussion. It is very possible that, on properly understanding them, the statistics of vacua will look rather different. For example, it might turn out that complex vacua (on the simplicity–complexity axis of Section 1) are much harder to stabilize than simple ones, disfavoring them.[4]

The third class of criticism [48] is deep and starts from the postulate that the holographic dual, whose nature depends crucially on the large scale structure of space-time and thus on the sign of the cosmological constant, is more fundamental than the approximate descriptions of string compactification which we work with now. In this view, because the duals of AdS and dS vacua are so different, trying to relate the two types of construction is misguided. This idea, which contradicts very basic tenets of nongravitational physics, can only be properly judged when we have a more complete understanding of string/M theory.

In my opinion, a convincing construction of metastable supersymmetry breaking de Sitter vacua and a good understanding of their physics must be founded on a controlled computation of the effective potential. Although major advances will be required, there is no known obstacle to someday computing the exact superpotential for a wide range of supersymmetric string compactifcations. The same cannot be said (yet) for the Kähler potential, the other necessary ingredient to determine the effective potential (the final ingredient, the D terms, are determined by the symmetries of the Kähler potential). Most work, both in string compactification and in the study of four-dimensional gauge theory, uses perturbative approximations to the Kähler potential. This often can be justified by other physical arguments, but it is a weak point. For some gauge theories in the large N limit, one can get the effective potential using gauge–gravity duality. This at least points the way to other techniques, and I believe there will be progress on controlled calculations of the Kähler potential and thus the full effective potential, though perhaps not soon.

3. Eternal Inflation and Measure Factor

We will assume that the reader is familiar with the theory of eternal inflation, if not he or she is directed to one of the many reviews such as [64,65]. Although there is controversy about the predictions of this theory, I generally follow the analysis given in [27] and Section 2 of [66] has a brief review from this point of view. To summarize the results of this theory, if we assume that cosmological dynamics is dominated by eternal inflation and semiclassical tunneling between metastable vacua with positive c.c.,

---

[4] There are several arguments that complex vacua might be disfavored; see [63].

we can show that the stochastic evolution of the multiverse is governed by a linear master equation for $N_i(t)$, the number of pocket universes of type i at time t. This equation takes the form

$$\frac{d}{dt}N_i(t) = \gamma_i N_i(t) + \sum_j \Gamma_{j \to i} N_j(t) - \sum_j \Gamma_{i \to j} N_i(t), \tag{9}$$

where γ_i represents the exponential growth due to inflation, and $\Gamma_{i \to j}$ is the tunneling rate from universe type i to type j. Usually, one uses the Coleman–DeLuccia (CDL) formula for these rates [67], calculated by describing the tunneling in terms of a domain wall solution and applying semiclassical quantum gravity. We will not quote this formula except to comment that the rates are typically exponentially small (as usual for tunneling rates) and even doubly exponentially small (as we explain below).

The right-hand side of Equation (9) can be rewritten as a matrix acting on $N_i(t)$, so that the equation can be solved by diagonalizing this matrix. In the large t limit, except for a set of initial conditions of measure zero, the solution will approach the dominant eigenvector; in other words, the eigenvector of the matrix with the largest eigenvalue. The i'th component of this eigenvector is then the factor $U(i)$ in Equation (1). Note, however, that the time required to reach this limit can be extremely long, a point we will return to below.

The structure of this dominant eigenvector depends very much on the tunneling rates Γ. We can illustrate this by considering two extreme cases. In the first, each rate $\Gamma_{i \to j}$ is either zero (tunneling is forbidden) or a constant rate ϵ. In this case, the landscape is completely described by an undirected graph (the rates are expected to be symmetric), and techniques of graph/network theory can be useful. A simplified landscape of this type of F theory vacua was studied in [29], and it was found that the high dimensionality of the landscape led to exponentially large ratios between the limiting $U(i)$'s.

The other extreme case is when one of the metastable vacua is far longer lived than the others, as then the dominant eigenvector will be highly concentrated on this vacuum. This leads to the following prescription for the measure factor: one must find the longest-lived metastable vacuum, call this V_0. Then (up to subleading factors depending on γ_i and an irrelevant overall factor), we have

$$U(i) \sim \Gamma_{0 \to i} \quad \forall i \neq 0. \tag{10}$$

Now, it is very plausible that V_0 is not anthropically acceptable ($A(0) = 0$), so the upshot is that the favored vacua are "close to V_0" in the sense that they can be easily reached from it by tunneling.

An analysis of the cosmology of a Bousso–Polchinski landscape along these lines was made in [28]. They argued that the hypothesis of a single longest-lived metastable vacuum will be generic under the following assumptions. First, we only consider vacua with $\Lambda > 0$; all tunnelings to vacua with $\Lambda \leq 0$ are ignored as they do not lead to inflating bubbles. Second, we use the Coleman–DeLuccia tunneling rates for $\Gamma_{i \to j}$. These include a factor of $\exp - (M_{Planck}^4/\Lambda_i - M_{Planck}^4/\Lambda_j)$, so "uptunnelings" with $\Lambda_j > \Lambda_i \approx 0$ are highly suppressed. Indeed, a realistic landscape must contain vacua with $\Lambda_i \sim 10^{-120} M_{Planck}^4$ and, in this sense, these tunnelings are doubly exponentially suppressed.

How does one find the longest-lived metastable vacuum? From the above, it might sound like it is the vacuum with the smallest positive Λ, which would be bad news as finding this one is computationally intractable [68]. Fortunately, there are other equally important factors in the tunneling rate. In fact, one can make suggestive arguments that the longest lived metastable vacuum is the one with the smallest scale of supersymmetry breaking [69], and that this comes from the compactification with the most complex topology [21]. If so, this would narrow down the search tremendously and might well lead to testable predictions—for example, that we should expect many hidden sectors associated with the complex topology of the extra dimensions.

Actually deriving Equation (9) from semiclassical gravity and making these arguments precise requires defining a time coordinate t and understanding the long time limit, which is quite ambiguous and this leads to paradoxes. By now, many definitions of the time coordinate as well as variations on the argument have been studied (a review is in [65]). Although the topic is by no means settled, many of the definitions and alternate arguments (for example, that of [70]) do seem to lead to the claim that the term $U(i)$ in the measure factor is as we just discussed, the tunneling rate from the longest-lived metastable vacuum.

Problems with the Eternal Inflation Analysis

Let us summarize the claim of the previous subsection. Starting with very general assumptions, namely that the dynamics of the multiverse is dominated by the process of eternal inflation, and that this process runs long enough to reach an equilibrium, one can derive a well-defined cosmological measure factor $U(i)$ expressed in terms of quantities we can actually hope to compute, namely the set of metastable dS vacua, their cosmological constants and the tunneling rates between them. This would be very impressive were it not for some crucial gaps in the argument. These gaps are known and discussed in the literature, and we suggest [70] as a good reference.

A very serious gap is that the analysis leading to Equation (9) was all done in four-dimensional semiclassical gravity. While it should also hold for the string/M theory solutions we are discussing, with four large dimensions and the other spatial dimensions stabilized to a fixed small manifold, of course, this is a tiny part of the full configuration space, in which any subset of the dimensions could be large. The only excuse I know of for ignoring this major omission is that it is plausible that having four large dimensions is the only way to get eternal inflation. The physics called upon in the dS vacuum constructions we cited does not generalize to higher dimensions, while general relativity behaves very differently in fewer dimensions. Of course, this argument is not entirely convincing. Even if we grant it, it might be that the dominant tunneling processes between the four-dimensional vacua pass through the larger configuration space of higher dimensional or even non-geometric configurations.

Another important gap is in the treatment of AdS vacua. Now, in Section 2, we stressed that, from the point of view of local physics, there can be no essential difference between dS, AdS and Minkowski vacua; indeed, the uncertainty principle makes it impossible to distinguish between them without making measurements at lengths $L \gg |\Lambda|^{-1/4}$. On the other hand, the global structures of these different space-times are very different, and they enter into the arguments for Equation (9) very differently. We know there are many supersymmetric AdS and Minkowski vacua of string theory, as we discussed in Section 2. According to CDL and other general arguments, supersymmetric vacua are absolutely stable to tunneling, so their decay rates are $\Gamma = 0$ (they are "terminal vacua"). Granting this, it is straightforward to take them into account, and the claim for the measure factor remains true. However, while solutions describing tunneling from dS to Minkowski vacua are non-singular and this treatment seems sensible, the solutions describing tunneling from dS to AdS vacua are singular, ending in a "big crunch" [67]. While these transitions are usually left out, it could be that AdS vacua do somehow tunnel back to dS and must be taken into account to properly define Equation (9).

Another potential problem with the whole picture arises from the double exponentially suppressed nature of tunneling rates. It is that the ratios between probabilities $P(i)$ in Equation (1) can be so large that vacua with very small values of $A(i)$ become important. This was pointed out in [28] which studied Equation (9) in an explicit Bousso–Polchinski flux landscape (Equation (7)). They found that, because of the wide variation of tunneling rates, the factor $U(i)$ was very irregular, in a way one can model as a roughly uniform constant plus a much larger fluctuating random variable. Now, because in inflationary cosmology observables such as $\delta\rho/\rho$ have quantum fluctuations, a vacuum with Λ violating the anthropic bound of [6] does not have $A = 0$. Rather, there is an exponentially small probability for a density enhancement to allow structure formation in a small part of the universe. If this suppression were overwhelmed by a double exponential, it could turn out that most observers live in small overdense regions of an otherwise empty universe. Even worse, to decide whether this is the case and make any predictions at all, one must compute the $A(i)$ to the same accuracy as the $U(i)$. Given the difficulty of understanding $A(i)$, this is problematic.

Universes with very small but nonzero $A(i)$ were referred to as "marginally hospitable" in [66] and lead to problems for mulitverse cosmology. The most extreme is the famous "Boltzmann brain" paradox [71,72], which points out that there is a nonzero (though doubly exponentially small) probability to create a local region of space-time containing any specified matter distribution—say, a disembodied brain imagining our universe—as a random thermodynamic fluctuation. Granting the advances in biology and artificial intelligence since Boltzmann's time, we can imagine more likely ways that this could happen, through fluctuations which create a local region in which evolution and/or engineering produces observers. Although very far-fetched from any point of view, if the space-time volume of the multiverse is doubly exponentially large, these could be the most numerous observers.

The main point is that double exponential probabilities are very small and even the most unlikely-seeming scenarios could be more likely. I would go so far as to advocate the principle that any theory which is complex enough to describe universal computing devices and whose predictions depend on differences between probabilities less than $10^{-10^{10}}$ and zero cannot be believed because the analysis needed to rule out other comparably rare events is computationally intractable (and perhaps even undecidable). However, I do not know how to make this argument precise.

What is clear is that the relevance of these double exponentially small rates to quantum cosmology depends on the assumption that the dynamics of multiverse cosmology runs to equilibrium. This is another way of stating the assumption of the large time limit made in the paragraph following Equation (9). The time taken to reach equilibrium can be estimated as the spectral gap of the matrix in Equation (9), which is more or less the second longest lifetime of a metastable vacuum. This will also be a double exponential, at least $10^{10^{120}}$ Planck times and probably much longer. This is the context in which I invite the reader to reflect on the above doubts about double exponentials.

This observation brings us to what I consider to be the main philosophical objection to the multiverse.[5] It is that, in postulating a multiverse, we are postulating a far larger and more complex structure than any model of what we can actually observe. This does not make the multiverse idea wrong, but it does motivate looking for a formulation of a multiverse theory without this property, or at least one that minimizes the complexity of the multiverse. Clearly, a good first step in this direction would be to have some way to quantify the complexity of the multiverse.

This question was first raised in [66], which also proposed a way to answer it. The conceit suggested there is to imagine that the multiverse is being simulated by a powerful quantum supercomputer, which is searching for vacua that satisfy the conditions required for the emergence of structure and observers, or at least those conditions which can be tested within a fixed amount of time (we refer to these as "hospitable" vacua).[6] Following this idea leads to a definition of the complexity of the multiverse in terms of the computational resources (number of quantum gate operations) required to do the simulation, analyzing the intermediate results and controlling the search. At least in semiclassical quantum gravity, one can use the freedom to choose a time coordinate to express the choice of which parts of the multiverse to simulate. Given an algorithm for making this choice, the corresponding measure factor $U(i)$ is simply the probability with which the vacuum V_i will be the first hospitable vacuum found by the supercomputer. One can even argue that there is a preferred time coordinate, "action time," which (following a conjecture in [73]) counts the number of gate operations required for a simulation. According to this conjecture, and consistent with the earlier analysis of [74], the observable universe could be simulated with $\sim 10^{120}$ operations. An "economical" multiverse proposal would be one which does not require too many more operations than this.

---

[5] See also the article [8] in this issue.
[6] Our point was not to claim that the multiverse is a simulation, in fact we explicitly postulate that no local observer can tell that the multiverse is being simulated. Rather, in the spirit of computational complexity theory, the point is to define the complexity of the multiverse in terms of a "reduction" to another problem whose complexity we know how to define, namely that of running a specified quantum computer program.

On thinking about this, it is clear that the standard eternal inflation treatment based on Equation (9), with its double exponential time to reach equilibrium, is very far from an economical multiverse. One can easily improve its efficiency by making simple changes to the search algorithm. For example, after simulating a given vacuum for 10^{100} years, if no interesting tunneling or other events have been detected, the computer simply abandons simulating that branch of the multiverse and switches to another. This particular variation can be analyzed as a modification of Equation (9) and one finds drastically different dynamics, which arguably lead to a measure factor concentrated on the vacua which are close to (in the sense of tunneling rates) whatever preferred initial conditions string/M theory might provide for the multiverse. If the initial conditions are simple, this would favor the "simple" side of the simplicity–complexity axis of Section 1. Further analysis of the resulting predictions will have to wait until we know something about these initial conditions, which may require finding a new formulation of string/M theory. However, this is to be expected of any approach which abandons the assumption that the dynamics of the multiverse is at equilibrium.

One sees that our understanding of the string landscape is still in its early exploratory days. Let me conclude by repeating what I think is the most important question about the multiverse:

Is the multiverse in equilibrium or not?

Funding: This research received no external funding.

Acknowledgments: I thank many people for discussions on these topics, particularly Bobby Acharya, Nima Arkani-Hamed, Sujay Ashok, Tom Banks, Raphael Bousso, Frederik Denef, Michael Dine, Bogdan Florea, Antonella Grassi, Brian Greene, Alan Guth, Shamit Kachru, Renata Kallosh, Gordy Kane, Andrei Linde, Bernie Shiffman, Eva Silverstein, Wati Taylor, Cumrun Vafa, Alex Vilenkin, Steve Zelditch and Claire Zukowski.

Conflicts of Interest: The author declares no conflict of interest.

References

1. Candelas, P.; Horowitz, G.T.; Strominger, A.; Witten, E. Vacuum Configurations for Superstrings. *Nucl. Phys. B* **1985**, *258*, 46–74. [CrossRef]
2. Witten, E. String theory dynamics in various dimensions. *Nucl. Phys. B* **1995**, *443*, 85–126. [CrossRef]
3. Denef, F.; Douglas, M.R. Distributions of flux vacua. *J. High Energy Phys.* **2004**, *2004*, 72. [CrossRef]
4. Taylor, W.; Wang, Y.N. The F-theory geometry with most flux vacua. *J. High Energy Phys.* **2015**, *2015*, 164. [CrossRef]
5. Douglas, M.R.; Kachru, S. Flux compactification. *Rev. Mod. Phys.* **2007**, *79*, 733. [CrossRef]
6. Weinberg, S. The Cosmological Constant Problem. *Rev. Mod. Phys.* **1989**, *61*, 1. [CrossRef]
7. Polchinski, J. The Cosmological Constant and the String Landscape. *arXiv* **2006**, arXiv:hep-th/0603249.
8. Heller, M. Multiverse—Too Much or Not Enough? *Universe* **2019**, *5*, 113. [CrossRef]
9. Sandora, M. Multiverse Predictions for Habitability: The Number of Stars and Their Properties. *Universe* **2019**, *5*, 149. [CrossRef]
10. Douglas, M.R. The Statistics of string / M theory vacua. *J. High Energy Phys.* **2003**, *2003*, 46. [CrossRef]
11. Acharya, B.S.; Douglas, M.R. A Finite landscape? *arXiv* **2006**, arXiv:hep-th/0606212.
12. Bousso, R.; Polchinski, J. Quantization of four form fluxes and dynamical neutralization of the cosmological constant. *J. High Energy Phys.* **2000**, *2000*, 006. [CrossRef]
13. Denef, F.; Douglas, M.R. Distributions of nonsupersymmetric flux vacua. *J. High Energy Phys.* **2005**, *2005*, 61. [CrossRef]
14. Acharya, B.S.; Bobkov, K.; Kane, G.L.; Shao, J.; Kumar, P. The G(2)-MSSM: An M Theory motivated model of Particle Physics. *Phys. Rev. D* **2008**, *78*, 065038. [CrossRef]
15. Marsano, J.; Saulina, N.; Schafer-Nameki, S. Gauge Mediation in F-Theory GUT Models. *Phys. Rev. D* **2009**, *80*, 046006. [CrossRef]
16. Dundee, B.; Raby, S.; Westphal, A. Moduli stabilization and SUSY breaking in heterotic orbifold string models. *Phys. Rev. D* **2010**, *82*, 126002. [CrossRef]
17. Aparicio, L.; Cerdeno, D.G.; Ibanez, L.E. Modulus-dominated SUSY-breaking soft terms in F-theory and their test at LHC. *J. High Energy Phys.* **2008**, *2008*, 99. [CrossRef]

18. Ibanez, L.E. From Strings to the LHC: Les Houches Lectures on String Phenomenology. *arXiv* **2012**, arXiv:1204.5296.
19. Douglas, M.R. Statistical analysis of the supersymmetry breaking scale. *arXiv* **2004**, arXiv:hep-th/0405279.
20. Susskind, L. Supersymmetry breaking in the anthropic landscape. In *From Fields to Strings: Circumnavigating Theoretical Physics*; World Scientific: Singapore, 2005; Volume 3, pp. 1745–1749.
21. Douglas, M.R. The string landscape and low-energy supersymmetry. *Les Houches Lect. Notes* **2015**, *97*, 315.
22. Banks, T.; Kaplan, D.B.; Nelson, A.E. Cosmological implications of dynamical supersymmetry breaking. *Phys. Rev. D* **1994**, *49*, 779. [CrossRef] [PubMed]
23. de Carlos, B.; Casas, J.A.; Quevedo, F.; Roulet, E. Model independent properties and cosmological implications of the dilaton and moduli sectors of 4-d strings. *Phys. Lett. B* **1993**, *318*, 447–456. [CrossRef]
24. Acharya, B.S.; Kane, G.; Kuflik, E. String Theories with Moduli Stabilization Imply Non-Thermal Cosmological History, and Particular Dark Matter. *arXiv* **2010**, arXiv:1006.3272.
25. Svrcek, P.; Witten, E. Axions In String Theory. *J. High Energy Phys.* **2006**, *2006*, 51. [CrossRef]
26. Arvanitaki, A.; Dimopoulos, S.; Dubovsky, S.; Kaloper, N.; March-Russell, J. String Axiverse. *Phys. Rev. D* **2010**, *81*, 123530. [CrossRef]
27. Garriga, J.; Schwartz-Perlov, D.; Vilenkin, A.; Winitzki, S. Probabilities in the inflationary multiverse. *J. Cosmol. Astropart. Phys.* **2006**, *2006*, 17. [CrossRef]
28. Schwartz-Perlov, D.; Vilenkin, A. Probabilities in the Bousso-Polchinski multiverse. *J. Cosmol. Astropart. Phys.* **2006**, *606*, 10. [CrossRef]
29. Carifio, J.; Cunningham, W.J.; Halverson, J.; Krioukov, D.; Long, C.; Nelson, B.D. Vacuum Selection from Cosmology on Networks of String Geometries. *Phys. Rev. Lett.* **2018**, *121*, 101602. [CrossRef]
30. Denef, F. Les Houches Lectures on Constructing String Vacua. *arXiv* **2008**, arXiv:0803.1194.
31. Silverstein, E. TASI lectures on cosmological observables and string theory. In *New Frontiers in Fields and Strings*; World Scientific: Singapore, 2016.
32. Berger, M. *A Panoramic View of Riemannian Geometry*; Springer: Berlin, Germany, 2003.
33. Douglas, M.R.; Kallosh, R. Compactification on negatively curved manifolds. *J. High Energy Phys.* **2010**, *2010*, 004.
34. Candelas, P.; Perevalov, E.; Rajesh, G. Toric geometry and enhanced gauge symmetry of F theory/heterotic vacua. *Nucl. Phys. B* **1997**, *507*, 445–474. [CrossRef]
35. Banks, T.; Dixon, L.J. Constraints on String Vacua with Space-Time Supersymmetry. *Nucl. Phys. B* **1988**, *307*, 93–108. [CrossRef]
36. Arkani-Hamed, N.; Motl, L.; Nicolis, A.; Vafa, C. The String landscape, black holes and gravity as the weakest force. *J. High Energy Phys.* **2007**, *2007*, 60. [CrossRef]
37. Palti, E. The Swampland: Introduction and Review. *Fortschritte Phys.* **2019**, *67*, 1900037. [CrossRef]
38. Giddings, S.B.; Kachru, S.; Polchinski, J. Hierarchies from fluxes in string compactifications. *Phys. Rev. D* **2003**, *66*, 106006. [CrossRef]
39. Candelas, P.; De La Ossa, X.C.; Green, P.S.; Parkes, L. A Pair of Calabi–Yau manifolds as an exactly soluble superconformal theory. *Nucl. Phys. B* **1991**, *359*, 21–74. [AMS/IP Stud. Adv. Math. 1998, 9, 31–95.] [CrossRef]
40. Kachru, S.; Kallosh, R.; Linde, A.D.; Trivedi, S.P. De Sitter vacua in string theory. *Phys. Rev. D* **2003**, *68*, 046005. [CrossRef]
41. Balasubramanian, V.; Berglund, P.; Conlon, J.P.; Quevedo, F. Systematics of Moduli Stabilisation in Calabi-Yau Flux Compactifications. *J. High Energy Phys.* **2005**, *2005*, 7. [CrossRef]
42. Conlon, J.P.; Quevedo, F.; Suruliz, K. Large-volume flux compactifications: Moduli spectrum and D3/D7 soft supersymmetry breaking. *J. High Energy Phys.* **2005**, *2005*, 7. [CrossRef]
43. Becker, K.; Becker, M.; Haack, M.; Louis, J. Supersymmetry breaking and alpha-prime corrections to flux induced potentials. *J. High Energy Phys.* **2002**, *2002*, 60. [CrossRef]
44. Kachru, S.; Pearson, J.; Verlinde, H.L. Brane/flux annihilation and the string dual of a nonsupersymmetric field theory. *J. High Energy Phys.* **2002**, *2002*, 21. [CrossRef]
45. Marsh, D.; McAllister, L.; Wrase, T. The Wasteland of Random Supergravities. *J. High Energy Phys.* **2012**, *2012*, 102. [CrossRef]
46. Dine, M. Classical and Quantum Stability in Putative Landscapes. *J. High Energy Phys.* **2017**, *2017*, 82. [CrossRef]

47. Yamada, M.; Vilenkin, A. Hessian eigenvalue distribution in a random Gaussian landscape. *J. High Energy Phys.* **2018**, *2018*, 29. [CrossRef]
48. Banks, T.; Dine, M.; Gorbatov, E. Is there a string theory landscape? *J. High Energy Phys.* **2004**, *2004*, 58. [CrossRef]
49. Sethi, S. Supersymmetry Breaking by Fluxes. *arXiv* **2017**, arXiv:1709.03554.
50. Danielsson, U.H.; Riet, T.V. What if string theory has no de Sitter vacua? *Int. J. Mod. Phys. D* **2018**, *27*, 1830007. [CrossRef]
51. Obied, G.; Ooguri, H.; Spodyneiko, L.; Vafa, C. De Sitter Space and the Swampland. *arXiv* **2018**, arXiv:1806.08362.
52. Ooguri, H.; Palti, E.; Shiu, G.; Vafa, C. Distance and de Sitter Conjectures on the Swampland. *arXiv* **2018**, arXiv:1810.05506.
53. Conlon, J.P. The de Sitter swampland conjecture and supersymmetric AdS vacua. *arXiv* **2018**, arXiv:1808.05040.
54. Kachru, S.; Trivedi, S. A comment on effective field theories of flux vacua. *arXiv* **2018**, arXiv:1808.08971.
55. Akrami, Y.; Kallosh, R.; Linde, A.; Vardanyan, V. The landscape, the swampland and the era of precision cosmology. *arXiv* **2018**, arXiv:1808.09440.
56. Gibbons, G.W. Aspects Of Supergravity Theories. In Proceedings of the Three lectures Given at GIFT Seminar on Theoretical Physics, San Feliu de Guixols, Spain, 4–11 June 1984.
57. Gibbons, G.W. Thoughts on tachyon cosmology. *Class. Quant. Grav.* **2003**, *20*, S321. [CrossRef]
58. de Wit, B.; Smit, D.J.; Dass, N.D.H. Residual Supersymmetry of Compactified D = 10 Supergravity. *Nucl. Phys. B* **1987**, *283*, 165–191. [CrossRef]
59. Maldacena, J.M.; Nunez, C. Supergravity description of field theories on curved manifolds and a no go theorem. *Int. J. Mod. Phys. A* **2001**, *16*, 822. [CrossRef]
60. Moritz, J.; Retolaza, A.; Westphal, A. Toward de Sitter space from ten dimensions. *Phys. Rev. D* **2018**, *97*, 046010. [CrossRef]
61. Moritz, J.; Retolaza, A.; Westphal, A. On uplifts by warped anti-D3-branes. *arXiv* **2018**, arXiv:1809.06618.
62. Bena, I.; Dudas, E.; Graña, M.; Lüst, S. Uplifting Runaways. *arXiv* **2018**, arXiv:1809.06861.
63. Douglas, M.R. Landscape and Complexity Catastrophe. Talk Presented at the 2018 String_data Workshop in Munich. Available online: https://indico.mpp.mpg.de/event/5578/contribution/18/material/slides/0.pdf (accessed on 19 July 2019).
64. Guth, A.H. Eternal inflation and its implications. *J. Phys. A Math. Theor.* **2007**, *40*, 6811. [CrossRef]
65. Freivogel, B. Making predictions in the multiverse. *Class. Quantum Gravity* **2011**, *28*, 204007. [CrossRef]
66. Denef, F.; Douglas, M.R.; Greene, B.; Zukowski, C. Computational complexity of the landscape II—Cosmological considerations. *Ann. Phys.* **2018**, *392*, 93–127. [CrossRef]
67. Coleman, S.R.; Luccia, F.D. Gravitational Effects on and of Vacuum Decay. *Phys. Rev. D* **1980**, *21*, 3305. [CrossRef]
68. Denef, F.; Douglas, M.R. Computational complexity of the landscape I. *Ann. Phys.* **2007**, *322*, 1096–1142. [CrossRef]
69. Dine, M.; Festuccia, G.; Morisse, A.; van den Broek, K. Metastable Domains of the Landscape. *J. High Energy Phys.* **2008**, *2008*, 14. [CrossRef]
70. Garriga, J.; Vilenkin, A. Watchers of the multiverse. *J. High Energy Phys.* **2013**, *2013*, 37. [CrossRef]
71. Dyson, L.; Kleban, M.; Susskind, L. Disturbing implications of a cosmological constant. *J. High Energy Phys.* **2002**, *2002*, 11. [CrossRef]
72. Albrecht, A.; Sorbo, L. Can the universe afford inflation? *Phys. Rev. D* **2004**, *70*, 063528. [CrossRef]
73. Brown, A.R.; Roberts, D.A.; Susskind, L.; Swingle, B.; Zhao, Y. Complexity, action, and black holes. *Phys. Rev. D* **2016**, *93*, 086006. [CrossRef]
74. Lloyd, S. Computational capacity of the universe. *Phys. Rev. Lett.* **2002**, *88*, 237901. [CrossRef]

 © 2019 by the authors. Licensee MDPI, Basel, Switzerland. This article is an open access article distributed under the terms and conditions of the Creative Commons Attribution (CC BY) license (http://creativecommons.org/licenses/by/4.0/).

Review

Time Reversal Symmetry in Cosmology and the Creation of a Universe–Antiuniverse Pair

Salvador J. Robles-Pérez [1,2]

1. Estación Ecológica de Biocosmología, Pedro de Alvarado, 14, 06411 Medellín, Spain; salvador.robles@educa.madrid.org
2. Departamento de Matemáticas, IES Miguel Delibes, Miguel Hernández 2, 28991 Torrejón de la Calzada, Spain

Received: 14 April 2019; Accepted: 12 June 2019; Published: 13 June 2019

Abstract: The classical evolution of the universe can be seen as a parametrised worldline of the minisuperspace, with the time variable t being the parameter that parametrises the worldline. The time reversal symmetry of the field equations implies that for any positive oriented solution there can be a symmetric negative oriented one that, in terms of the same time variable, respectively represent an expanding and a contracting universe. However, the choice of the time variable induced by the correct value of the Schrödinger equation in the two universes makes it so that their physical time variables can be reversely related. In that case, the two universes would both be expanding universes from the perspective of their internal inhabitants, who identify matter with the particles that move in their spacetimes and antimatter with the particles that move in the time reversely symmetric universe. If the assumptions considered are consistent with a realistic scenario of our universe, the creation of a universe–antiuniverse pair might explain two main and related problems in cosmology: the time asymmetry and the primordial matter–antimatter asymmetry of our universe.

Keywords: quantum cosmology; origin of the universe; time reversal symmetry

PACS: 98.80.Qc; 98.80.Bp; 11.30.-j

---

1. Introduction

There is a formal analogy between the evolution of the universe in the minisuperspace and the trajectory of a test particle in a curved spacetime. The former is given, for a homogeneous and isotropic universe, by the solutions of the field equation $a(t)$ and $\vec{\varphi}(t) = (\varphi_1(t), \ldots, \varphi_n(t))$, where a is the scale factor and φ_i are n scalar fields that represent the matter content of the universe. The evolution of the universe can then be seen as a parametrised trajectory in the $n+1$ dimensional space formed by the coordinates a and $\vec{\varphi}$, which is called the minisuperspace. The trajectory is the worldline that extremizes the Einstein–Hilbert action, the time variable t is the parameter that parametrises the worldline, and the parametric coordinates along the worldline are the classical solutions, $(a(t), \vec{\varphi}(t))$.

From that point of view, the time reversal invariance of the laws of physics translates in the minisuperspace into the invariance that we have in running the worldline in the two possible directions, forward and backward, along the worldline. It is similar to what happens with the trajectory of a test particle in the spacetime. In particle physics, Feynman interpreted the time-forward and the time-backward solutions of the trajectory of a test particle as the trajectories of the particles and antiparticles of the Dirac's theory [1]. In cosmology, we can also assume that the two symmetric solutions may form a universe–antiuniverse pair. In the universe, however, a forward oriented trajectory with respect to the scale factor component, $\dot{a} > 0$, means an increasing value of the scale factor, so it represents an expanding universe. Similarly, a backward solution ($\dot{a} < 0$) represents a contracting universe. Therefore, the created pair contains, in terms of the same time variable, a contracting universe and an expanding universe.

However, the analysis of the emergence of the classical spacetime in quantum cosmology suggests that the time variables of the two universes should be reversely related [2,3], $t_1 = -t_2$. In that case, the matter that propagates in one of the universes can naturally be identified from the point of view of the symmetric universe with antimatter, and vice versa. Note that from the quantum cosmology perspective, the semiclassical picture of quantum matter fields propagating in a classical background spacetime is an emergent feature that appears, after some decoherence process, in the semiclassical regime [4,5]. In that case, we shall see in this paper that in order to obtain the correct value of the Schrödinger equation in the two universes their time variables would be reversely related. Then, the time variables measured by the internal observers in their particle physics experiments (i.e., the time variables that appear in the Schrödinger equation of their physical experiments) would be reversely related and, from that point of view, the matter that propagates in a hypothetical partner universe could naturally be identified with the primordial antimatter that is absent in the observer's universe. From the global perspective of the composite state, then, the apparent asymmetry between matter and antimatter would be restored.

However, a caveat should be made on the assumptions of homogeneity and isotropy for the initial spacetime manifold, which is in the basis of the time reversal symmetry of the cosmological field equations, and is therefore a condition for the creation of universes in pairs with reversely related time variables. From the point of view of a full quantum theory of gravity, the creation of all kinds of universes with all kinds of (even exotic) spacetime geometries is expected, so the creation of a homogeneous and isotropic universe should be considered as a particular case, and the consequent scenario of the creation of a universe–antiuniverse pair only as a plausible one. This is said even though the scenario might be rather realistic provided that we assume that the fluctuations of the spacetime are relatively small from the very onset of the universe. The observational data suggest that at least from a very remote past our universe essentially looks homogeneous and isotropic, with relatively small inhomogeneities and anisotropies compared with the energy of the homogeneous and isotropic background. Accordingly, we shall assume in this paper that the universe left the Euclidean gravitational vacuum and started inflating from an initial spatial hypersurface, $\Sigma(a_i)$, that is small but large enough to assume that the fluctuations of the spacetime are subdominant (i.e., $a_i \gg l_P$). This is not an unrealistic scenario. For instance, in the Higgs inflation scenario [6], $a_i \sim V^{-1/2} \propto \xi l_P$, where V is the potential of the Higgs in the initial slow roll regime and $\xi \gg 1$ is the strong coupling between the Higgs and gravity. We shall later on comment on the effect that the fluctuations of the spacetime would have on the breaking of the time reversal symmetry.

In this paper we review and gather the main results of previous works [2,3,7–9] and extend the hypothesis presented in [10] for the restoration of the primordial matter–antimatter asymmetry to the more general scenario of two homogeneous and isotropic pieces of the spacetime whose time variables, according to the time reversal symmetry of the Einstein–Hilbert action, are reversely related. This will prepare the arena for more detailed future developments on the subject. In Section 2, we present the analogy between the classical evolution of the universe in the minisuperspace and the trajectory of a test particle in a curved spacetime. It is shown that the time reversal symmetry of the action and the conservation of the total momentum in the process of creation of the universe would imply that the universes should be created in pairs with opposite values of their momenta so that the total momentum is zero. Section 3 analyses the correlations between the spacetimes that emerge from the two wave functions that are associated to the opposite values of the momenta conjugated to the scale factor. It is determined that in order to obtain the correct value of the Schrödinger equation in the two universes, their physical time variables must be reversely related. Thus, the particles that propagate in one of the universes are naturally identified with the antiparticles that are left in the partner universe. In Section 4, we summarise and make some conclusions.

2. Time Reversal Symmetry in Classical Cosmology

Let us consider a homogeneous and isotropic spacetime and a scalar field φ, which represents the matter content of the universe, that propagates minimally coupled to gravity under the action of the potential, $V(\varphi)$. The spacetime is then foliated in homogeneous and isotropic slices, with a total line element given by

$$ds^2 = -N^2(t)dt^2 + a^2(t)d\Omega_3^2, \qquad (1)$$

where $a(t)$ is the scale factor and $N(t)$ is the lapse function that parametrises the time variable ($N = 1$ corresponds to cosmic time), and the homogeneous mode of the scalar field is $\varphi(t)$. Small inhomogeneities around this homogeneous and isotropic background can also be considered [11,12], but as far as the inhomogeneities remain small, the dynamics of the background essentially depends on the values of the scale factor and the homogeneous mode of the scalar field, $a(t)$ and $\varphi(t)$. From this point of view, the evolution of the universe is determined by the functions $a(t)$ and $\varphi(t)$ that extremise the Hilbert–Einstein action, which for the present case can be written as [13]:

$$S = \int dt N \left(\frac{1}{2N^2} G_{AB} \frac{dq^A}{dt} \frac{dq^B}{dt} - \mathcal{V}(q) \right), \qquad (2)$$

where $q^A = \{a, \varphi\}$ are the coordinates of the configuration space[1], G_{AB} is given by

$$G_{AB} = \text{diag}(-a, a^3), \qquad (3)$$

and $\mathcal{V}(q)$ contains all the potential terms of the spacetime and the scalar field

$$\mathcal{V}(q) = \frac{1}{2} \left(-\kappa a + a^3 V(\varphi) \right), \qquad (4)$$

where $\kappa = 0, \pm 1$ for flat, closed and open spatial slices of the whole spacetime. An explicit term for a cosmological constant is implicitly included in the case of a constant value of the potential, $V(\varphi) = \frac{\Lambda}{3}$. The Euler–Lagrange equations derived from the variation of the action (2) are [13]:

$$\frac{\ddot{a}}{a} + \frac{\dot{a}^2}{2a^2} + \frac{1}{2a^2} = -3\left(\frac{1}{2}\dot{\varphi}^2 - V\right), \quad \ddot{\varphi} + 3\frac{\dot{a}}{a}\dot{\varphi} + \frac{\partial V}{\partial \varphi} = 0. \qquad (5)$$

The Friedmann equation [14]

$$\left(\frac{\dot{a}}{a}\right)^2 + \frac{k}{a^2} = \frac{1}{2}\dot{\varphi}^2 + V(\varphi) \qquad (6)$$

turns out to be the Hamiltonian constraint that appears in quantum cosmology from the invariance of the Hilbert–Einstein action (2) under time reparametrisation, $\frac{\delta S}{\delta N} = 0$. The field Equations (5) and (6) can generally be difficult to solve analytically, but the exact or the approximate solutions of the field equations basically give the evolution of the universe. It is easy to see that these equations are invariant under the reversal change in the time variable, $t \rightarrow -t$. This means that for any given solution $a(t)$ and $\varphi(t)$ one may also consider the symmetric solution, $a(-t)$ and $\varphi(-t)$.

The action (2) and the minisupermetric (3) clearly reveal the geometric character of the configuration space, which is called the minisuperspace[2], where the scale factor would formally

---

[1] For convenience, the initial scalar field φ was rescaled according to $\varphi \rightarrow \frac{1}{\sqrt{2}}\varphi$.
[2] Generally speaking, we call superspace the space of all possible geometries, modulo diffeomorphisms, and all the matter field configurations that can be fitted in those spacetime [13,15]. However, when we restrict the degrees of freedom by the assumption of some symmetries, like the homogeneity and isotropy that we are considering here, then it is called a minisuperspace.

play the role of the time-like variable and the scalar field would formally play the role of the space-like variable[3]. Therefore, an alternative but equivalent point of view for the evolution of the universe considers that the time-dependent solutions of the scale factor and the scalar field, $a(t)$ and $\varphi(t)$, are the parametric equations of a trajectory in the minisuperspace, where the time variable t acts as the (non-affine) parameter in terms of which the trajectory of a "test universe" is described (Figure 1). The Euler–Lagrange equations associated with the action (2), given by (5), can be rewritten as the equations of the non-geodesic curves

$$\ddot{q}^A + \Gamma^A_{BC}\dot{q}^B\dot{q}^C = -G^{AB}\frac{\partial V(q)}{\partial q^B}, \tag{7}$$

where G^{AB} is the inverse of the minisupermetric G_{AB}. The momentum conjugated to the minisuperspace variables can be directly obtained from (2),

$$p_a = -\frac{a\dot{a}}{N}, \quad p_\varphi = \frac{a^3\dot{\varphi}}{N}, \tag{8}$$

and the Hamiltonian constraint, $\frac{\delta H}{\delta N} = 0$, then reads

$$G^{AB}p_A p_B + m^2_{\text{eff}}(q) = 0, \tag{9}$$

where for convenience we have defined[4] $m^2_{\text{eff}}(q) = 2V(q)$. In the present case, it yields

$$-\frac{1}{a}p_a^2 + \frac{1}{a^3}p_\varphi^2 + m^2_{\text{eff}}(q) = 0, \tag{10}$$

which is the Friedmann Equation (6) expressed in terms of the momenta instead of in terms of the time derivatives of the minisuperspace variables. As pointed out above, the geodesic Equation (7) and the momentum constraint (9) and (10) are invariant under a reversal change of the time variable. This means that the solutions may come in pairs with opposite values of the associated momenta (note that the momenta given in (8) are not invariant under the same change). From (8) and (10), it is easy to see that in terms of the cosmological time ($N = 1$), the two symmetric solutions are

$$a\frac{da}{dt} = -p_a = \pm\sqrt{\frac{1}{a^2}p_\varphi^2 + am^2_{\text{eff}}(q)}. \tag{11}$$

This is clearly reminiscent of the solutions of the trajectory of a test particle moving in spacetime [16]. For instance, in Minkowski spacetime[5], the time component of the geodesics satisfies

$$\frac{dt}{d\tau} = -p_t = \pm\sqrt{\vec{p}^2 + m^2}, \tag{12}$$

where τ is an affine parameter and $p_t = \pm E$ is the energy of the test particle. In the spacetime, the two signs in (12) represent the opposite values of the time component of the tangent vector to the geodesic—that is, the two ways in which the geodesic can be run: forward in time and backward in

---

[3] Let us recall, however, that this is just a formal analogy, and let us also notice that in the case of considering n scalar field minimally coupled to gravity, the line element of the minisuperspace would be

$$ds^2 = -ada^2 + a^3\delta_{ij}d\varphi^i d\varphi^j,$$

so the scalar fields would parallel the role of n spatial variables in a $n+1$ dimensional spacetime.

[4] Written in this way, the resemblance between the description of the trajectory in the minisuperspace and the description of a trajectory in the spacetime is quite evident.

[5] A similar procedure can be followed in a curved spacetime.

time. This was used by Feynman [1] to interpret the trajectories of an electron and a positron as the trajectory of one single electron bouncing from backwards to forwards in time (Figure 2 Top).

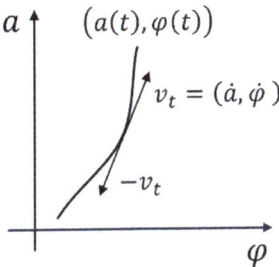

Figure 1. The evolution of the universe can be seen as a parametrised trajectory in the minisuperspace, whose parametric coordinates are given by the solutions of the field equation, $a(t)$ and $\varphi(t)$.

In the case of the universe, the two solutions given in (11) also represent two universes: one universe moving forward in the scale factor component and the other moving backward in the scale factor component. In the minisuperspace, however, moving forward in the scale factor component entails an increasing value of the scale factor, so the associated solution represents an expanding universe; and moving backward in the scale factor component entails a decreasing value of the scale factor, so the symmetric solution represents a contracting universe. Therefore, the two symmetric solutions form an expanding–contracting pair of universes (Figure 2 Bottom-Left). The total momentum conjugated to minisuperspace variables is conserved because the values of the momenta associated to the two symmetric solutions are reversely related. However, let us notice that the field Equation (5) and the Friedmann Equation (6) are invariant under a reversal change in the time variable, $t \to -t$, so from a theoretical point of view we could have chosen $-t$ as the time variable and then the solutions that represent an expanding and a contracting universe would have been interchanged. In this paper we are interested in the creation of the universe from the spacetime foam [17], so we shall interpret a contracting–expanding pair of symmetric solutions as the trajectories in the minisuperspace of two newborn universes, both expanding in terms of their reversely related time variables, $t_1 = -t_2$ (Figure 2 Bottom-Right).

Therefore, we shall assume that the universes are created in pairs, both expanding in terms of their internal, reversely related time variables. In terms of the same time variable, however, one of the universes is an expanding universe and the other is a contracting universe. For instance, for an inhabitant of one of the universes, say Alice, her universe is the expanding one and the partner universe (that she does not see) would be the contracting one. However, it is not contracting for Bob, an inhabitant of the partner universe, for whom things are the other way around, it is his universe the one that is expanding (in terms of his time variable) and Alice's universe, from his point of view, the one that is contracting. Thus, the particles that move in the two universes look like they were propagating backward and forward in time, depending on the observer's point of view. Assuming the CPT theorem, the particles that propagate in the disconnected pieces of the spacetime have consequently opposite values of their charge and parity, so they can be identified in the quantum theory of the composite system with particles and antiparticles. The inhabitants of the two universes can only see the particles that propagate in their own spacetimes, but if they would find any signature of the existence of a time reversely related universe, then they could infer that at the onset primordial antimatter was mainly created in the partner universe, and thus they could conclude that the matter–antimatter asymmetry that they observe in their universes is only an apparent asymmetry that becomes restored in the composite picture of the two symmetric universes.

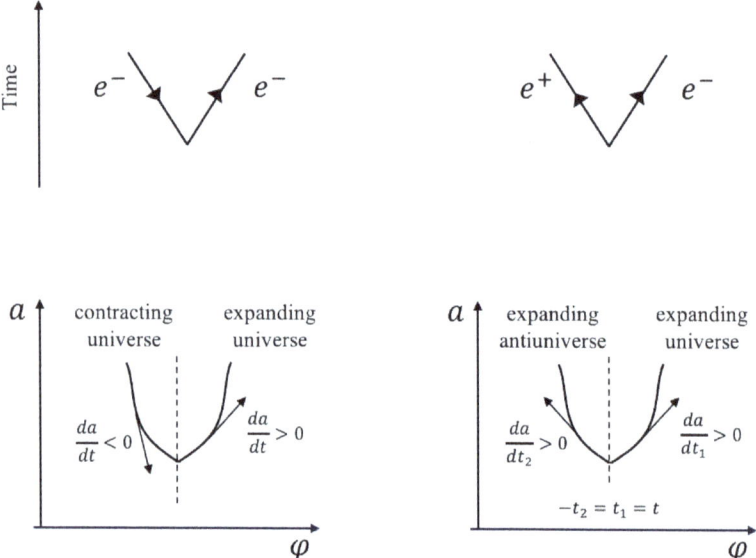

Figure 2. Top: the creation of an electron-positron pair (**Right**) can equivalently be seen as the trajectory of an electron bouncing from backwards to forwards in time (**Left**). Bottom: two symmetric cosmological solutions can represent, in terms of the same time variable t, a contracting and an expanding universe (**Left**). In terms of the reversely related time variables, $t_1 = -t_2$, the two symmetric solutions can represent a universe-antiuniverse pair (**Right**).

3. Quantum Cosmology and the Creation of Universes

One would expect that the creation of the universe should have a quantum origin. Therefore, let us analyse the creation of a pair of time reversely related universes in quantum cosmology. The quantum state of the universe is described by a wave function that depends on the metric components of the spacetime as well as on the degrees of freedom of the matter fields. It is the solution of the Wheeler–DeWitt equation, which is essentially the canonically quantised version of the classical Hamiltonian constraint. This is generally a very complicated function. However, as was pointed out in Section 1, we are assuming that the universe emerges from the gravitational vacuum and starts inflating from an initial spatial hypersurface $\Sigma(a_i)$ that is small but large enough to consider that the fluctuations of the spacetime are subdominant. In that case, we can model the universe as weakly coupled fields propagating in an essentially homogeneous and isotropic background spacetime. Then, the Hamiltonian of the whole universe can be split into the Hamiltonian of the background and the Hamiltonian of the matter fields [11,12]

$$(\hat{H}_{bg} + \hat{H}_m)\phi = 0, \tag{13}$$

where the Hamiltonian of the background spacetime, H_{bg}, is given by the quantum version of the classical Hamiltonian (10)

$$\hat{H}_{bg} = \frac{1}{2a}\left(\hbar^2 \frac{\partial^2}{\partial a^2} + \frac{\hbar^2}{a}\frac{\partial}{\partial a} - \frac{\hbar^2}{a^2}\frac{\partial^2}{\partial \phi^2} + a^4 V(\varphi) - a^2\right), \tag{14}$$

and H_m contains the Hamiltonian of all the matter fields and their interactions. The wave function, $\phi = \phi(a, \varphi; x_m)$, where x_m are the variables of the local matter fields, can then be expressed in the semiclassical regime as a linear combination of WKB solutions, that is [4,5]:

$$\phi = \sum \phi_+ + \phi_- = \sum C e^{\frac{i}{\hbar}S} \psi_+ + C e^{-\frac{i}{\hbar}S} \psi_-, \qquad (15)$$

where $C = C(a, \varphi)$ is a real slow-varying function of the background variables, $S = S(a, \varphi)$ is the action of the background spacetime, $\psi = \psi(a, \varphi; x_m)$ is a complex wave function that contains all the dependence on the matter degrees of freedom, with $\psi_- = \psi_+^*$, and the sum in (15) extends to all possible classical configurations. A relevant feature to be noticed here is that, because of the Hermitian character of the Hamiltonian (13)—which in turn is rooted in the time reversal symmetry of the classical Hamiltonian constraint (10)—the general solution of the Wheeler–DeWitt equation can always be expressed in terms of the two complex conjugated, independent solutions that correspond to the two possible signs in the exponentials of (15). We shall now see that these two wave functions represent similar universes with essentially the same evolution of their spacetimes and similar matter fields propagating therein. However, because the momenta conjugated to the scale factor associated to the two complex conjugated solutions in (15) are reversely related, the time variables of their spacetimes are also reversely related.

In order to see how the wave functions ϕ_+ and ϕ_- of (15) represent a particular universe, one can insert them into the Wheeler–DeWitt Equation (13). After some decoherence process between the two wave functions, which is a guarantee for the smallness of the fluctuations of the spacetime [5,18], one can expect that the Wheeler–DeWitt equation must be satisfied order by order in an expansion in $\hbar$. At order $\hbar^0$ one obtains the following Hamilton–Jacobi equation [3,12]:

$$-\left(\frac{\partial S}{\partial a}\right)^2 + \frac{1}{a^2}\left(\frac{\partial S}{\partial \varphi}\right)^2 + a^4 V(\varphi) - a^2 = 0. \qquad (16)$$

This equation contains the dynamics of the background spacetime. It can be converted into the Friedmann equation by defining the WKB time variable given by [12]

$$\frac{\partial}{\partial t} = \pm \nabla S \cdot \nabla \equiv \pm \left(-\frac{1}{a}\frac{\partial S}{\partial a}\frac{\partial}{\partial a} + \frac{1}{a^3}\frac{\partial S}{\partial \varphi}\frac{\partial}{\partial \varphi}\right), \qquad (17)$$

where ∇ is the gradient of the minisuperspace [12]. In terms of the WKB time variable,

$$\dot{a}^2 = \frac{1}{a^2}\left(\frac{\partial S}{\partial a}\right)^2, \quad \dot{\varphi}^2 = \frac{1}{a^6}\left(\frac{\partial S}{\partial \varphi}\right)^2, \qquad (18)$$

so that the Hamilton–Jacobi Equation (16) turns out to be the Friedmann Equation (6). It thus describes the evolution of the background spacetime. Furthermore, let us notice that at order $\hbar^0$ the momentum conjugated to the scale factor associated to the wave functions ϕ_+ and ϕ_- is given by

$$p_a = -i\hbar \frac{\partial \phi_\pm}{\partial a} = \pm \frac{\partial S}{\partial a}, \qquad (19)$$

where the plus sign corresponds to ϕ_+ and the minus sign to ϕ_-. They are thus reversely related and the total momentum associated to the creation of the two universes represented by ϕ_+ and ϕ_- is zero.

Furthermore, at first order in $\hbar$ in the expansion of the Wheeler–DeWitt equation, one obtains [3,12]:

$$\mp i\hbar \left(-\frac{1}{a}\frac{\partial S}{\partial a}\frac{\partial}{\partial a} + \frac{1}{a^3}\frac{\partial S}{\partial \varphi}\frac{\partial}{\partial \varphi}\right)\psi = H_m \psi, \qquad (20)$$

where the positive and negative signs correspond to ϕ_- and ϕ_+ (15), respectively. The term in brackets in (20) is actually the WKB time variable defined in the background spacetime, given by (17), meaning that (20) is essentially the Schrödinger equation for the matter fields that propagate in the classical background spacetime represented by ϕ_+ and ϕ_-. We then recover the semiclassical picture of quantum matter fields propagating in a classical background. However, in order to have the proper sign in the Schrödinger equation in each single universe, we need to choose the positive sign in the definition of the time variable t in (17) for the wave function ϕ_- and the negative sign of the time variable for the wave function ϕ_+. If we assume that the time variable involved in the Schrödinger equation is the *physical* time variable in the sense that it is the time measured by the observers in their particle physics experiments (so it is the time variable measured by actual clocks, which are eventually made of matter), then the physical time variables of the two universes are reversely related. Note that the eventual inhabitants of the universes will only see the matter of their respective universes and therefore cannot observe the antimatter (from their point of view) that propagates in the symmetric universe. There is also a Euclidean gap between the two universes that prevents matter and antimatter from collapse (Figure 3). Therefore, from the point of view of an individual observer there is nothing in principle that makes him suspect the existence of a time reversely symmetric universe except perhaps the occurrence of an asymmetry that is hard to explain within the single universe scenario. In principle, it is only from a symmetry consideration that the observer can pose the existence of another universe that justifies the apparent primordial asymmetry between matter and antimatter[6].

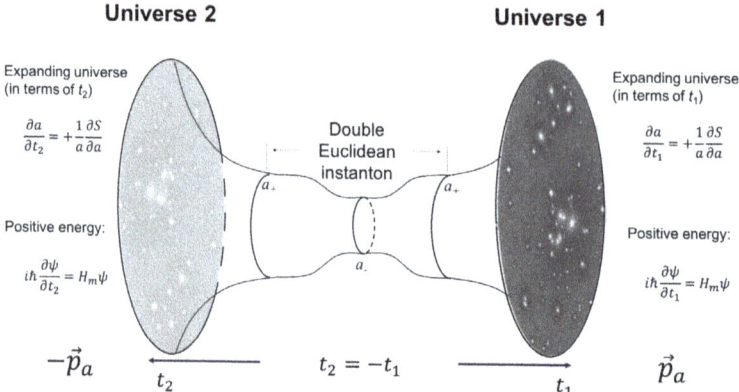

Figure 3. The creation of universes in entangled pairs [3]. The time variables of the two symmetric universes are reversely related. This provides us with the correct value of the Schrödinger equation in the two universes. At the onset, primordial matter would be created in the observer's universe and antimatter in the symmetric one. Particles and antiparticles do not collapse because of the Euclidean gap that exists between the two newborn universes [3,10].

The creation of particles and antiparticles would follow a similar procedure to that customary considered in the context of a single universe (e.g., [19]). After the slow roll regime the inflaton field eventually approaches a minimum of the potential and starts oscillating. In each oscillation it decays through different channels and in subsequent stages into the particles of the standard model. The interaction with the inflaton field produces a series of consecutive, non-adiabatic changes of the vacuum of the matter fields. The associated particle production can then be derived from the

---

[6] That does not exclude the possibility that other mechanisms of matter–antimatter asymmetry given in the context of a single universe can also contribute to the total asymmetry observed, but it does assume that the main mechanism would be the creation of matter and antimatter in two separated symmetric universes, provided that the former are not yet fully satisfactory within the Standard Model of particle physics.

Bogolyubov transformation that relates the vacuum states before and after the non-adiabatic change, with a particle production given by

$$|0\bar{0}\rangle = \prod_k \frac{1}{|\alpha_k|^{\frac{1}{2}}} \left(\sum_n \left(\frac{\beta_k}{2\alpha_k} \right)^n |n_k, \bar{n}_{-k}\rangle \right), \tag{21}$$

where $|0\bar{0}\rangle$ is the vacuum state of the corresponding matter field before the oscillation and $|n_k, \bar{n}_{-k}\rangle$ is the state representing the number of particles and antiparticles created with momentum k after the interaction. The functions α_k and β_k are the coefficients of the Bogolyubov transformation that relates the wave functions of the modes before and after the interaction, and they contain the effect of the interaction between the inflaton and the matter fields. In the single-universe scenario, because the symmetries of the background, particles and antiparticles are created in perfectly correlated pairs $|n_k, \bar{n}_{-k}\rangle$ so the number of them is exactly balanced, at least in principle. Different mechanisms that would produce some asymmetry in the creation of matter and antimatter are invoked [19]. However, they usually consider some modification or extension of the standard model of particle physics.

In the scenario of the twin universes with reversely related time variables presented here, the state of the matter field would still be given by (21). However, matter and antimatter would be created in different universes, so (21) should be rewritten as:

$$|0^{(1)}0^{(2)}\rangle = \prod_k \frac{1}{|\alpha_k|^{\frac{1}{2}}} \left(\sum_n \left(\frac{\beta_k}{2\alpha_k} \right)^n |n_k^{(1)}, n_{-k}^{(2)}\rangle \right), \tag{22}$$

where $n_k^{(1)}$ are the particles in the mode k created in one of the universes, labeled as 1, and $n_k^{(2)}$ are the particles in the mode k created in the other universe, labeled as 2. For the inhabitants of the two universes, the primordial antimatter is essentially created in the partner universe, that is, for an observer in universe 1, $n_k^{(2)}$ are the antiparticles that she misses in her universe and, analogously, $n_k^{(1)}$ are the antiparticles in mode k that are left from the point of view of an observer of universe 2. Particles and antiparticles cannot interact, and therefore cannot annihilate each other, because of the Euclidean gap (a quantum barrier) that separates the universes (see Figure 3 and Refs. [3,10]).

Let us now make a comment about the influence that the fluctuations of the spacetime could have in the global picture of the creation of the universes in symmetric pairs and the subsequent creation of matter and antimatter separately. The assumption of homogeneity and isotropy made from the very beginning is determining for the time reversal symmetry of the field equations, and eventually for the creation of the universes in symmetric pairs. This means that the scenario presented here is at most a plausible scenario. However, one would expect that as long as the deviation from the homogeneous and isotropic background is relatively small, the picture would essentially be rather similar. The small fluctuations of the spacetime would cause the symmetry to stop being an exact symmetry to become an approximate symmetry. It would be expected that the creation of particles and antiparticles would cease to occur in perfectly correlated pairs and that (22) would only be approximate, with corrections at different levels. However, one might expect that this mechanism of creating matter and antimatter separately could still be dominant, as long as the deviations from the homogeneous and isotropic spacetime were small. In addition, other mechanisms that produce matter and antimatter asymmetry can still be considered, and they could work jointly to produce the observed amount of asymmetry. Perhaps the creation of universes in (imperfectly) symmetric pairs could relax the tensions or the anomalous fine tuning that might exist between the theoretical models and the observational data. It might also entail other observable effects [20]. Therefore, we believe that it is an appealing scenario that deserves further analysis and a deeper understanding.

4. Conclusions

The evolution of the universe can be seen as a worldline of the minisuperspace formed by the scale factor, which formally plays the role of a time-like variable, and the scalar fields, which formally play the role of the spatial components. From that point of view, the time reversal symmetry of the field equations becomes equivalent to the invariance of the geodesics of the minisuperspace under a reversal parametrisation of their non-affine parameter.

Positively oriented paths with respect to the scale factor component in the minisuperspace entail an increasing value of the scale factor, so they represent expanding universes. On the contrary, negatively oriented worldlines with respect to the scale factor component represent contracting universes. However, because of the time reversal invariance of the Lagrangian of a homogeneous and isotropic spacetime, the terms "expansion" and "contraction" can be interchanged so we end up with four symmetric solutions—two oriented forward and two oriented backwards. The former represent two expanding universes with their time variables reversely related and the latter represent two virtual universes that rapidly return to the gravitational vacuum from which they emerged.

Quantum mechanically, the solutions of the Wheeler–DeWitt equation can always be given in complex conjugated pairs because of the Hermitian character of the Hamiltonian constraint. Two complex conjugated solutions entail opposite values of the momentum conjugated to their scale factors, so the creation of universes in pairs whose wave functions are complex conjugated has an associated total zero momentum. Furthermore, the analysis of the emergence of the semiclassical spacetime in the universes that they represent suggests that the physical time variables of their spacetimes should be reversely related. Then, the inhabitants of the two universes would naturally identify matter with the particles that propagate in their spacetimes and the unobserved primordial antimatter with the particles that would propagate in the symmetric universe.

The consideration of a homogeneous and isotropic spacetime makes the scenario presented here at most a plausible one. However, it might be realistic if our universe was created and started inflating from an initial hypersurface that is small but large enough to assume that the fluctuations of the spacetime are subdominant. In that case, small deviations from the homogeneous and isotropy of the spacetime would transform the exact time reversal symmetry into an approximate one. Nevertheless, as long as the deviations are relatively small, one would expect that the global picture would not differ very much from the one depicted here. The creation of matter and antimatter in separated universes might still be the dominant one, or at least it could help or enhance other mechanisms that are already considered in the context of a single scenario.

Finally, this work sets the arena for a deeper and more detailed study, which might eventually unveil whether the universe is actually part of a universe–antiuniverse pair.

Funding: This research received no external funding.

Conflicts of Interest: The author declares no conflict of interest.

References

1. Feynman, R.P. The theory of positrons. *Phys. Rev.* **1949**, *76*, 749. [CrossRef]
2. Robles-Pérez, S. Quantum cosmology of a conformal multiverse. *Phys. Rev. D* **2017**, *96*, 063511. [CrossRef]
3. Robles-Pérez, S.J. Cosmological perturbations in the entangled inflationary universe. *Phys. Rev. D* **2018**, *97*, 066018. [CrossRef]
4. Hartle, J.B. The quantum mechanics of cosmology. In *Quantum Cosmology and Baby Universes*; Coleman, S.; Hartle, J.B.; Piran, T.; Weinberg, S., Eds.; World Scientific: London, UK, 1990; Volume 7.
5. Kiefer, C. Decoherence in quantum electrodynamics and quantum gravity. *Phys. Rev. D* **1992**, *46*, 1658–1670. [CrossRef]
6. Bezrukov, F.L.; Shaposhnikov, M.E. The Standard Model Higgs boson as the inflaton. *Phys. Lett. B* **2008**, *659*, 703. [CrossRef]

7. Robles-Pérez, S.; González-Díaz, P.F. Quantum entanglement in the multiverse. *J. Exp. Theor. Phys.* **2014**, *118*, 34–53. [CrossRef]
8. Robles-Pérez, S.J. Creation of entangled universes avoids the Big Bang singularity. *J. Gravity* **2014**, *2014*, 382675. [CrossRef]
9. Robles-Pérez, S.; Balcerzak, A.; Dąbrowski, M.P.; Krämer, M. Inter-universal entanglement in a cyclic multiverse. *Phys. Rev. D* **2017**, *95*, 083505. [CrossRef]
10. Robles-Pérez, S.J. Restoration of matter-antimatter symmetry in the multiverse. *arXiv* **2017**, arXiv:1706.06304.
11. Halliwell, J.J.; Hawking, S.W. Origin of structure in the Universe. *Phys. Rev. D* **1985**, *31*, 1777–1791. [CrossRef]
12. Kiefer, C. Continuous measurement of mini-superspace variables by higher multipoles. *Class. Quant. Grav.* **1987**, *4*, 1369–1382. [CrossRef]
13. Kiefer, C. *Quantum Gravity*; Oxford University Press: Oxford, UK, 2007.
14. Linde, A. Particle physics and inflationary cosmology. In *Contemporary Concepts in Physics*; Harwood Academic Publishers: Chur, Switzerland, 1993; Volume 5.
15. Wiltshire, D.L. An Introduction to Quantum Cosmology. In *Cosmology: The Physics of the Universe*; Robson, B., Visvanathan, N., Woolcock, W., Eds.; World Scientific: Singapore, 1996; pp. 473–531.
16. Garay, I.; Robles-Pérez, S. Classical geodesics from the canonical quantisation of spacetime coordinates. *arXiv* **2019**, arXiv:1901.05171.
17. Hawking, S.W. Spacetime foam. *Nucl. Phys. B* **1978**, *144*, 349–362. [CrossRef]
18. Halliwell, J.J. Decoherence in quantum cosmology. *Phys. Rev. D* **1989**, *39*, 2912–2923. [CrossRef]
19. Mukhanov, V.F. *Physical Foundations of Cosmology*; Cambridge University Press: Cambridge, UK, 2008.
20. Boyle, L.; Finn, K.; Turok, N. CPT-Symmetric universe. *Phys. Rev. Lett.* **2018**, *121*, 251301. [CrossRef]

© 2019 by the authors. Licensee MDPI, Basel, Switzerland. This article is an open access article distributed under the terms and conditions of the Creative Commons Attribution (CC BY) license (http://creativecommons.org/licenses/by/4.0/).

Article

Possible Origins and Properties of an Expanding, Dark Energy Providing *Dark Multiverse*

Eckhard Rebhan

Institut für Theoretische Physik, Heinrich-Heine-Universität, D-40225 Düsseldorf, Germany;
rebhan@thphy.uni-duesseldorf.de

Received: 31 May 2019; Accepted:17 July 2019; Published: 24 July 2019

Abstract: The model of a multiverse is advanced, which endows subuniverses like ours with space and time and imparts to their matter all information about the physical laws. It expands driven by dark energy (DE), which is felt in our Universe (U) by mass input and expansion–acceleration. This *dark multiverse* (DM) owes its origin to a creatio ex nihilo, described in previous work by a tunneling process in quasi-classical approximation. Here, this origin is treated again in the context of quantum gravity (QG) by solving a Wheeler de Witt (WdW) equation. Different than usual, the minisuperspace employed is not spanned by the expansion parameter a but by the volume $2\pi^2 a^3$. This not only modifies the WdW-equation, but also probabilities and solution properties. A "soft entry" can serve the same purpose as a tunneling process. Sections of solutions are identified, which show qualitative features of a volume-quantisation, albeit without a stringent quantitative definition. A timeless, spatially four-dimensional primordial state is also treated, modifying a state proposed by Hartle and Hawking (HH). For the later classical evolution, elaborated in earlier papers, a wave function is calculated and linked to the solutions for the quantum regime (QR). It is interpreted to mean that the expansion of the DM proceeds in submicroscopic leaps. Further results are also derived for the classical solutions.

Keywords: multiverse; dark energy; creation from nothing; soft entry; quantum gravity; Wheeler-de Witt-equation; Bohm-like interpretation; volume-quantisation; space atoms; information storage and transfer

1. Introduction

The multiverse conception is currently in a state which is comparable to that of black holes before the proof of their existence. Because one does not know for sure if something like a multiverse exists, and since all the more nothing is known about its properties, the occupation with multiverses has the touch of the exotic. In addition, it is not foreseeable that one comes to solid conclusions about their existence and characteristics as fast as with black holes. It is therefore not surprising that there are very different concepts for them. It is certainly advantageous, if a concept sticks to proven equations: it has often turned out that previously unknown phenomena are actually realized in nature, if they are in harmony with approved physics. In addition, it would speak for a specific concept if detectable influences on U would result from it. However, it cannot be ruled out that a concept, based on modified basic equations, will finally prevail. If some day a particular concept comes out on top, the others become more or less obsolete, which could of course also happen to the current concept. However, that does not necessarily mean that the others were completely worthless because they could have paved the way for the ultimate winner.

The present concept was initiated in Ref. [1], refined in Ref. [2] and is further developed in this paper. The most important results of the earlier work are briefly summarized in Section 2. Section 3 deals with two scenarios for the origin of the DM, the first with a tunneling process as proposed by

Vilenkin [3,4] and Linde [5] (VL), and the second with a modification of a timeless primordial state proposed by Hartle and Hawking [6]. Both scenarios are dealt with by using two methods, 1. relatively short with a quasi-classical, approximate treatment of the QR, and 2. much more detailed in the context of QG by solving an unusual WdW-equation, which we consider appropriate and deduce in the Appendix. Furthermore, an equation for the classical regime (CR) is derived, whose solution is smoothly connected to that for the QR and interpreted in a special way. In doing so, we encounter a solution that represents an alternative to the tunneling solution and can be called a "soft entry". In Section 4.1, we critically reflect on the basic idea of our DM-concept, the storage and transmission of information about the laws of physics. In this context, we point to difficulties arising from loop theory of QG (LQG) and suggest ways in which they may be overcome. The identification of solution sections exhibiting properties of a spatial volume quantisation provides this subsection with a particular significance. In Section 5, properties of the classic solution are elaborated, which were only hinted at or not dealt with in Ref. [2]. At first, the conditions are calculated, which, allowing for primordial matter, enable an (unstable) equilibrium between this and the DE at the beginning of the DM-expansion.

Then, the minimal the age of the DM is calculated, which is compatible with the maximum curvature allowed by corresponding measurements in U. It is also clarified how it comes about that the irreversible solution of the cosmological equations with a friction term in the equation for the expansion acceleration, $\ddot{a}(t)$, solves as well the reversible equations for a scalar field driven expansion. Finally, it is investigated how the DM-expansion behaves in comparison to the expansion of U.

For the readers' convenience, all abbreviations are listed before Appendix A.

2. Summary of Previous Work on the DM-Concept

Essence and determination of the considered DM are that the latter provides space and time for subuniverses with ours among them, i.e., the latter arise in an already prefabricated space. Thus, it was not generated along with them but long before. Thus far, we assumed that it originated at a finite time in the past by a creation from nothing, emerging continuously from nothingness through quantum mechanical tunneling. For this, it must be a closed and expanding multiverse of positive curvature, if topologically more complex situations (three-torus) are excluded, and its expansion must start at zero velocity immediately after the tunneling. The maximum spatial curvature that would be compatible with relevant measurements of negative result is so small that the space, in which U lives, extends far beyond its particle horizon. As a consequence, large areas exist outside that are causally not connected with the interior, are thus structurally quite different, and can be considered as parts of a multiverse embracing U.

The ingredient forming the backbone of the DM or the space-time spanned by it, respectively, is assumed to be dark energy (DE). It is no additional extra, but the medium which is generating space and time. Subuniverses like ours are assumed to originate from fluctuations of independent inflation fields. Since the DE of the DM also fills U, it can be considered as a fingerprint of the DM on U. It causes not only a continuously accelerated expansion of the DM but is also responsible for the accelerated expansion presently observed in U, if its mass density ϱ_Λ is properly adjusted. If, as in inflation theories, its initial value is chosen close to the Planck density, then ϱ_Λ must decay by a factor of about 10^{-120} so it can be identified with the DE observed in U. The decay factor 10^{-120} agrees approximately with the factor, by which the presently observed mass density ϱ_0 of the cosmological constant differs from the value predicted by elementary particle theory. This suggested the assumption of a mechanism, by which, in the course of the increasing expansion of the DM, the effect of a huge and unchangeable cosmological constant is continuously reduced down to its present much lower efficiency. It turned out that this mechanism can be intimately linked to a proposed solution for another, so far largely ignored conundrum of physics, namely the question of how it comes about that in each point of space-time the physical laws are obeyed.

For this purpose, going far beyond what was set out above, we assumed that the space-time of the DM is of a very special kind in that it is encoded with all the information about the physical laws, to which all material components occurring in the subuniverses must obey. This could be achieved in a geometric way, as is the case with the laws of gravitation according to the general theory of relativity, or with the laws of electrodynamics, using an additional spatial dimension as in the Kaluza–Klein theory. Alternatively, the physical laws could as well be stored in subatomic structures of a granular space-time in a similar way as the laws of biological growth are encoded in the DNA. According to our former assumptions, the information about all physical laws, and the physical agents to whom they relate, is transmitted through the creational tunneling process to the subsequent initial state of the DM. This means that the primordial nothingness is understood as a state of pure information without space and time.

As stated above, we assume that ϱ_Λ keeps its large initial value during the whole evolution of the DM. Without reduction it would cause an extremely accelerated expansion. The space added by this must be equipped with all information listed above via transfer from already existing space. This will take time that is not available, if the spatial expansion proceeds too fast. Therefore, we assumed that this transfer impedes the spatial expansion, and that this can be cumulatively represented by a friction term in the cosmological equations. Without it, the Friedman–Lemaître (FL) equation would be $\dot{a}^2(t) = (8\pi G/3)\varrho_\Lambda a^2 - c^2$ with the immediate consequence $\ddot{a}(t) = (8\pi G/3)\varrho_\Lambda a$. Introducing a linear friction term, the latter equation becomes

$$\ddot{a}(t) = -f\dot{a}(t) + \frac{8\pi G}{3}\varrho_\Lambda a, \tag{1}$$

where $f > 0$ is a constant. Multiplication with $\dot{a}(t)$ and integration with respect to t yields

$$\dot{a}^2(t) = \frac{8\pi G}{3}\varrho(t)a^2 - c^2 \quad \text{with} \quad \varrho = \varrho_\Lambda + \varrho_f \quad \varrho_f = -\frac{3f}{4\pi G a^2}\int_0^t \dot{a}^2(t')\,dt'. \tag{2}$$

The integration constant was chosen such that, at $t=0$, the time immediately after the tunneling process, the DM starts with $\dot{x}(t)_{t=0} = 0$ and $\varrho = \varrho_\Lambda$, where

$$\varrho_\Lambda = \frac{\varrho_*}{x_i^2} \quad \text{with} \quad \varrho_* = \frac{3}{8\pi}\varrho_P = \frac{3c^2}{8\pi G\, l_P^2} \tag{3}$$

($a_i = a(0)$ and $x_i = a_i/l_P$ with $l_P = \sqrt{\hbar G/c^3}$ = Planck length, ϱ_P = Planck density), and evolves according to the usual FL-Equation (2). It is therefore possible to say that our model leads to solutions of the standard theory which are only re-interpreted in an unusual way. In Ref. [2], it was shown that even the cosmological equations for a scalar field Φ, driven by a potential $V(\Phi)$, are satisfied. As a result, the interaction of the huge cosmological constant with energy density $\varrho_\Lambda c^2$ and the friction force $-f\dot{a}(t)$ of Equation (1) has exactly the same effect as a dark energy (exerting an anti-gravitational force and providing mass in our universe), which is represented by a scalar field Φ with energy density $\varrho c^2 = (\varrho_\Lambda + \varrho_f)c^2$, and is therefore also referred to as DE. Its origin is inseparably linked to the emergence of space-time because, according to the above, the latter is generated by it. Because of the associated information about properties like mass or charge, in principle, particles should also be incorporated in the FL-equation at least in the form of a cumulative mass density. Since this would not disclose anything new as compared to the results of Ref [1], as in Ref [2], this is relinquished here.

In the dimensionless quantities,

$$x = \frac{a}{l_P} \quad \tau = \frac{t}{t_P} \quad \rho = \frac{\varrho}{\varrho_\Lambda} \tag{4}$$

($t_P = \sqrt{\hbar G/c^5}$ = Planck time), Equations (1) and (2) become[1]

$$\ddot{x}(\tau) + 2\sigma \dot{x}(\tau) - x/x_i^2 = 0 \quad \text{with} \quad \sigma = \frac{f t_P}{2} \tag{5}$$

and

$$\dot{x}^2(\tau) = \rho \left(x/x_i \right)^2 - 1. \tag{6}$$

The solution of the last equation for the initial conditions $x = x_i$ and $\dot{x}(\tau) = 0$ at $\tau = 0$ is

$$x(\tau) = \frac{e^{\gamma \tau} + \gamma^2 x_i^2 \, e^{-\tau/(\gamma x_i^2)}}{1 + \gamma^2 x_i^2} x_i \quad \text{with} \quad \gamma = \sqrt{1/x_i^2 + \sigma^2} - \sigma. \tag{7}$$

Employing the Wick rotation $\tau = -iu$ for $\tau < 0$, Equation (6) is converted into

$$\dot{x}^2(u) = 1 - \rho \, x^2/x_i^2. \tag{8}$$

For $\rho \equiv 1$ and $x = 1$ plus $\dot{x}(u) = 0$ at $u = 0$, the solution $x(u) = \cos u$, shown in Figure 1, is obtained. Since γ and γx_i are both extremely small,

$$x(\tau) = x_i \, e^{\gamma \tau} \quad \text{for all} \quad \tau \geq 0 \tag{9}$$

is an excellent approximation. Equation (6) is satisfied by calculating from it the mass density

$$\rho(\tau) = \left(1 + \dot{x}^2(\tau)\right) x_i^2/x^2. \tag{10}$$

Inserting in it solution (9), using Equation (4) and the last of Equations (3), and cutting out x_i^2, we obtain for the present state $x = x_0$ by solving for γ

$$\gamma = \sqrt{\frac{\varrho_0}{\varrho_*} - \frac{1}{x_0^2}} \approx \sqrt{\frac{\varrho_0}{\varrho_*}} \approx 10^{-61}, \tag{11}$$

where ϱ_0 is the value of the DE-density measured at present in U, and $\tau_0 \gtrsim 175 \, \tau_{u0}$ (with τ_{u0} = present age of U) is the condition that allows the neglect of $1/x_0^2$ according to Ref. [2]. Because γ is so small, $\rho = x_i^2/x^2$ holds according to Equation (10), and the volume $v = V/(2\pi^2 l_P^3)$ is given by $v = x^3$ according to Equation (A10) of the Appendix. The total energy of M is therefore $E = \varrho c^2 V \sim x_i^2 x$, so it grows with increasing expansion $x(t)$. Consequently, the friction force $-f\dot{a}(t)$ does not, as usual, convert potential energy into heat; instead, it only slows down the energy increase of M and, simultaneously, the decrease of (negative) gravitational energy.

Annotation: We consider the case $x_i \neq 1$ only in Section 4.1, everywhere else we assume $x_i = 1$.

---

[1] For later purposes (Section 4.1), we consider here a slightly more general case. Unfortunately, when Ref. [2] was written down for printing, in the derivation of Equation (26) there (Equation (7) here), two different types of calculation were mixed up, which resulted in two sign errors. For correction, the following substitutions must be made there: $e^{\gamma_2} \to e^{-\gamma_2}$ in Equation (22), and $\alpha \gamma_1 + \beta \gamma_2 \to \alpha \gamma_1 - \beta \gamma_2$ in the unnumbered equation immediately thereafter. The resulting Equation (26) is, however, correct.

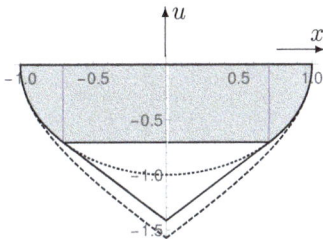

Figure 1. Approximate solutions for the QR, discussed from the bottom up. 1. dashed curve: $u = \arccos x$, 2. full curve: solution (13) continued by straight lines, 3. dotted curve: semicircle, 4. curve with gray filling: HH-like solution.

3. Two Scenarios for the Origin of the Dark Multiverse

In Refs. [1,2], it was assumed that the DM comes about by a creation from nothing, conveyed by a quantum mechanical tunneling process as introduced into cosmology by Vilenkin [3,4] and Linde [5]. At about the same time, Hartle and Hawking presented a quite different proposal, according to which the primordial condition of the universe is a timeless quantum state in four spatial dimensions on equal footing, filling essentially a 4D-sphere. The VL-approach employs four spatial coordinates as well; however, one of them comes out by a Wick rotation of the time, a mathematical tool, by which the effectively time-dependent tunneling process is described in a particularly easy, albeit approximate manner. In the HH-approach, on the other hand, the positively curved space is truly four-dimensional; due to symmetry requirements (no boundary condition), no point is distinguished from the others, which in any way could be interpreted as a starting point. Nevertheless, time is still implicitly contained in that the wave function of the timeless state is linked to a three-dimensional probability density and depends on the metric coefficients of a 3D-space, while matter fields involve a 3D-mass-density. No simple approximation of the quantum behavior is possible as in the VL-approach, rather a true quantum theory of the gravitational field must be invoked. For this purpose, the WdW-equation is chosen, according to which the states of the multiverse are generally timeless.

In this paper, two separate approaches are undertaken, the first modifying the VL-approach, and the second the HH-approach.

3.1. Quasiclassical Approximation of the Quantum Regime

1. *Creation from nothing through tunneling.* It is interesting to look into the question of whether a tunneling solution of Equation (8) is also possible when $x(u)$, as in the HH-approach, forms a semicircle (page 86 of Ref. [7]) so that no point is distinguished. Accordingly, we demand the validity of the equation

$$x(u) = -\sqrt{1 - u^2} \qquad (12)$$

and determine ρ from Equation (8) in such a way that Equation (12) is satisfied. From the latter follows $\dot{x}^2(u) = u^2/(1-u^2)$. Inserting this in Equation (8), eliminating u by use of Equation (12), and resolving with respect to ρ yields

$$\rho(x) = \frac{2x^2 - 1}{x^4} \begin{cases} = 0 & \text{for} \quad x = 1/\sqrt{2}, \\ < 0 & \text{for} \quad x < 1/\sqrt{2}, \\ \to -\infty & \text{for} \quad x \to 0. \end{cases} \qquad (13)$$

This treatment of the QR is an approximation which is getting worse the deeper one gets into it. In that sense, the negative values of the density ρ may perhaps not be taken seriously. However, for $x \to 0$, a singularity appears that is just what should be avoided in a quantum treatment.

There are two ways to get out of this situation with a viable solution.

1. the solution (13), yielding $\rho(x) \geq 0$ in the range $1/\sqrt{2} \leq x \leq 1$, is continued into the range $1/\sqrt{2} \geq x \geq 0$ by a straight line with the slope $\dot{x}(u) = \pm 1$ following from Equation (8) for $\rho = 0$.
2. The solution (13) is cut off at $x = 1/\sqrt{2}$.

Both possibilities are shown in Figure 1. The truncated HH-like solution, supplemented by two straight lines, has a peak which constitutes a distinguished point and must therefore be regarded as a tunneling solution of the VL-type. It represents an alternative to the solution (45) of Ref. [2], $x(u) = \cos u$ or $u = \arccos x$, but does not comply with the HH-concept.

2. *HH-like primordial state.* In the case of the solution cut off at $x = 1/\sqrt{2}$ with $\rho = 0$, the situation changes. If the two end points are connected by a horizontal straight line and the resulting corners rounded, one arrives at a solution, whose boundary, although not having a constant curvature, does not contain any point which could be interpreted as a temporal beginning. Although this solution results from the effort to construct a VL-like tunneling solution, it makes sense to accept it as a solution compatible with the HH-intention, so that the fourth spatial coordinate (which resulted from time by a Wick Rotation) is on equal footing to the others. It is obvious that the treatment of this model with the WdW-equation in the context of QG is much more appropriate because of its time independence.

3.2. Employing Quantum Gravity with Use of a Wheeler-De Witt Equation

In this section, the solution of the WdW-equation [8,9], applied to a properly chosen minisuperspace, is used to investigate the tunneling process considered in Ref. [2] (and ending up in a modification of the latter), and a modified HH-concept. Despite the 50 years that have passed since the introduction of this equation, and despite the many papers and book contributions in which it was treated, it is still not as straightforward as in ordinary quantum mechanics to find the right solution for a given concept and to interpret it appropriately (see e.g., [10], quoted in pages 206–207 of Ref. [11], or page 135 of Ref. [12]). This is partly because there are two types of probability interpretation, either by transition probabilities or by the square of the wave function. In addition, problems arise with respect to the Hermiteness of operators, setting proper boundary conditions, and the role of time. Considerable effort was made to find analytic solutions to the WdW-equation (see e.g., [13]) and study such conditions with them.

The ambition of this paper goes in a different direction. We want to use the square of the wave function for calculating probabilities. However, regarding its normalization and the associated probability interpretation, we try a way that deviates from the usual. In doing so, we find it useful to base the WdW-equation on a different minisuperspace for the following reason: in several places, the three-dimensionality of space enters the relevant WdW-equation in a decisive way. This can be seen in its derivation, which is therefore carried out in the Appendix A with reference to the critical points. Furthermore, densities contained in it like ϱ relate to three dimensions. Finally, the three-dimensionality is also reflected in solutions. On the one hand, the wave function for the CR depends on a^3, if it is calculated in the usual way (see item 3 in Section 3.2.1). On the other hand, its square, the probability density, decreases over time instead of increasing as expected. All of this has prompted us to choose $V = 2\pi^2 a^3$ instead of a as the variable spanning the minisuperspace. This leads not only to more plausible probability densities, but also to a modified WdW-equation.

3.2.1. Creation from Nothing

1. *Solution in the QR.* In finding the right solution for a creation from nothing, we restrict ourselves to the option $\rho \equiv 1^2$, for which we must solve Equation (A9) (with $v_i = 1$),

$$\psi''(v) + \left(\frac{\pi}{2}\right)^2 \left(1 - v^{-2/3}\right)\psi = 0. \qquad (14)$$

According to pages 203–206 of Ref. [11], a solution is considered the best, for which $\psi^2(x)$ is largest at $x = 0$ and decreases fast with increasing x because it roughly corresponds to the tunneling through the potential barrier given by the factor $\sim x^4 - x^2$ near ψ in Equation (A8). We take the position that probability statements make sense only within an already existing multiverse, while conclusions based on such statements about the start of the tunneling are external, that is, they come from nowhere and must be regarded as meaningless. In the selection of a solution for an equivalent process, being suggested as a viable alternative on page 204 of Ref. [11] with reference to Ref. [10], we rely entirely on the probability interpretation of ψ^2 and choose $\psi(v)$ so that with increasing v its square permanently increases. The solution shown in Figure 2 has this property; in the place where it is linked with the wave function for the CR, it satisfies the further constraint $\psi'(v) = 0$, which is explained further down. In the next paragraph, another argument is developed which supports our selection. Although derived with the goal of a tunneling process, this solution has nothing in common with the latter and therefore should be named differently, e.g., "soft entry". Equation (14) has, of course, also a genuine tunneling solution, which can be treated in exactly the same way as our "soft entry" solution.

2. *Interpretation.* Contributions to the literature on QG such as [14–16] give the impression that too little information is available for the assignment and interpretation of the WdW-equation. In Ref. [16], the interesting suggestion is made to use the de Broglie–Bohm interpretation for this purpose. According to this, from the representation $\Psi(\vec{x}, t) = e^{iS(\vec{x},t)/\hbar}$ of the wave function, particle trajectories satisfying $\dot{\vec{x}}(t) = \nabla S$ are derived. Unfortunately, this is not possible in our case because the solutions of Equation (14) are real whence $S \equiv 0$. Therefore, we develop an alternative, which, like the de Broglie–Bohm approach, leaves quantum mechanics completely unchanged and concerns only its interpretation. Under application to the solutions of Equation (14), it consists essentially in reversing the step $p_V \to (\hbar/i)\partial/\partial V$ that leads from the classical to the quantum-mechanical description. Specifically, for the spatial density $\Psi \hat{p}_V \Psi$ of the quantum mechanical momentum, we make the ansatz

$$i\Psi \hat{p}_V \Psi = \frac{p_V}{V} \quad \text{or} \quad \hbar \Psi \Psi'(V) = \frac{c^2 \dot{V}(t)}{12\pi G V^2} \qquad (15)$$

($p_V = \partial L/\partial \dot{V}$ is obtained from Equations (A2)–(A4)) by which it is identified with the classical momentum. The multiplication of the left side by i constitutes the return to classical values and can be interpreted as reverse Wick rotation. With Equations (4), $l_P/t_P = c$, $l_P = \sqrt{\hbar G/c^3}$, $\psi = \psi/l_P^{3/2}$ and $V = 2\pi^2 l_P^3 v$ going over to dimensionless quantities, and resolving with respect to $\dot{v}(\tau)$, the last equation becomes

$$\dot{v}(\tau) = 12\pi v^2 \psi \psi'(v) = 6\pi v^2 d\psi^2/dv. \qquad (16)$$

This means that the DM moves to larger values of ψ^2 until it terminates the entry process, if everywhere $\psi'(v) \geq 0$. The velocity $\dot{v}(\tau)$ and $\psi^2(v)$, up to a normalization factor the probability

---

[2] Note that by this assumption or $\varrho \equiv \varrho_\Lambda$ resp. the term ϱ_f in Equation (2) drops out, i.e., the friction term plays a role only in the CR.

density of our "soft entry" solution, are depicted in Figure 2. From that, it becomes particularly clear that our solution is suitable for describing a creation from nothing. Regarding the dynamics, one could even say, at least for this solution that in the QR the role of time is taken over by the probability density $dP/dv \sim \psi^2(v)$.

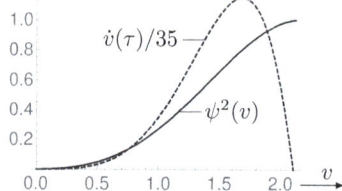

Figure 2. QR-solution of the WdW-equation Equation (14) to the square, and velocity $\dot{v}(\tau)$ from Equation (16) (divided by 35 for combinability with ψ^2) for the VL-type "soft entry".

3. *Wave function of the the CR.* We want to connect the wave function of the QR with that of the CR and have to determine the latter. For this, it is advantageous to use the interpretation of the classical solution based on Equations (5) and (6) with a friction term because then the minisuperspace contains only the one variable v. Furthermore, we can use the fact that we already know the classical solution (9), which satisfies the equation

$$\dot{x}^2(\tau) - \gamma^2 x^2 = 0. \tag{17}$$

We first consider the quantisation of the latter in the minisuperspace spanned by a or x resp. This can be done in the same way as in the derivation of Equation (A7). With $x = a/l_P$, Equation (17) becomes $\dot{a}^2(t) - (\gamma c/l_P)^2 a^2 = 0$. The equation corresponding to Equation (A2) is obtained by setting $\Phi(t) = 0$, $U_a = 0$ and $U_\Phi = 3\gamma^2 c^4/(8\pi G l_P^2)$. With this, one can go directly to Equation (A6) and from there to dimensionless quantities, ending up at the WdW-equation

$$\psi''(x) + \left(\frac{3\pi\gamma}{2}\right)^2 x^4 \psi = 0. \tag{18}$$

Its general solution is

$$\psi_c(x) = \sqrt{u}\left[a\,\Gamma(5/6)\,J_{-\frac{1}{6}}(u^3/3) + b\,\Gamma(7/6)\,J_{\frac{1}{6}}(u^3/3)\right] \quad \text{with} \quad u = \left(\frac{3\pi\gamma}{2}\right)^{1/3} x, \tag{19}$$

where $J_{\pm\frac{1}{6}}(u^3/3)$ are Bessel functions, and a and b are integration constants. As mentioned in the introduction to Section 3.2, the solution depends on the volume $v = x^3$. Furthermore, for $u \gg 1$, the envelope of the Bessel functions is $u^{-3/2}$ whence, due to the factor $\sqrt{u}$ in front of ψ_c, the envelope of $\psi_c^2(x)$ decays with u as $1/u^2$. This solution would actually lead the DM-expansion to ever smaller probabilities, so absolutely not what one would reasonably expect.

The situation turns for the better, if we employ the minisuperspace spanned by v. The corresponding WdW-equation can be easily deduced from Equation (18) by replacing in it dx^2 by $(dv^2/9x^4)$ according to Equation (A10). The resulting equation is

$$\psi''(v) + \left(\frac{\pi\gamma}{2}\right)^2 \psi = 0 \tag{20}$$

and has the solution

$$\psi_c(v) = \sin[\omega(v-\delta)] \quad \text{with} \quad \omega = \frac{\pi\gamma}{2}, \tag{21}$$

where δ is an integration constant. This solution must be continuously and with continuous derivative be connected to the solution for the QR, that of Equation (14). For the latter as well as for Equation (20), it holds that together with $\psi(v)$ also $C\psi(v)$ is a solution, where C is a freely selectable constant. This allows us to freely set the value of one of the two solutions at the junction and to continuously join the other. Due to the extremely small value of γ, the derivative $\psi'_c(v) = \omega \cos[\omega(v-\delta)]$ of the solution (21) for the CR is essentially equal to zero for all v. It follows that the derivative of the solution for the QR must be virtually zero at the junction. For the sake of simplicity, we therefore demand that the derivative of both solutions is exactly zero at the junction. With the boundary condition $\psi'(v_i) = 0$ for the QR-solution at the border to the CR, it follows from Equation (16) that there also $\dot{v}(\tau) = 0$ and thus $\dot{x}(\tau) = 0$ must apply. Thus, QG requires the same boundary condition as the quasi-classical approximation.

Figure 3 shows the solutions thus obtained. The joining point is $v_i = 2.103$ or $x_i = v_i^{1/3} = 1.28$ and not $x_i = 1$ as in Section 3.1. This is due to the requirement $\psi'(v_i) = 0$ and the fact that Equation (14) does not allow a solution with $\psi(0) = 0$ and $\psi'(1) = 0$. The classical solution (21) is entered twice in the figure, once over $\omega(v-\delta)$ (horizontal straight line), and once over $(4/\omega)\omega(v-\delta)$ (dashed curve).

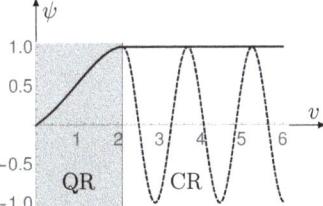

Figure 3. The full curve is a joint representation of the "soft entry" solution $\psi(v)$ for the QR (gray filling) and $\psi_c(v)$ for the CR. The dashed curve shows $\psi_c(v)$ once again, but highly compressed in the horizontal direction in order to visualize the oscillations.

4. *Normalization of $\psi(v)$ and corresponding probabilities.* The freely selectable factor C in the solutions $C\psi(x)$ of Equations (14) and (20) allows a particularly simple fulfillment of the normalization condition (A11), for which only

$$C = \left(2\pi^2 \int_r \psi^2 dv\right)^{-1/2}$$

must be chosen. The normalized solution is then $C\psi(v)$. This does not only apply to the solution for the QR, but also to combined solutions for the QR and the CR. For comparing states that have been traversed up to a certain time τ (corresponding volume $v(\tau)$), it appears reasonable to extend the normalization range from $v = 0$ up to $v(\tau)$. Over time, the upper boundary $v(\tau)$ becomes larger and larger, with the result that the probability density of fixed and duly normalized intermediate states becomes ever smaller. Therefore, it is obviously useless to compare intermediate states of different normalization. In other words, it only makes sense to compare states with the same normalization, which amounts to determining the ratio of their probabilities. However, this is just as well obtained from a wave function without normalization, which is valid for the entire range. (This is the reason why our figures are not based on normalized wave functions.)

For interpreting the dynamics of the system described by the wave function (21), the probability density $\psi_c^2(v)$ can not be used as in the case of the "soft entry" solution. The reason is that it has many local maxima, and the system would remain with one of them, once it has reached it. This problem can be solved by considering the set of ψ_c values between two successive zeros as a single quantum state of the classical system. This range has the size Δv following from $\gamma \pi \Delta v/2 = \pi$ or $\Delta v = 2/\gamma$ and corresponds to the x-range $\Delta x = (\Delta v)^{1/3}$ or

$$\Delta a = 2.71 \cdot 10^{20}\, l_p = 3.14 \cdot 10^{-15}\, \text{m} = 5.94 \cdot 10^{-5}\, r_B, \qquad (22)$$

where $r_B = 5.29 \cdot 10^{-11}\,\text{m} =$ Bohr radius. Regarding the dynamics, we assume that the system jumps from one state to the next in leaps of the length $\Delta a = 2.71 \cdot 10^{20}\, l_p$. In this interpretation, up to an uninteresting factor, the probability density of the various system states is given by the average of the local density ψ_c^2 over a half period, that is, by

$$\left\langle \frac{dP}{dv} \right\rangle \sim \left\langle \psi_c^2(v) \right\rangle = \left\langle \sin^2\left[\gamma\pi(v-\delta)/2\right] \right\rangle = \frac{1}{2}.$$

With our solution, the probability of the different states does not decrease as with the solution (19) based on the usual minisuperspace. However, it does not increase as desired either. This can, however, be improved by taking advantage of the fact that the classical dynamic is an evolution in time, and by relating the probability density not to volume but to time (using $dP/d\tau$ instead of dP/dv). Setting

$$\left\langle \frac{dP}{d\tau} \right\rangle = \left\langle \frac{dP}{dv} \right\rangle \dot{v}(\tau) \sim \dot{v}(\tau)/2$$

for $\dot{v}(\tau)$ we employ the classical solution: from $v = x^3$ and use of Equation (9), we get $\dot{v}(\tau) = 3x^2 \dot{x}(\tau) = 3\gamma x^3 = 3\gamma v$ and finally

$$\left\langle \frac{dP}{d\tau} \right\rangle \sim \frac{3\gamma}{2} v, \qquad (23)$$

where v does the jumps of height $\Delta v = 2/\gamma$ described above. Plotted above v, $<dP/d\tau>$ thus follows a staircase-like curve of the slope $3\gamma/2$. Once again, the incremental probability of the individual quantum states can be considered as a substitute for time.

3.2.2. Modified HH-Approach

1. *Depictive model.* In this section, we investigate an alternative to the HH-approach which differs comparatively more from the latter than the approach in Section 3.1. We now assume that the timeless primordial state is a 4D-sphere of radius $R \approx l_P$ in a 4D-space (coordinates x, y, z, u) of constant positive curvature, uniformly filled with DE of the 3D-density $\varrho = \varrho_\Lambda$. It can be represented by the surface

$$x^2 + y^2 + z^2 + u^2 + v^2 = R^2 \qquad (24)$$

of a 5D-sphere in a Euclidean 5D-space. Assuming that the 3D-subspace spanned by x, y, z is homogeneous and isotropic, its metric belongs to the class of metrics with line element

$$ds^2 = -a^2\left(\chi^2 + \sin^2\chi\, d\Omega\right) \quad \text{with} \quad d\Omega = d\vartheta^2 + \sin^2\vartheta\, d\varphi^2 \quad \text{and} \quad \sin\chi = \frac{r}{R}, \qquad (25)$$

where $r = \sqrt{x^2 + y^2 + z^2}$ and $-\pi \leq \chi \leq \pi$. For both $0 \leq \chi \leq \pi$ and $-\pi \leq \chi \leq 0$, a 3D-space of the same metric is transversed, so one can say that, for given a, two identical homogeneous, isotropic and closed 3D-spaces of Volume ${}^3V(a) = 2\pi^2 a^3$ exist. Due to the factor a^2, not present in the representation (24) of the surface of the initial 5D-sphere, the metric of its homogeneous and isotropic 3D-sub-spaces is a subclass of the metrics of Equation (24). We must therefore find out which restrictions result for the expansion parameter a. According to Equation (24), $x = y = z = 0$ for $u^2 + v^2 = R^2$, from which it follows that ${}^3V = 2\pi^2 a^3 = 0$ and $a = 0$. A maximum value $\bar{a}$ of a follows from the fact that the total volume $2\int_0^{\bar{a}} {}^3V(a)\, da = 4\pi^2 \int_0^{\bar{a}} a^3\, da = \pi^2 \bar{a}^4$ of 3D-sub-spaces, belonging to the range $0 \leq a \leq \bar{a}$ of permissible a values, may not be larger than the surface $^4S = (8/3)\pi^2 R^4$ of the initially given sphere in the Euclidean 5D-space, and is given by

$$\bar{a} = (8/3)^{1/4} R \approx 1.28\, R\,. \tag{26}$$

The surface of the Euclidean 5D-sphere is invariant under rotations $x, y, z, u, v \rightarrow x', y', z', u', v'$, each of which leads to another set of homogeneous and isotropic 3D-sub-spaces, which for given $a \leq \bar{a}$ all have the same metric and shape. All in all, we get a number[3] of similar 3D-sub-spaces, which, based on their abundance, are equally probable. We assume that each of them can serve as a timeless primordial state of the DM. (An important difference to the HH-model is that the latter has only one coordinate which can turn from space-like to time-like; furthermore, the initial state is represented only by the lower part of the 4D-sphere, while the upper part is reserved for the evolution in time, see pages 80–83 of Ref. [7].)

2. *Treatment in the framework of quantum gravity.* For the above model, which is partly based on classical ideas, a suitable solution of the WdW-equation for the QR is to be found. Because we again want to have a continuous connection between the solutions for the QR and the CR, once again we impose the boundary conditions $\psi(v_i) = \psi_c(v_i)$ and $\psi'(v_i) = \psi'_c(v_i)$. In order to implement our concept from above as accurately as possible, we look for a solution in the QR whose probability density $dP/dv \sim \psi^2(v)$ is as constant as possible. This condition can only roughly be satisfied (see Figure 4). Furthermore, it appears reasonable to require for this timeless state that its total momentum disappears, i.e.,

$$\int_{QR} \psi(v)\, \hat{p}_V\, \psi(v)\, dv = \frac{\hbar}{i} \int_{QR} \psi(v)\, \psi'(v)\, dv = \frac{\hbar}{i}\left(\psi^2(v_i) - \psi^2(0)\right) = 0\,. \tag{27}$$

Therefore, $\psi^2(v_i) = \psi^2(0)$ must apply what can be easily satisfied. (Note that zero is the only value for which the total momentum is not imaginary.) In Figure 4, a solution satisfying the above requirements is entered together with the properly adapted classical solution $\psi_c(v)$, Equation (21).

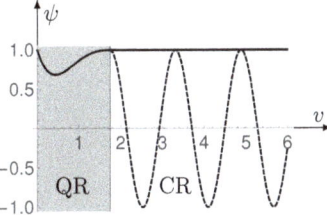

Figure 4. Joint representation of the HH-like QR-solution $\psi(v)$ and the CR-solution $\psi_c(v)$, the latter being shown a second time in dashed style and with strong horizontal compression.

Both in the HH-concept, used as a model, and in the present concept, the transition from the time-independent primordial state in four spatial dimensions to a time-dependent evolution in three spatial dimensions represents a critical point. (In the original version of the HH-concept [6], it appears somewhat non-transparent [17] by being accomplished via transition probabilities. In a later contribution—see pages 85–86 of Ref. [7]—the transition is explicitly performed by linking the purely spatial solution for the QR with the time-dependent solution for the CR, the same as done by us in this and in Section 3.2.1.)

---

[3] Due to space-related quantum effects, this number could be quite small.

A look at the averaged probability density in the CR does not reveal why the DM should make this transition. In our case, it could slightly be favored by the fact that each of the four spatial coordinates and, in addition, coordinates resulting from them by a rotation (if allowed by LQG) are eligible for the transition to a time coordinate. However, another means can help: because a transition from space to time is concerned, it offers itself to relate the probability density in the CR to the time interval $d\tau$ as in Section 3.2.1. According to Equation (23), due to the prefactor γ, $dP/d\tau$ starts with an extremely small value which is smaller than dP/dx by many orders of magnitude. However, since it is a probability assessment of the same state, it must be considered as equivalent. In this way, the transition from space to time is associated with a corresponding change in relating the probability densities. Thus, the transition from purely spatial to time-dependent states finally leads to an evolution with increasing probabilities.

4. Critical Comments on the Quantum Results

4.1. Notes on the Storage and Transmission of Information

In Ref. [2], different possibilities have been indicated regarding how the information about the physical laws and their physical implementation could be stored in the DM. A storage in structured arrangements of space grains as examined in the LQG would seem most attractive. Unfortunately, due to the size of the smallest volume allowed by the LQG, i.e., the volume

$$V_{LQG} = \sqrt{(8\pi\tilde{\gamma})^3/(6\sqrt{3})}\, l_P^3 = 39.1\, l_P^3 \quad \text{for} \quad \tilde{\gamma} = 1 \tag{28}$$

of the space atoms according to page 30 or Ref. [18], this does not seem possible. In addition, the limits for the storage and transmission of information, determined by Bekenstein [19,20] and Bremermann [21], would also be infringed, provided that they were valid in the QR. Our "soft entry" process may serve as an example. The volume of the initial state after the entry is $v \approx 2$ (see Figure 3) or

$$V \approx 2\pi^2\, l_P^3 = 39.5\, l_P^3. \tag{29}$$

This corresponds fairly closely to the smallest volume (28) of the LQG, and, in that, according to the latter, substructures are not possible in which huge amounts of information could be stored. The mass of the DE contained in this volume is approximately $2\pi^2 l_P^3 \varrho_\Lambda = 2.36\, m_P$ where $m_P = 2.18 \cdot 10^{-8}$ kg is the Planck mass. According to Bremermann, the upper limit for the information transmission speed is $\approx 1.36 \cdot 10^{50}$ bits/(kg s). With an initial mass $\approx 2.4\, m_P$ of the DM (reached after the "soft entry") and a entry duration of $\approx t_P$, the maximum transferable information is

$$I = 1.36 \cdot 10^{50}\, \frac{2.36\, m_P\, t_P}{\text{kg s}}\, \text{bits} = 0.38\, \text{bits},$$

which is far too little.

Despite the limitations of the LQG, let us at this point tentatively have a look at the possibility that, still undetected, much finer substructures below the LQG structures exist. (This is of course a rather speculative assumption.) After all, at least the WdW-Equation (A9),

$$\psi''(v) + \left(\frac{\pi}{2}\right)^2 \left(v_i^{-2/3} - v^{-2/3}\right)\psi = 0 \tag{30}$$

allows such solutions. (In order to stay within the scope of valid physics, one would have to assume that the information concerned is not bound to matter. Otherwise, Bronstein's black hole argument would be violated, according to which anything smaller than l_P is "hidden inside its own mini black hole", see pages 7–8 of Ref. [18]). A corresponding solution for the HH-like case with $\varrho \equiv \varrho_*$ and, once again, $\int_{QR} \psi\, \hat{p}_v \psi\, dv = 0$ is shown in Figure 5. (The discontinuous transition from short to extremely long waves is due to the fact that, in the transition from the QR to the CR, $\rho'(x)$ jumps from zero to $-2x_i^2/x^3$, see Equation (10); it could be remedied by prescribing a continuous transition of $\rho'(x)$.) Like the solution for the CR, it contains many zeros of $\psi(v)$, if $v_i \ll 1$. As for the CR-solution, we interpret the regions between neighboring zeros as single states. Except for those with small v, the distance between adjacent zeros is pretty much $\Delta v = 2 v_i^{1/3}$ or $\Delta V = 4 x_i \pi^2 l_P^3$, which can be derived from Equation (30): for $v \gg v_i$, the latter reduces to

$$\psi''(v) = -(\pi v_i^{-1/3}/2)^2 \psi, \tag{31}$$

whence

$$\psi(v) = \sin[(\pi v_i^{-1/3}/2)(v-\delta)] \tag{32}$$

and $\Delta v = 2 v_i^{1/3}$ or

$$\Delta V = 4 x_i \pi^2 l_P^3 = 39.5 \, x_i \, l_P^3 \tag{33}$$

exactly. For $x_i = v_i = 1$, these are quanta of almost the same size as provided by the LQG according to Equation (28) with $\tilde{\gamma} = 1$. (A spectrum of quanta can not be deduced here.) Equations (28) and (33) yield even the same volume quanta exactly, i.e., $\Delta V = V_{LQG}$, if x_i is chosen according to

$$x_i = \frac{4 \tilde{\gamma}^{3/2}}{3^{3/4} \sqrt{\pi}} = 0.99 \, \tilde{\gamma}^{3/2} . \tag{34}$$

Based on Equations (3) and (10), the same applies if $\dot{x}(\tau) = 0$ for $x = x_i$, while ρ_i and ϱ_i are chosen according to

$$\rho_i = 1 \quad \text{and} \quad \varrho_i = \varrho_*/x_i^2 . \tag{35}$$

After Ref. [18], $\tilde{\gamma}$ is a quantity of $\mathcal{O}(1)$, from which it follows that $\varrho_i = \mathcal{O}(\varrho_*)$. The latter is a conclusion that does not emerge from our investigation.

Concerning our information problem, there are still other options to be considered. As already suggested in Ref. [2], information might also be stored in hidden higher dimensions, which would be an issue for string theory. Another possibility would be that both string and loop theory are involved. In the Discussion, Section 6, the urgency of the information problem is emphasized once again.

Another way of solving this problem, perhaps even the best one, would be to allow a much larger volume of the initial state. Even with an initial volume of the order $\Delta v = 1/\gamma = 10^{61}$ (magnitude of the volume jumps of the classical solution), one would still be far in the range of quantum physics, albeit outside the range where QG is absolutely required. However, nothing seems to speak against using it there. This case of a much larger initial volume is addressed more closely in Section 4.3.

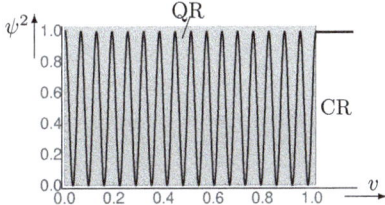

Figure 5. QR-solution of Equation (30) with $\int_{QR} \psi \, \hat{p}_v \psi \, dv = 0$ and substructures obtained by choosing $v_i \gg 1$. For the reason of presentability, only a small value, $v_i = 10^{-4.5}$, was used. The short horizontal line at the top right is the beginning of $\psi_c^2(v)$.

4.2. Notes on the Primordial States

According to Figures 3 and 4, the width of the QR, resulting from solutions of Equation (14), is $\Delta v \approx 2$, essentially the same as found above for the volume quanta. This means that, in its primordial state, the DM consists of nothing more than a single space atom. If so, then the question arises as to whether it makes any sense to distinguish between solutions of different shape, e.g., a tunneling and a "soft entry" solution. This could at best serve to characterize space atoms of different shapes, but then should be done in a different way than usual. Furthermore, the question arises of whether the initial state for the classical evolution would not have to be composed of many space atoms. This possibility is discussed in the next subsection.

4.3. Primordial State with Large Volume

Figure 6 shows what the probability density of our "soft entry" process looks like when the QR is much further extended. As already indicated above, the size $\Delta v = 1/\gamma = 10^{61}$ of the volume jumps performed by the classical solution appears to be a reasonable choice for the volume of the QR. The latter is then made up of about 10^{60} space atoms, which could be enough to accommodate the amount of information required by our DE model. For not having to choose a much smaller initial density $\varrho_\Lambda = \varrho_*/x_i^2$ (see Equation (3)) because of $x_i^2 = v_i^{2/3} \gg 1$, we give up the condition $\dot{x}(\tau) = 0$ for $x = x_i$ leading to the latter, and must replace the consequential boundary condition $\psi'(v_i) = 0$. This means that we can use Equation (14) and must only continue the "soft entry" solution underlying Figure 3 to larger values of v. For Figure 6, the position of the upper QR-boundary was chosen so that Equation (27) is satisfied. The transition from QR to CR would be best accomplished by passing over $\rho(x)$ with continuous derivative $\rho'(x)$ to the decreasing density of the CR, and then looking for a common solution for the transition region, a lengthy procedure not accomplished here.

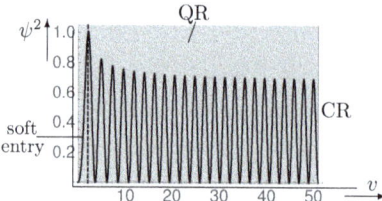

Figure 6. Square of our "soft entry" solution, extending from $v = 0$ to the dashed line, and at constant $\rho = 1$ continued over a larger v-interval. The upper boundary value ≈ 50, selected for visualization, should in reality be much larger.

As can be seen from Figure 6, our "soft entry" solution fits very well with our interpretation of the ψ^2-values between adjacent zeroes as quantum states. Obviously, the quantum states with small v-values are somewhat higher and wider than the others. This can be explained by the fact that they are subjected to a particularly strong spatial curvature. With the correspondingly continued HH-like state of Figure 4, the fit is not nearly as good. However, the extended "soft entry" state of Figure 6 can also be reinterpreted as HH-like. (For this purpose, Equation (27) was taken into account). For our purposes, this interpretation appears even more appropriate because it is difficult to imagine how large amounts of information will be accommodated, if the volume quanta, which are supposed to store them, arise one after the other.

The rather large volume of the HH-like primordial state thus obtained may even settle a never resolved discrepancy between the VL- and the HH-approach. According to Ref. [17], the VL-approach favors a universe (multiverse in our case) with an initial state of the smallest possible size, whereas the HH-approach favors one of the largest imaginable size. As a sort of compromise, our DM-model favors an HH-like primordial state of medium size between the smallest and largest possible.

5. Properties of the CR-Solution

5.1. Initial Equilibrium between DE and Matter

In Ref. [2], it was suggested that the initial state of the expansion phase could be an – albeit unstable – equilibrium between DE and matter. Because of the importance of this, the related calculation is made up here. For the sake of simplicity, it is limited to case $a_i = l_p$. The additional matter term $\rho_m = \rho_{m_i}/x^n$ converts Equation (6) for $x_i = 1$ into

$$\dot{x}^2(\tau) = (\rho + \rho_{m_i}/x^n)\, x^2 - 1. \tag{36}$$

With $\dot{x}(\tau) = 0$ for $x_i = x(0) = 1$, from this, we get

$$\rho_i + \rho_{m_i} = 1. \tag{37}$$

With $\dot{\rho}(\tau)/\dot{x}(\tau) = \rho'(x)$, the time derivative of Equation (36) results in

$$\ddot{x}(\tau) = \left(\rho'(x) - \frac{n\rho_{m_i}}{x^{n+1}}\right)\frac{x^2}{2} + \left(\rho(x) + \frac{\rho_{m_i}}{x^n}\right)x.$$

In the case of the friction-involving interpretation, for $\rho_{m_i} \to 0$ and $\rho(x) \to 1$, this equation must reduce to Equation (5), i.e., $x^2 \rho'(x)/2 + x\rho \to -2\sigma\dot{x} + x$. In consequence, in the last equation, $x^2 \rho'_\phi(x)/2$ must be replaced by $-2\sigma\dot{x}$. With the equilibrium conditions $\ddot{x}(\tau) = \dot{x}(\tau) = 0$ for $x = x_i = 1$, we get from the last equation

$$\rho_i + (2-n)\rho_{mi}/2 = 0. \tag{38}$$

Resolving Equations (37) and (38) with respect to ρ_i and ρ_{mi} yields the results shown in Table 1.

Table 1. Equilibrium values of ρ_i and ρ_{mi}.

	Generally	$n = 3$	$n = 4$
ρ_{mi}	$2/n$	$2/3$	$1/2$
ρ_i	$(n-2)/n$	$1/3$	$1/2$

Because in Ref. [1] the influence of an initial (unstable) equilibrium on the solutions $x(\tau)$ has already been dealt with in detail, and because it is unlikely that, in the present case, significant changes will occur (due to the extremely slow temporal change of $x(\tau)$, the situation is already very equilibrium-like), and also for reasons of clarity, a correspondingly extended treatment is waved here.

5.2. Minimum Age of the Dark Multiverse

In Ref. [2], the age of the DM was evaluated on the simplifying assumption that its size exceeds that of U by a factor $\zeta = 100$ or more. This is far beyond the lower limit of $\zeta \approx 8.4$, posed by measurements of the spatial curvature. Therefore, in the following, the age limit is determined. For this purpose, the factor γ in the solution (9) for $x(\tau)$ must be taken from the full Equation (36) of Ref. [2], $\varrho/\varrho_0 = \gamma^2 \varrho_\Lambda/\varrho_0 + 0.522/(\zeta^2 X^2)$, where $X = a/a_0$ with $a_0 = R\zeta$ = presents a value of a and $R = 2.22 \cdot 10^{26}$ m = metric radius of the observable boundary of U. Putting $\varrho = \varrho_0$ and $X = 1$, we obtain from this equation

$$\gamma = \sqrt{(1 - 0.522/\zeta^2)\varrho_0/\varrho_\Lambda}. \tag{39}$$

Inserting this, $x = R\zeta X/l_p$, $\tau = t_{H0}\mathcal{T}/t_p$ with $\mathcal{T} = t/t_{H0}$, and $t_{H0} = 4.41 \cdot 10^{17}$ s = present Hubble time, in Equation (9) and resolving the latter with respect to $\mathcal{T}$ yields its present value

$$\mathcal{T}_0(\zeta) = \frac{170.4 + 1.21\,\ln\zeta}{\sqrt{1 - 0.522/\zeta^2}}. \tag{40}$$

In Figure 7, the function $T_0(\zeta)$ is shown together with the aforementioned lower limit $\zeta \approx 8.4$.

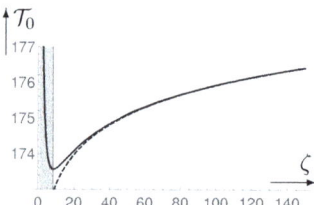

Figure 7. Age T_0 of the DM as a function of ζ. Due to inadmissible curvature values, the shaded area $\zeta < 8.4$ must be excluded. The dashed curve represents the approximation $170.4 + 1.21 \ln \zeta$.

5.3. Irreversibility of the Friction-Involving Interpretation

By including the friction term $-f\dot{a}(t)$ in Equation (1), for particularized reasons (time savings for the transmission of information), an irreversible element was deliberately introduced into the basic cosmological equations derived from general relativity. It is surprising that the irreversible solutions thus obtained are also solutions of the usual equations involving a scalar field Φ. It is worthwhile to investigate how this comes about. For this, we first consider the equations with a scalar field Φ,

$$\dot{a}^2(t) = \frac{8\pi G}{3} \varrho\, a^2 - c^2 \quad \text{with} \quad \varrho = \frac{\hbar^2 \dot{\Phi}^2(t)}{2\mu c^4} + \frac{V(\Phi)}{c^2}, \tag{41}$$

$$\ddot{\Phi}(t) + 3H\dot{\Phi}(t) + \frac{\mu c^2}{\hbar^2} V'(\Phi) = 0 \quad \text{with} \quad H = \frac{\dot{a}(t)}{a}. \tag{42}$$

All of them are invariant with respect to time reversal because dt appears in them only quadratically. ($3H\dot{\Phi}(t) \sim \dot{a}\,\dot{\Phi}$ is invariant because $\dot{a}$ and $\dot{\Phi}$ change their sign simultaneously with t.)

Let us now turn to the irreversible interpretation involving friction. In this, the FL-Equation (2) is only seemingly reversible, since $\varrho_f \to -\varrho_f$ for $t \to -t$. What was shown in Ref. [2] is that the solutions of this equation coincide with the solutions of the system of Equations (41) and (42) only as time is moving forward; for $t \to -t$, this is no longer true. (This does not immediately become apparent from Ref. [2] because the solutions of the system (41) and (42) were only calculated forward in time. For example, Equation (60) from there, $\dot{\varphi}(\tau) = \sqrt{-x\,\varrho'(x)/(3\,\varrho_\Lambda)}$, used to calculate the solution, would have to be replaced by $\dot{\varphi}(\tau) = -\sqrt{-x\,\varrho'(x)/(3\,\varrho_\Lambda)}$ for $t \to -t$.) The resolution of the seeming contradiction is therefore that the fulfillment of both the reversible and the irreversible equations by our solution applies only to advancing time.

5.4. Behavior of the DE in Our Universe

It was shown in Ref. [1] that the rest mass density of the DE, measured in M and U, matches for $t = t_* + \vartheta$, where ϑ is the proper time in U, and t_* is the time of the origin of U as judged from M. Our model assumes that in M the DE is at rest with respect to the coordinates t, χ, etc. This does not mean that it is at rest as well in U. It is shown below that it does so at least approximately.

In the following, we consider two states of U.

State 1: U contains only the DE of the DM, and, omitting the angular contributions, the square of the line element is

$$ds_M^2 = c^2 dt^2 - a^2(t)\, d\chi^2. \tag{43}$$

State 2: is the real present state of U with

$$ds_U^2 = c^2 d\vartheta^2 - a_U^2(\vartheta)\, d\eta^2, \tag{44}$$

where the subscript $_U$ denotes the affiliation to U, η is the radial, and ϑ the time-coordinate. We assume that the origins of the radial coordinates χ and η coincide, and search for a transformation $t = t(\eta, \vartheta), \chi = \chi(\eta, \vartheta)$ that converts ds_M^2 into a form commensurable with ds_U^2. In order to keep an unchanged rest mass density, we set

$$t = t_* + \vartheta \qquad \chi = \chi(\eta, \vartheta). \tag{45}$$

With this (specifically $dt = d\vartheta$), from Equation (43), we get

$$ds_M^2 = \left(c^2 - a^2(t)\chi_\vartheta^2\right) d\vartheta^2 - 2a^2(t)\chi_\eta \chi_\vartheta \, d\eta \, d\vartheta - a^2(t)\chi_\eta^2 \, d\eta^2,$$

where χ_η and χ_ϑ are the partial derivatives of $\chi(\eta, \vartheta)$ with respect to η and ϑ resp. To determine how the position $\chi = \text{const}$ is judged in U, following from $d\chi = \chi_\eta d\eta + \chi_\vartheta d\vartheta = 0$, we insert in this $d\eta = -\chi_\vartheta \, d\vartheta / \chi_\eta$ whence $2a^2(t)\chi_\eta \chi_\vartheta \, d\eta \, d\vartheta = -2a^2(t)\chi_\vartheta^2 \, d\vartheta^2$, and obtain

$$ds_M^2 = \left(c^2 + a^2(t)\chi_\vartheta^2\right) d\vartheta^2 - a^2 \chi_\eta^2 \, d\eta^2. \tag{46}$$

According to Equations (43) and (46), the differential contribution of an element $d\chi$ to the distance $d = a(t)\chi$ of the position χ under consideration, expressed in terms of the radial coordinate η of state 2, is $a \, d\chi = a \chi_\eta d\eta$. This is true because $dt = 0$ for length measurements, and $dt = d\vartheta = 0$ according to Equation (45). With this, from Equations (44) and (46), we get $a \chi_\eta d\eta = a_U d\eta$ or

$$\chi = (a_U/a)\eta \quad \text{and} \quad d(t) = a(t)\chi = a_U(\vartheta)\eta = d_U(\vartheta), \tag{47}$$

where $d(t)$ or $d_U(\vartheta)$ is the distance from the origin of the DE-element at χ in situation 1 or 2 resp. According to this, the expansion velocity of the DE in U is

$$\dot{d}_U(\vartheta) = \dot{d}(t) = \dot{a}(t)\chi = \frac{\gamma}{t_P} a \chi = \frac{\gamma}{t_P} d_U, \tag{48}$$

where $\dot{a}(t) = l_P \dot{x}(\tau) d\tau/dt = l_P \gamma x/t_P = \gamma a/t_P$ was used. The expansion velocity of the cosmic substrate in U is

$$\dot{d}_{Ucs}(\vartheta) = \dot{a}_U(\vartheta)\eta = \frac{\dot{a}_U(\vartheta)}{a_U} d_u = \frac{d_U}{t_H}, \tag{49}$$

where $\dot{a}_U(\vartheta)/\dot{a}_U = H(\vartheta) = 1/t_H$ was used, with H and t_H being the Hubble parameter and the Hubble time resp. Inserting for t_H the present value $t_{H0} = 4.41 \cdot 10^{17}$ s, with use of Equations (11), (48) and (49), and $t_P = 5.39 \cdot 10^{-44}$ s, the ratio of the two expansion velocities becomes

$$\frac{\dot{d}_U(\vartheta)}{\dot{d}_{Ucs}(\vartheta)} = \frac{\gamma \, t_{H0}}{t_P} = 0.83. \tag{50}$$

In view of uncertainties regarding the Hubble parameter and the huge numbers involved, it spoils nothing to say that the two velocities are essentially the same. Even a slightly larger expansion velocity of the DE cannot be excluded. In this case, the following scenario could come into effect: Since the observed expansion-acceleration of the material components of U is attributed to the DE, according to the principle actio = reactio, this should conversely delay the DE-expansion. The mass density of the DE could then even be slightly below the usual value because some of the acceleration caused by it would emanate from its kinetic energy. However, if the numerical values obtained above are correct, just the opposite will happen: a decelerating effect on the matter must be compensated by a slightly higher mass density of the DE.

6. Discussion

The concept of a timeless, spatially four-dimensional HH-like state developed in Sections 3.1 and 3.2.2 as a primordial state represents one of the most important supplements to our model of the DM. In contrast to the tunneling or the "soft entry" concept, the information about the physical laws and the tools needed for their implementation are bound to matter from the very beginning. This can certainly be regarded as an advantage over the other two concepts, in which the tunneling or entry is preceded by a further state of pure information without integration into matter, a state that cannot be described within the framework of valid physical laws. According to Hawking, the HH-state could never have been created due to its timelessness because its creation would require a temporal before. The same argument would apply as well to our HH-like state. However, we cannot agree with this interpretation. It is certainly true that in many, perhaps most cases, cause and effect follow each other in time, which is usually associated with the use of these words. However, there are counterexamples in mathematics and logics, where, instead of cause and effect, the word pair precondition and consequence is used. In this case, the first implicates the second and not vice versa, with time being irrelevant. This is exactly what is also conceivable for the HH- or our HH-like state.

Because of its timelessness, like the HH-model, our HH-like model can only be treated within the framework of a QG-theory. Our unusual choice of a minisuperspace, using the volume as basis of the WdW-equation, was due to the fact that the usual procedure yields very implausible probabilities in the CR. Our approach leads not only to more plausible probabilities, but also to the identification of single states, which can be interpreted as spatial volume quanta.

With regard to the most important feature of our DM model, the storage and transmission of information about the physical laws, the investigations in Section 4.1 have shown how difficult it is to integrate deeper details beyond the friction term into common physics. On the other hand, it seems to be an urgent demand to open up new ways in this respect because the problem is palpable: elementary particles such as the electron have no receiving device and no brain, with which they could receive commands and implement them into action, and the properties with which they are described (e.g., rest mass, charge and spin) are not sufficient to act and react as required by physical laws. In this respect, at present, questions and speculations are instead the order of the day. As seen in Section 4.1, QG without loop theory would allow substructures below the LQG-structures. Would this be possible, or are the limits posed by LQG too restrictive? What role does the DE play in this game? At least the HH-like primordial state with large volume, treated in Section 4.3, seems to provide a useful approach.

Regarding newly identified properties of the solution for the CR, we point out the behavior of the DE in U, examined in Section 5.4. It is understood that DE causes the acceleration of galaxy expansion, and there must be a counteraction to this action. We suggested that the latter might consist in slowing down a somewhat higher expansion velocity of the DE, but this could not be conclusively demonstrated in consideration of very narrow numerical relationships and possible inaccuracies. Thus, after all, that too has to be ranked as a speculation. On the other hand, another, potentially disturbing problem with the friction term introduced into the equation for the DM-expansion-acceleration $\ddot{a}(t)$ could be adequately solved.

As indicated at the beginning of the Introduction, it is currently a huge problem to prove beyond doubt that there is a multiverse at all and, even more so, to verify the validity of assumptions that enter into a multiverse concept. In our case, one possibility would be to examine more closely the footprint of the multiverse in our universe, specifically the time dependence of the DE density $\varrho(t) c^2$ given by Equations (9) and (10). It would be even more difficult, but also more interesting, to unveil their decomposition into the two opposing components $\varrho_\Lambda(t)c^2$ and $\varrho_f(t)c^2$ given by Equation (2).

Funding: This research received no external funding.

Conflicts of Interest: The author declares no conflict of interest

Abbreviations

The following abbreviations are used in this manuscript:

M	multiverse
U	our universe
DE	dark energy
DM	dark multiverse
QR	quantum regime
CR	classical regime
VL	Vilenkin–Linde
HH	Hartle–Hawking
QG	quantum gravity
LQG	loop quantum gravity
WdW	Wheeler–de Witt
FL	Friedmann–Lemaître

Appendix A

Inserting

$$\varrho_\Phi = \frac{\hbar^2 \dot\Phi^2(t)}{2\mu c^4} + \frac{U_\Phi(\Phi)}{c^2} \tag{A1}$$

in Equation (2) and multiplying the latter with $3c^2/(8\pi G a^2)$ yields

$$\mathcal{H} := \frac{3c^2}{8\pi G a^2}\dot a^2(t) - \frac{\hbar^2 \dot\Phi^2(t)}{2\mu c^2} + U_a(a) - U_\Phi(\Phi) = 0 \quad \text{with} \quad U_a = \frac{3c^4}{8\pi G a^2}. \tag{A2}$$

The case of constant density $\varrho_\Phi \equiv \varrho_\Lambda$, examined closer in the main body, is included by setting $\dot\Phi(t) \to 0$ and $U_\Phi/c^2 \to \varrho_\Lambda$. Having the same dimension as $\varrho_\Phi c^2$, $\mathcal{H}$ can be interpreted as an energy density. Integrating it at given a over the total volume V of the DM yields the quantity

$$H = \mathcal{H} V \quad \text{with} \quad V = 2\pi^2 a^3. \tag{A3}$$

Note that, by way of the multiplication by $V \sim a^3$, the three-dimensionality of space enters decisively into the derivation. This is also expressed by the fact that ϱ_Φ and $\mathcal{H}$ are 3D-densities. Interpreting H as the Hamiltonian of the system,

$$L = U(\mathcal{H} - 2U) \quad \text{with} \quad U = U_a(a) - U_\Phi(\Phi) \tag{A4}$$

is the corresponding Lagrangian. The associated momenta are

$$p_a = \frac{\partial L}{\partial \dot a} = \frac{3\pi c^2 a \dot a}{2G} \qquad p_\Phi = \frac{\partial L}{\partial \dot\Phi} = -\frac{2\pi^2 \hbar^2 a^3 \dot\Phi}{\mu c^2}. \tag{A5}$$

(Since $\dot p_a(t) = \partial L/\partial a$ and $\dot p_\Phi(t) = \partial L/\partial \Phi$ result in Equations (41) and (42), our above interpretations of H and L are justified.) Resolving the last two equations with respect to $\dot a$ and $\dot\Phi$ resp. and inserting the results into Equation (A3) with (A2) yields

$$H = \frac{G}{3\pi c^2 a} p_a^2 - \frac{\mu c^2}{4\pi^2 \hbar^2 a^3} p_\Phi^2 + 2\pi^2 a^3 (U_a(a) - U_\Phi(\Phi)) = 0.$$

Now, employing the quantisation rules $p_a \to (\hbar/i)\partial/\partial a$ and $p_\Phi \to (\hbar/i)\partial/\partial\Phi$, after multiplication with $3\pi c^2 a/(G\hbar^2)$, we obtain the equation

$$\hat H \Psi = \left(-\frac{\partial^2}{\partial a^2} + \frac{3\mu c^4}{4\pi G \hbar^2 a^2} \frac{\partial^2}{\partial \Phi^2} + \frac{9\pi^2 c^6}{4G^2 \hbar^2} a^2 - \frac{6\pi^3 c^2 a^4}{G\hbar^2} U_\Phi \right) \Psi = 0, \tag{A6}$$

where $\Psi(a,\Phi)$ is the wave function of the DM in the so-called minisuperspace spanned by the variables a and Φ. Inserting $a=xl_p$ with $l_p=\sqrt{\hbar G/c^3}$, $\Psi=\psi/l_p^{3/2}$ with dimensionless ψ, $U_\Phi(\Phi)=\varrho_\Lambda c^2 u_\varphi(\varphi)$, $\varphi=(\hbar/c^2)\sqrt{8\pi G/(3\mu)}\,\Phi$, and Equation (3) yields

$$\left(-\frac{\partial^2}{\partial x^2}+\frac{2}{x^2}\frac{\partial^2}{\partial\phi^2}+\frac{9\pi^2}{4}\left[x^2-\frac{x^4 u_\phi(\phi)}{x_i^2}\right]\right)\psi=0. \tag{A7}$$

In the case $\varrho_\Phi\equiv\varrho_\Lambda$ or $\rho\equiv 1$, the substitutions $\dot\Phi(t)\to 0$ and $U_\Phi/c^2\to\varrho_\Lambda$ translate into $\partial^2/\partial\phi^2\to 0$ and $u_\phi(\phi)\to 1$ by what Equation (A7) reduces to the particularly simple and for $x_i=1$ frequently studied case

$$\psi''(x)+\left(\frac{3\pi}{2}\right)^2\left(x^4/x_i^2-x^2\right)\psi=0. \tag{A8}$$

For the reasons depicted in Section 3.2, we do not employ this equation. Instead, we apply the equation which is obtained, if, in the above derivation, from Equation (A5) onward, a is eliminated by using $V=2\pi^2 a^3$, $p_V=\partial L/\partial V$ in place of p_a, and a dimensionless volume $v=V/(2\pi^2 l_p^3)$. (The initial volume is $V_i=2\pi^2 a_i^3=2\pi^2 l_P^3 x_i^3$, and the corresponding dimensionless volume is $v_i=V_i/(2\pi^2 l_P^3)=x_i^3$.) The result of this procedure is

$$\psi''(v)+\left(\frac{\pi}{2}\right)^2(v_i^{-2/3}-v^{-2/3})\psi=0. \tag{A9}$$

It can be read directly from Equation (A8) by eliminating x through use of

$$v=x^3 \quad\text{and}\quad dv=3x^2 dx \tag{A10}$$

and by replacing $\psi''(x)$ with $9x^4\psi''(v)$. Note that Equation (A9) does not result from Equation (A8) simply through transformation and therefore has different solutions.

We are only dealing with real solutions Ψ in this paper, and for them the (also real) Hamiltonian $\hat H$ is obviously Hermitian, no matter how the scalar product $(\Psi,\hat H\psi)$ is defined. Ψ^2 is the 3D-density of the probability P, i.e., $\Psi^2=dP/dV$ so that its volume integral over the full range R of possible V values yields 1, i.e., we have

$$\frac{dP}{dV}=\Psi^2 \qquad \int_R dP=\int_R \Psi^2\,dV=2\pi^2\int_r \psi^2\,dv=1. \tag{A11}$$

References

1. Rebhan, E. Model of a multiverse providing the dark energy of our universe. *Int. J. Mod. Phys. A* **2017**, *32*, 1750149. [CrossRef]
2. Rebhan, E. Acceleration of cosmic expansion through huge cosmological constant progressively reduced by submicroscopic information transfer. *Int. J. Mod. Phys. A* **2018**, *33*, 1850137. [CrossRef]
3. Vilenkin, A. Creation of universes from nothing. *Phys. Lett. B* **1982**, *117*, 25–28. [CrossRef]
4. Vilenkin, A. Quantum origin of the universe. *Nucl. Phys. B* **1985**, *252*, 141–152. [CrossRef]
5. Linde, A.D. Quantum creation of the inflationary universe. *Lett. Nuovo Cim.* **1984**, *39*, 401–405. [CrossRef]
6. Hartle, J.B.; Hawking, S.W. Wave function of the universe. *Phys. Rev. B* **1983**, *28*, 2960–2975. [CrossRef]
7. Hawking, S.; Penrose, R. *The Nature of Space and Time*; Princeton University Press: Princeton, NJ, USA, 2000.
8. DeWitt, C.M.; Wheeler, J.A. (Eds.) Lectures in Mathematics and Physics. In *Batelle Rencontres*; Benjamin: New York, NY, USA, 1968.
9. DeWitt, B.S. Quantum Theory of Gravity. I. The Canonical Theory. *Phys. Rev.* **1967**, *160*, 1113–1148. [CrossRef]
10. Coleman, S. Why there is nothing rather than something: A theory of the cosmological constant. *Nucl. Phys. B* **1988**, *310*, 643–668. [CrossRef]

11. Linde, A. *Particle Physics and Inflationary Cosmology*; Contemporary Concepts Physics 5; Harwood Academic Publishers: Reading , UK, 1990; pp. 1–362.
12. Rovelli, C. *Reality Is Not What It Seems: The Journey to Quantum Gravity*; Penguin Publishing Group: London, UK, 2017.
13. Vilenkin, A.; Yamada, M. Tunneling wave function of the universe. *arXiv* **2018**, arXiv:1808.02032v2.
14. Peres, A. Critique of the Wheeler-DeWitt Equation. In *On Einstein's Path. Essays in Honor of Engelbert Schucking*; Harvey, A., Ed.; Springer: New York, NY, USA; 1999; pp. 367–379
15. Patrick, P. Using Trajectories in Quantum Cosmology. *arXiv* **2018**, arXiv:1902.00796v1.
16. Pinto-Neto, N.; Fabris, J.C. Quantum cosmology from the de Broglie–Bohm perspective. *Class. Quantum Gravity* **2013**, *30*, 143001. [CrossRef]
17. Vilenkin, A. *Many Worlds in One*; Hill and Wang: New York, NY, USA, 2006; pp. 190–191.
18. Rovelli, C.; Vidotto, F. *Covariant Loop Quantum Gravity: An Elementary Introduction to Quantum Gravity and Spinfoam Theory*; Cambridge University Press: Cambridge, UK, 2014. Available online: http://www.cpt.univ-mrs.fr/~rovelli/IntroductionLQG.pdf (accessed on 27 May 2019).
19. Bekenstein, J.D. Universal upper bound on the entropy-to-energy ratio for bounded systems. *Phys. Rev. D* **1981**, *23*, 287. [CrossRef]
20. Bekenstein, J.D. Energy cost of information transfer. *Phys. Rev. Lett.* **1981**, *46*, 623–626. [CrossRef]
21. Bremermann, H.J. Optimization through evolution and recombination. In *Self-Organizing Systems 1962*; Yovits, M.C., Jacobi, G.T., Goldstein, G.D., Eds.; Spartan Books: Washington, DC, USA, 1962; pp. 93–106.

© 2019 by the author. Licensee MDPI, Basel, Switzerland. This article is an open access article distributed under the terms and conditions of the Creative Commons Attribution (CC BY) license (http://creativecommons.org/licenses/by/4.0/).

Article

Multiverse Predictions for Habitability: The Number of Stars and Their Properties

McCullen Sandora [1,2]

1. Institute of Cosmology, Department of Physics and Astronomy, Tufts University, Medford, MA 02155, USA; mccullen.sandora@gmail.com
2. Center for Particle Cosmology, Department of Physics and Astronomy, University of Pennsylvania, Philadelphia, PA 19104, USA

Received: 14 May 2019; Accepted: 10 June 2019; Published: 13 June 2019

Abstract: In a multiverse setting, we expect to be situated in a universe that is exceptionally good at producing life. Though the conditions for what life needs to arise and thrive are currently unknown, many will be tested in the coming decades. Here we investigate several different habitability criteria, and their influence on multiverse expectations: Does complex life need photosynthesis? Is there a minimum timescale necessary for development? Can life arise on tidally locked planets? Are convective stars habitable? Variously adopting different stances on each of these criteria can alter whether our observed values of the fine structure constant, the electron to proton mass ratio, and the strength of gravity are typical to high significance. This serves as a way of generating predictions for the requirements of life that can be tested with future observations, any of which could falsify the multiverse scenario.

Keywords: multiverse; habitability; stars

1. Introduction

Science is beginning to embrace the idea of a multiverse—that is, that the laws of physics have the potential of being different elsewhere. In this framework, some of the parameters in our standard models of particle physics and cosmology vary from universe to universe, and are not capable of being explained mechanistically. This does not necessarily mean that there is no explanatory power of these theories; however, One of the requirements for a physical theory becomes that it allows sufficient complexity to give rise to what are termed observers, of which we as humans are presumably representative. These privileged, information processing-rich arrangements of matter are fragile, and so are extremely sensitive to the types of environments the underlying physics is capable of producing. A universe without our panoply of atomic states, for example, is expected to be devoid of sufficient complexity to give rise to these observers.

The mode of explanation we can hope for in such a scenario is to determine the chances of observing the laws of physics to be what they are, subject to the condition that we are typical observers. This style of reasoning usually goes by the name 'the principle of mediocrity' [1]. In order to employ it, it becomes necessary to try to quantify how many observers a universe with a given set of physical parameters is likely to host. Traditionally, cosmologists have shied away from the detailed criteria necessary for life and then intelligent life to emerge, largely because the specifics of the requirements are very uncertain at our current state of knowledge. Thus, attention has focused on the predictions for cosmological observables like the cosmological constant and density contrast [2–5], as opposed to quantities that may affect mesoscopic properties of observers.

In this context, a useful proxy for this complicated task has been to simply take the fraction of baryons that ended up in galaxies above a certain threshold mass, with the understanding that this is necessary in order for heavy elements to be synthesized and recycled into another generation of stars and planets. However, this crude method, while useful for determining preferred values of cosmological parameters, has really only resulted in a rather limited number of mostly postdictions, such as the need to live in a universe which is big, old, empty, and cold.

While many anthropic boundaries regarding microscopic physical parameters such as the proton mass, electron mass, fine structure constant, and strength of gravity have been delineated [6–8] (for a recent review see [9]), comparatively little attention has been paid to these in the context of the principle of mediocrity. However, these largely determine many of the properties of our mesoscopic world, and so the details of the microphysical parameters will ultimately dictate quite strongly which universes will be capable of supporting observers. Placing this further level of realism on our estimations, however, requires a refined understanding of habitability. While still a major open issue, over the past few decades science has made amazing progress in understanding this question: We now have a much clearer view of the architecture of other planetary systems [10], we have discovered the ubiquity of preorganic chemical complexes throughout the galaxy [11], the outer reaches of our own solar system have revealed remarkable complexity [12,13], we understand the formation of planetary systems to unprecedented levels, and we now have atmospheric spectroscopy of nearly a dozen extrasolar planets (albeit mostly hot Jupiters) [14]. As amazing as this progress has been, the coming decades are slated to exhibit an even more immense growth of knowledge of the galaxy and its components: Projects like TESS and CHEOPS will find a slew of new exoplanets [15,16], with sensitivity pushing into the Earthlike regime. Experiments like the James Webb Space Telescope [17] are expected to directly characterize the atmospheres of several Earth-sized planets [18], the disequilibrium of which will make it possible to infer the presence or absence of biospheres [19]. In addition, further afield, when the next generation of telescopes such as TMT, PLATO, HabEx, and LUVOIR will be able to deliver a large enough population to do meaningful statistics on atmospheric properties [20], we will be able to characterize the ubiquity of microbial life, as well as which environmental factors its presence correlates with [21].

Rather than wait for the findings of these missions to further our understanding of the conditions for habitability, our position now represents a unique opportunity: We can test various habitability criteria for their compatibility within the multiverse framework. If a certain criterion is incompatible in the sense that it would make the values of the fundamental constants that we observe highly improbable among typical observers, then we can make the prediction that this criterion does not accurately reflect the habitability properties of our universe. When we are finally able to measure the distribution of life in our galaxy, if we indeed confirm this habitability criterion, then we will have strong evidence that the multiverse hypothesis is wrong. Likewise, if a criterion is necessary in order for our observations to be typical, but is later found to be wrong, this would be evidence against the multiverse as well. Put more succinctly, if our universe is good at something, we expect that to be important for life, and if it is bad at it, we expect it to not be important for life. Here, by saying that 'our universe is good at something', we really mean that by adopting the habitability criterion in question, our presence in this universe is probable, and, equally importantly, by not adopting it, our presence in this universe is improbable. Because it is easier to determine what our universe is good at than what life needs, the former can be done first, and used as a prediction for the latter. This logic is displayed in Figure 1.

A simple example will illustrate this approach: Suppose we take the hypothesis that the probability of the emergence of intelligent life around a star is proportional to its total lifetime. Then, in the multiverse setting, we would expect to live in a universe where the lifetime of stars is as long as possible. This is not the case, as we will show in [22]; therefore, the multiverse necessarily predicts that stellar lifetime cannot have a large impact on habitability. If, once we detect several biospheres, we find a correlation between the age of the star and the presence of life, we will have falsified this prediction the multiverse has made. Upcoming experiments aimed at characterizing the atmospheres of exoplanets will make this task feasible.

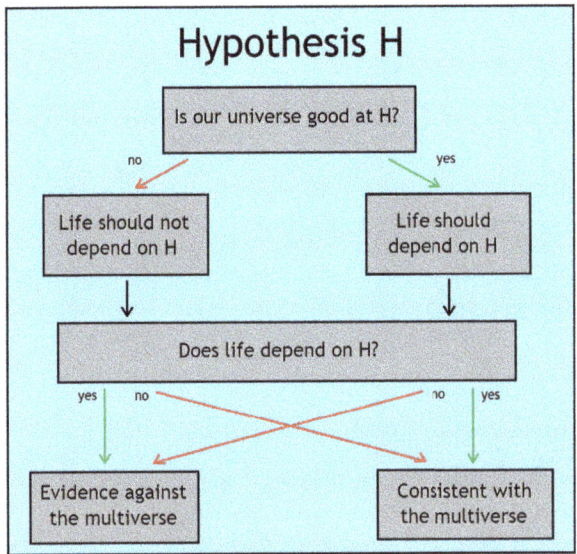

Figure 1. The underlying logic behind this task. We consider separate habitability hypotheses 'H', and determine whether our universe is good at H, in the sense that adopting this notion of habitability makes our presence in this universe probable (and equally importantly, that not adopting this notion makes our presence in this universe improbable). This yields a prediction for whether or not H is a good requirement for habitability that can then be tested against upcoming observations. There are dozens of proposed habitability criteria in the literature, and though not all of them will have a significant influence on our likelihood, many will. If we do live in a multiverse, we expect compatibility with all of these tests; if just one yields incompatible results, we will be able to rule out the multiverse hypothesis to a potentially very high degree of confidence.

This logic can be repeated for a number of different proposed habitability criteria: Estimates for the habitability of our universe change drastically depending on whether life can or cannot exist: On tidally locked planets; around dwarf stars; without the aid of photosynthesis; off of Earth-mass planets; outside the temperate zone; without plate tectonics; and in the presence of dangers such as comets, supernovae, and gamma ray bursts, among many others. Considering the impact of each of these can yield a separate prediction for what life requires, and consequently where we should expect to find it in upcoming surveys. Incorporating each will alter the distribution of observers throughout the purported multiverse, to differing levels of importance. Considering the gamut will include dozens of considerations for which should be crucial in the multiverse context, and promises to yield several strong predictions for where life should be found, sometimes up to a statistical significance of 6σ. This undertaking will take some effort, but it promises to elevate the multiverse hypothesis to standard, falsifiable, scientific theory.

Let us further stress that each habitability criterion acts as a quasi-independent test, which can greatly strengthen our conclusions of whether the multiverse exists or not. There are arguments both for and against every criterion we consider here, and since the issue is not likely to be settled through logical argument, in general it would pay to remain agnostic toward which are true. The multiverse, however, makes specific predictions about which criteria are right. If even one of these predictions fails to be in accord with observations, we will have strong evidence against the multiverse hypothesis. Conversely, if all of the predictions we make are shown to be true, this would be strong evidence for the multiverse.

We initiate this task in this paper by considering only the simplest criteria for habitability, based off counting the number of potentially habitable stars. Section 2 is devoted to outlining the formalism and detailing the observational facts that need to be reconciled. Section 3 presents the simplest possible estimate of habitability, where every star is taken as potentially life bearing with equal capability. We improve upon this in Section 4, where we compare multiverse expectations with the following proposed stellar habitability criteria: Is photosynthesis necessary for complex life? Is there a minimum timescale for developing intelligence? Are tidally locked planets habitable? Are convective stars habitable? Though most of the criteria we consider in isolation fail to yield a satisfactory account of our observed values, we finally display one that does, based off the total entropy produced throughout a star's lifetime, which serves as a minimal working model in terms of habitability criteria. We conclude in Section 5, and demonstrate that including multiple criteria simultaneously can lead to 'epistatic' effects in the probability of observing our parameters. Several concrete predictions for the distribution of life throughout our universe are made based off the success of each hypothesis, and adding further refinements to the scenarios considered here are capable of yielding multiple more.

2. Preliminaries

2.1. Properties and Probabilities of Our Universe

What are the properties of the world we are trying to explain with this approach? We focus here on three dimensionless constants, the fine structure constant $\alpha = e^2/(4\pi)$, where e is the charge of the electron, the ratio of the electron mass to the proton mass $\beta = m_e/m_p$, and the strength of gravity $\gamma = m_p/M_{pl}$, where M_{pl} is the reduced Planck mass. The values of these three quantities determine a great deal of the macroscopic characteristics of the universe [23]. In this approach, it is necessary to ask what values these parameters may take in order for the universe to be compatible with life, and where our observed values are situated within this allowed region. The different positions of these three variables will guide how sensitive we expect the criteria for habitability to be on these variables, which in turn will translate into an expectation for the types of environments that life is capable of thriving in.

The electron to proton mass ratio can be a factor of 2.15 larger than its current value before the processes involved in stellar fusion stop being operational [6], though detractors of this requirement [24,25] state that other fusion processes would take place. A factor of 4.9 and hydrogen would become unstable, creating a universe filled entirely with neutrons, incapable of complex chemistry (though even this scenario has been argued to be capable of producing life [26], highlighting the extreme degree of contention any speculative statements in this subject bring). Throughout this work, we will take the first, more stringent bound, as the border of the anthropically viable region, but relaxing this could readily be incorporated into our framework.

The fine structure constant also affects the stability of hydrogen, and will cause it to decay if it were 2.07 times larger. (It was found in [27] that a few percent increase would also preclude sufficient carbon production in stars, but this can be compensated by altering the pion mass). While this upper limit depends on the value of the electron to proton mass ratio, this can be compensated by the difference between the down and up quark masses, so that in effect the two limits are independent. A lower bound on the fine structure constant of about 1/5 the observed value was found in [5,28], based on the requirement for galactic cooling, which we will need to use as several scenarios we encounter favor small α.

Unlike the other two quantities, the strength of gravity can be several orders of magnitude stronger without any adverse effect toward life. As we will show in Section 4, it can be roughly 134 times stronger before the lifetime of stars becomes shorter than biological evolutionary timescales. There are lower limits to this, as with all other quantities, as well, which we will delay discussing until [29], but the relevant point here is that none of these quantities are very closely situated to their minimum allowed values. Let us also comment on the bound found in [30], which appears to place a stronger upper bound on the strength of gravity than the one we use. The quantity they consider is M_{pl}/GeV, keeping particle physics fixed: As such, it relates to various cosmological processes that depend on the Planck mass. They determine that the rate of close encounters between star systems is too frequent if this quantity is about ten times smaller, depending on the model for fluctuations, baryogenesis, and dark matter. While this is an important anthropic boundary that we will return to in the future, it can be alleviated by altering the other constants we consider, and turns out to be weaker than the stellar lifetime bound if α and β are allowed to vary.

For a habitability criterion to be compatible with the multiverse, our measured values must be compatible with what a typical observer would find. In particular, the fact that the bounds on α and β are only a few times larger than their observed values indicates that a relatively weak dependence on these quantities is preferred. In contrast, the habitability condition must impose a restriction that strongly favors weak gravitational strength in order to counteract the preference for larger values of γ.

Before we continue on, let us make this notion of typicality more quantitative. For each observable quantity $x = \alpha, \beta, \gamma$, we define the typicality $\mathbb{P}(x_{\text{obs}})$ as the cumulative probability that a value more extreme than ours is observed. Then, we have

$$\mathbb{P}(x_{\text{obs}}) = \min\left\{ \mathrm{P}(x > x_{\text{obs}}), \mathrm{P}(x < x_{\text{obs}}) \right\} \tag{1}$$

In this definition, we integrate over all other variables not under direct consideration, so that if there is a large portion of universes that have a different value of multiple parameters simultaneously, our observation is penalized. This combats the tendency for degeneracies in parameter space that we will encounter, which lead to misleading statistics if the other variables were to be held fixed. In addition, note that we take the minimum of the cumulative distribution function and its complement, to disfavor the scenarios where our observed values are anomalously close to the boundary of a region, though they may lie in a heavily favored location. With this definition, the maximal value of this quantity is $1/2$.

It is also possible to define a global typicality, which counts the fraction of observers that would find themselves in universes less typical than ours: This is not quite reconstructible from the quantities above, but in all cases considered it yields no additional insights, so is not put to use here. Then, our statistic for whether a notion of habitability is compatible with the multiverse is the combination of the probabilities for the three parameters parameters we consider, α, β, and γ.

2.2. Drake Parameters

Now that the expectations for the definition of habitability have been outlined, we must find a way to estimate the relative number of observers within a universe, and then extract how this quantity depends on the underlying physical parameters. Fortunately, the technology for doing this has been developed some time ago, as the well known Drake equation. Here, we make use of a slight modification of the 'archaeological form' of the Drake equation outlined in [31]: Our expression for the habitability of a given universe is equal to the expected number of observers that are produced throughout the course of its evolution. This can be broken down into the following product of factors:

$$\mathbb{H} = N_\star \times f_\text{p} \times n_\text{e} \times f_\text{bio} \times f_\text{int} \times N_\text{obs} \tag{2}$$

where here $N_\star$ is the number of habitable stars in the universe, f_p is the fraction of star systems that contain planets, n_e is the average number of habitable planets in systems that do possess planets, f_bio is the fraction of habitable planets on which life emerges, f_int is the fraction of life bearing planets that ultimately develop intelligent organisms, and N_obs is the number of observers per intelligent species. The first application of the Drake equation to multiverse reasoning was in [32].

As always, the point of this equation is meant to be organizational: It marshals the great variety of factors that dictate the emergence of intelligence into largely factorizable subproblems, and allows us to cleanly isolate the assumptions that go into each. While, as usual, the overall normalization remains highly uncertain, use of this equation will allow us to directly compare the relative numbers of observers for any two universes, given that we state our assumptions on how each of these parameters depends on the laws of physics.

Each factor in this equation deserves a fair amount of attention in its own right, and consequently, rather than overload the reader with every consideration that goes into this analysis at once, we will split this estimation into multiple separate papers. Our strategy will be to work our way through the Drake factors one at a time, starting from the left and moving our way to the right. It is important to note that although the final conclusions can only be made once all of these factors are considered in a unified picture, care is taken to report only those results that carry through once this synthesis takes place, and to explicitly state when conclusions will be altered in the full analysis. A completely satisfactory account of all three of our observed values will only be achieved once $f_{\rm bio}$ is taken into the fold.

We begin, then, in this paper, by considering how the number of habitable stars depends on the laws of physics, effectively taking the simplified ansatz that the number of observers will be directly proportional to the number of stars, independent of any other properties of the universe. We will first define precisely what is meant by this quantity in an infinite universe, then use straightforward estimates for the average size of stars to arrive at our first, simplest potential definition of habitability. This will be shown to be incompatible with the multiverse hypothesis, which motivates searching for refinements to improve. Next, we discuss additional proposed criteria for stellar habitability, and compare how these criteria fare.

Subsequent papers will deal with the other factors. In [29] we will discuss the two relating to the formation of planets. We will confront which factors are necessary for a star to produce planets, their resultant properties, and what the conditions on the parameters are required in order to achieve this. We will find that these considerations do not alter the probability distributions themselves much, but that they do place strong bounds on the allowed parameter space, many of which are the strongest lower bounds that can be found in the literature.

Next, in [22] we discuss what planetary characteristics may possibly influence the advent of simple life. We will consider a slew of possibilities here, and find that most of them are incompatible with the multiverse hypothesis, leading to clear predictions for where life should be found in our universe.

In [33] we tackle the question of how often intelligence emerges from simple life. Our approach will be to determine the rate of suppression of intelligent life, such as the mass extinctions that have plagued our planet over the course of geological time, and how the rates of these depend on the physical constants. Throughout, we will emphasize the testable predictions the multiverse hypothesis offers, and suggest the quickest ways to falsify them.

Before we begin, we need to relate the habitability to the probability of measuring particular values of the observables, because they need not be exactly equal: If there is an underlying prior distribution of the space of variables, presumably set by the ultimate physical theory, this must be taken into account as well, so that the probability of finding oneself in a given universe will be proportional to the habitability of that universe multiplied by that universe's chances of occurring:

$$\mathrm{P}(\alpha,\beta,\gamma) \propto \mathbb{H}(\alpha,\beta,\gamma)\, p_{\rm prior}(\alpha,\beta,\gamma) \tag{3}$$

While the precise form of the prior may need to await a fuller understanding of the ultimate theory of nature, we can make a plausible ansatz for each of the variables we are concerned with here. As we will find, the habitability often depends on the parameters much more strongly than the prior anyway, and so adopting a mildly different one will not appreciably alter any of our results.

We expect the prior on the fine structure constant α to be nearly flat, without any strong features or special values that would skew the distribution too much in their favor. Again, if the reader has reason to believe in some other prior, the details will not change much. The ratio of proton mass to Planck mass, on the other hand, is taken to be scale invariant, or flat in logarithmic space: $p_{\rm prior}(\gamma) \propto 1/\gamma$. The reasoning behind this is that the proton mass is dictated by the scale at which the strong force becomes confining, which through renormalization group analysis is given by $m_p \sim M e^{C - 2\pi/(9\alpha_s)}$, where M is some large mass scale, α_s is the strength of the strong force at high energies, and C is a coefficient that depends on the heavy quark masses. If this is taken to be roughly uniform at that scale, then the distribution for γ will be scale invariant (up to unimportant logarithmic corrections, which anyway depend on the precise distribution for the coupling). Similarly, the ratio of electron mass to proton mass will also be taken to be scale invariant, since not only is the proton mass scale invariant, but there is also reason to suspect that the Yukawa couplings of the standard model follow a scale invariant distribution as well [34]. Then, our final result for the probability of measuring a particular value of the parameters will be:

$$P(\alpha, \beta, \gamma) \propto \frac{\mathbb{H}(\alpha, \beta, \gamma)}{\beta \gamma} \quad (4)$$

Note that adopting this distribution, based off the plausibility of high energy physics priors, assumes that these are uncorrelated with the properties which affect how likely a particular universe is to arise, such as the reheating temperature, or the nucleation rate in the context of false vacuum inflation. This is a plausible assumption, as these properties of the theory are presumably set by the inflaton sector, that is insensitive to the masses and couplings of light states, but if the reader has reason to suspect an alternative scenario they should feel free to adopt their own prior. In addition, note that in this work we assume that the multiverse context accommodates all three of these quantities as variable and uncorrelated. This also seems plausible, especially since two of them are composed of multiple factors, but an alternative view could readily be incorporated in this formalism as well. We do not consider universes that are radically different than our own, such as having different types of particles, forces, or number of dimensions. We regard the comparison with potential habitats in those universes as too speculative to make immediate progress, and anyway not amenable to the type of reasoning we employ here.

With this, we are now ready to make our simplest appraisal of the overall habitability of our universe.

3. Number of Stars in the Universe $N_\star$

3.1. What Is Meant by this Quantity?

In this section we elaborate on what we mean by the number of stars in a given universe. If universes are infinite, this is an ill-defined concept, and comparing the relative number of stars in two different universes is ambiguous, which is a manifestation of the measure problem [35,36].

The fact that there is no obvious unique choice for this comparison has plagued cosmologists since the early days of multiverse reasoning. Many proposals have been made for how to regulate the infinities one must deal with, in order to be able to compare two finite numbers. Most proposals immediately run into drastic conflict with observation, which helps to winnow down the possibilities to a smaller subset. Encouragingly, of the few that remain, several have been shown to be equivalent to each other, even though the starting points were radically different [37,38]. However, as it stands there are multiple existing measures, with no obvious way of specifying which is correct. Thankfully for our purposes, much of this ambiguity only affects the distribution of cosmological parameters, leaving the microphysical parameters that we focus on relatively independent of the measure.

Here we make use of the scale factor cutoff measure [39], which states that the probability of an observer arising in a particular universe should be simply proportional to the number of baryons that have found their way into a suitably large galaxy cluster by the time the universe has reached a certain size. This cutoff size is arbitrary, and will not affect probabilities as long as it is taken to be longer than the time of peak galactic assembly. Additionally, the total number of baryons is infinite in an infinite universe, motivating the need to regulate by truncating to a finite region of space: In practice this can be accomplished by merely noting that the ratio of two probabilities is then equal to the ratio of baryon densities, again independent of cutoff. (Here, care must be taken to regularize in an unbiased way, since densities change in expanding universes, but as noted this will only have an affect cosmological parameters).

While the requirement on cluster mass is usually used to penalize universes that do not produce large enough halos (for instance, if the cosmological constant is too large [2] or the density contrast too small [3,5]), the reason for this has to do with the formation with planets, and so will not concern us here. We will return to this subject in [29], where we investigate what sets the minimum mass of a halo in terms of fundamental parameters. For now, we restrict our attention to universes where the majority of baryons falls into star forming regions, and instead add a layer of sophistication to the criteria, that the number of observers produced will be proportional to the number of stars.

To begin, we make the simplification that the efficiency of star formation, that is, the total amount of matter that ultimately becomes stars, is independent of the halo mass, and equal to $\epsilon_\star = 0.03$ [40]. This is in fact not a good approximation, as this quantity is known to be affected by many feedback processes that can lower the efficiency for both small mass and large mass halos [41]. We will refine our prescription to take these effects into account in a future publication, but expect that they should play more of a role for cosmological observables, rather than the ones we focus on here. Additionally, the efficiency is not taken to be a strong function of microphysical parameters. (Though it does in principle: For example, if the fine structure constant is too low, cooling will be so inefficient that gas will never fragment into collapsing clouds [5].)

3.2. Is Habitability Simply Proportional to the Number of Stars?

We are now ready to make our simplest estimate of the habitability of a universe, that the number of observers is proportional to the number of stars. This makes the quite unreasonable assumption that every star is equally habitable, yet it will serve as the calculational substrate, on top of which further refinements can be added. Modifications of this basic framework are the subject of the later sections of this paper, as well as the forthcoming sequels. Then, it stands to reason that the number of stars will be inversely proportional to how large they are: A universe where stars are smaller would ultimately be able to make more of them with the same amount of initial material. If each star is an independent opportunity to develop intelligent life, then universes that produce the smallest possible stars would have the greatest number of observers.

How large are stars, then? As is well known, the typical stellar mass scale is given by the quantity $M_0 = (8\pi)^{3/2} M_{pl}^3 / m_p^2 = 1.8 M_\odot$ (e.g., [42]). However, the average stellar mass is considerably lighter than this, and exhibits additional dependence on the physical constants. To estimate the average, we need to know the distribution of stellar masses. This is given by the initial mass function (IMF); for the majority of this paper we take this to be of the classic power law form, $p_{IMF}(\lambda) \propto \lambda^{-\beta_{IMF}}$ [43], which defines the dimensionless quantity $\lambda = M_\star/M_0$. The quantity $\beta_{IMF} = 2.35$ is referred to as the Salpeter slope, and, as usual with power law exponents, is set by universal processes that do not depend on physical parameters [44]. More realistic treatments instead use a broken power law, lognormal distribution, or some other form [45,46], which accurately reflects the details of the feedback mechanisms accompanying star formation, but this is an unnecessary complication that obfuscates but does not appreciably change our results. We will incorporate a more sophisticated IMF into this formalism in Section 5. The most relevant feature of this distribution is that it is very steep, making the vast majority of stars born rather close to the minimal possible mass, and larger stars extremely rare. As such, the average stellar mass is simply proportional to the minimum, and so it will be essential to estimate this quantity.

The minimum stellar mass can be determined based off the requirement that its central temperature must be high enough to ignite hydrogen fusion. Particular attention was paid to the dependence of this minimum mass on physical parameters in [42], where the central temperature of a star was determined to be $T \approx \lambda^{4/3} m_e$. This must be compared to the required temperature, which in [47] was found to be given by the Gamow energy, the threshold above which thermal fluctuations can routinely instigate tunneling through the repulsive barrier between two protons, $T_G \sim \alpha^2 m_p$. Demanding the central temperature be greater than this gives $\lambda_{min} = 0.22 \alpha^{3/2} \beta^{-3/4}$. This same scaling was also found in [23], where they demanded the essentially equivalent requirement that the scattering time is shorter than the Kelvin-Helmholtz time.

The average stellar mass can be computed from the initial mass function we employ as $\langle \lambda \rangle = (\beta_{\text{IMF}} - 1)/(\beta_{\text{IMF}} - 2)\lambda_{\min}$, but this actually underestimates the average stellar mass. The origin of this discrepancy come from the fact that the initial mass function deviates from a power law for small masses, reflecting feedback in the star formation process [44]. However, the normalization of this value is not important for our analysis, since it will only enter into the probability multiplicatively, and so we can simply take $\langle \lambda \rangle \propto \lambda_{\min}$. This relation holds in more realistic treatments as well.

A maximum stellar mass also exists, based on the criterion that a star must be gravitationally stable. This was found to be $\lambda_{\max} = 56$ in [42], independent of any constants. This cutoff may easily be included in our analysis, but it would considerably complicate the final expressions for the probability. Due to the extreme rarity of stars larger than this mass, we neglect this cutoff here, which does not alter any of the numbers we find to the precision we report.

The expected habitability of a universe with given constants is then:

$$\mathbb{H}_\star \propto N_\star \propto \frac{\epsilon_\star m_p}{\langle \lambda \rangle M_0} \propto \frac{\beta^{3/4} \gamma^3}{\alpha^{3/2}} \tag{5}$$

This, along with the measure from Equation (4), determines the probability distribution for observing values of the three parameters under consideration. Bearing in mind the total range for these values, we may calculate the probabilities of observing ours to be:[1]

$$\mathbb{P}(\alpha_{\text{obs}}) = 0.20, \quad \mathbb{P}(\beta_{\text{obs}}) = 0.44, \quad \mathbb{P}(\gamma_{\text{obs}}) = 4.2 \times 10^{-7} \tag{6}$$

The number of habitable stars is plotted in three different subplanes of the parameter space in Figure 2[2].

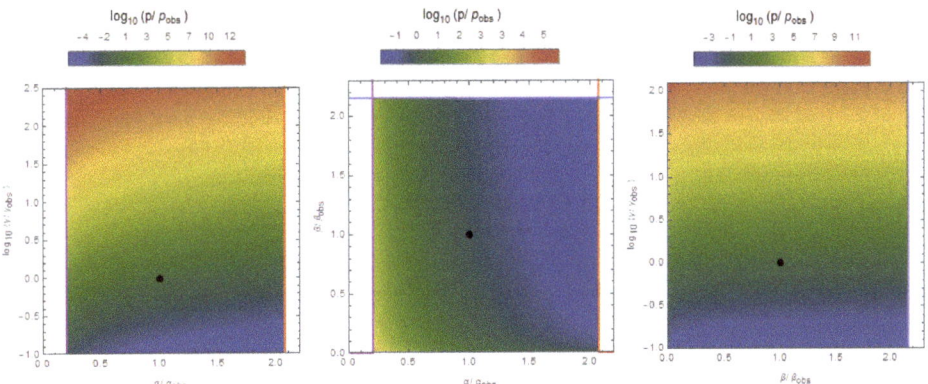

Figure 2. The distribution of stars throughout the multiverse. Each point represents a universe with a given set of parameters; the black dot represents our values. A strong preference for large γ, weak preference for small β, and a slightly stronger preference for small α value can be seen. Note that the γ axes are logarithmic, as well as the color display for the probability, which spans 16 order of magnitude.

---

[1] The code to compute all probabilities discussed in the text is made available at https://github.com/mccsandora/Multiverse-Habitability-Handler.
[2] The above used the log-uniform prior for β and γ, as discussed above. If instead we use a uniform prior, we find $\mathbb{P}(\alpha_{\text{obs}}) = 0.20$, $\mathbb{P}(\beta_{\text{obs}}) = 0.26$, and $\mathbb{P}(\gamma_{\text{obs}}) = 3.1 \times 10^{-9}$. We see that the probabilities for α and β are affected only slightly, and the probability for γ is decreased by two orders of magnitude. This is a fairly typical result.

As can be seen, there are allowed values of the parameters that contain many more stars than in our universe. Most notable is the preferences for larger gravitational strength, which represents a 5.1σ deviation from typicality. Based off this consideration alone, we conclude that this habitability criterion is incompatible with the premise that we are typical observers in a multiverse ensemble. This allows us to make our first testable prediction, that not all stars in our universe should be equally habitable. This should strike the reader as somewhat of an underwhelming prediction, since there are few that would wager against this statement, but it demonstrates a first possible test of the multiverse hypothesis. Including subsequent layers of realism will yield concomitantly sophisticated predictions.

4. Habitability Dependent on Stellar Properties

Up to this point, we have treated every stellar mass as potentially habitable, and postulated that the universe should maximize the total number of stars, irrespective of their properties. This is not justified, as there are a number of criteria that could render a star incapable of supporting life-bearing planets, and so now we develop the tools to reflect that in our calculations. Here, we focus on stellar characteristics that are a function of the mass only, though other aspects, such as metallicity, will be dealt with in [29]. Even further aspects, such as rotation, composition, environment, etc. would be interesting to investigate in the future. We first consider each additional criterion in isolation in this section, and then in Section 5 we consider various combinations of criteria to investigate their joint effects.

4.1. Is Photosynthesis Necessary for Complex Life?

Photosynthesis, the process by which photons are harvested by life for the purposes of creating chemical energy, was one of the absolutely key innovations in the history of life on our planet. By using the energy imparted on a specific molecule, this mechanism makes it possible to strip off electrons, which can then be used to process carbon dioxide into sugar. While other sources of energy may be exploited for this task [48], the sheer magnitude of available energy coming from the sun makes the harvesting of this source unrivaled, 3 orders of magnitude above any other potential source [49]. Today, there are many different molecular bases for anoxygenic photochemical systems, suggesting that it arose independently many times [50] and that it arose quite early in the history of the planet, perhaps 3.5 Ga [51], or even 3.8 Ga [52]. Photosynthesis provides the basis for the organic material for essentially the entire biosphere today (even at hydrothermal vents, who utilize organic material precipitated from above [48]). It was argued in [48] to inevitably arise in any situation where organic carbon is scarce and light energy is available.

Oxygenic photosynthesis is even more crucial for life on Earth. The ability to harvest the hydrogen atoms from water molecules allowed the process to yield 18 times the amount of energy as anoxygenic photosynthesis [53,54], provided a much more abundant supply chain, and, subsequently, oxidized the entire atmosphere. The Cambrian explosion occurred only after this event, and many complex organisms, including ourselves, require high levels of oxygen to perform the necessary level of metabolic activity [55]. Additionally, the atmospheric oxygen content was essential for the development of an ozone shield, which allowed subsequent colonization of the land surface.

Photosynthesis is by no means automatic, however. It relies on a coincidence where the energy of photons produced by the sun is roughly coincident with the energy required to ionize common molecules. This fact, that starlight is right at the molecular bond threshold, is one of the most remarkable anthropic coincidences. Originally pointed out by [56] on the basis that the stellar temperature be such that molecular bonds may form a partially convective outer layer, this bound was reinterpreted by [23] by noting that the energy may be harvested for chemical purposes. The requirements on the fundamental parameters can be found by equating the surface temperature of a star with the Rydberg energy. Using formulas in the appendix, this can be seen to occur for sunlike stars only if:

$$\alpha^6 \left(\frac{m_e}{m_p}\right)^2 \approx \frac{m_p}{\sqrt{8\pi}\, M_{pl}} \tag{7}$$

In [57] the precise details of the star were taken into account more carefully, altering the form of this expression slightly depending on the type of scattering that occurs. It is striking how well this equality holds in our universe, where the two are equal up to a factor of 1.7. Here, the temperature dependence on the size of the star is not taken into account in this expression because the spread in stellar temperatures is actually quite small. However, this makes the degree of tuning somewhat obtuse, and so we improve upon this standard analysis in what follows. This motivates our second ansatz for the habitability of a universe, that the number of observers is proportional to the number of stars capable of eliciting photosynthesis.

This begs the question, of what the allowable range for photosynthesis actually is. Though evidence across many different lineages suggests that photosystems have evolved to utilize the wavelengths with the most number of photons, (subject to some additional considerations) [54], there are hard physical limits for which wavelength photons are potentially photosynthetically useful. A lower bound often quoted is 400 nm [58] as below this photodissociation of most molecules occurs (though fluorescent pigments may potentially circumvent this bound [59]). An upper bound of 1100 nm was deduced in [60] on the basis that below this energy photons are indistinguishable from vibrational modes of molecules. A species of purple bacteria has been found that utilizes 1020 nm photons, though to split electrons from ferrous iron, which requires less energy [54]. The longest wavelength used for splitting water was recently found to be 750 nm [61]. In the following, we will refer to the maximally optimistic wavelength range, between 400–1100 nm, as the 'photosynthesis criterion', and the range 600–750 nm as the 'yellow criterion'. In this section we will stick to the former, and in the following subsection vary these bounds, commenting on the implications for what locales photosynthesis should be found around throughout our universe.

With this condition, the habitability of a universe becomes $\mathbb{H} = N_\star f_{\text{photo}}$, with:

$$f_{\text{photo}} = \int_{\lambda_{\text{fizzle}}}^{\lambda_{\text{fry}}} d\lambda\, p_{\text{IMF}}(\lambda) \tag{8}$$

Here λ_{fizzle} is the stellar mass with spectral temperature too weak for photosynthesis, and λ_{fry} the mass which is too hot. These both depend on the values taken for the limiting wavelengths, as illustrated in Figure 3. This should give the reader some idea of the width of allowed values of the parameters, which was not reported in the original treatments of this coincidence.

This leads to the following estimate for the habitability of the universe:

$$\mathbb{H}_{\text{photo}} \propto \alpha^{-3/2} \beta^{3/4} \gamma^3 \left(\min\left\{1, 0.45 \frac{L_{\text{fizzle}}}{1100\text{ nm}} Y^{1/4}\right\}^{2.84} - \min\left\{1, 0.16 \frac{L_{\text{fry}}}{400\text{ nm}} Y^{1/4}\right\}^{2.84} \right) \tag{9}$$

With L_{fizzle} and L_{fry} being the longest and shortest suitable wavelengths, respectively, and:

$$Y = 3.19 \frac{\gamma}{\alpha^{63/20} \beta^{137/40}} \tag{10}$$

which is normalized to 1 for our observed values. As can be seen, this quantity, which controls the fraction of photosynthetic stars, differs from the expectation given by Equation (7). This discrepancy is due to the fact that the original analysis restricted attention to sunlike stars, whereas we have considered the entire range of stars as potentially photosynthetic. This criterion leads to probabilities:

$$\mathbb{P}(\alpha_{\text{obs}}) = 0.32, \quad \mathbb{P}(\beta_{\text{obs}}) = 0.23, \quad \mathbb{P}(\gamma_{\text{obs}}) = 5.2 \times 10^{-7} \tag{11}$$

When compared with Equation (6), the typicality of our observed electron to proton mass ratio is worse by about a factor of 1.9, the fine structure constant better by 1.6, and the strength of gravity better by 1.3. So far, this hypothesis does not seem to add much to the discussion. However, it is premature to dismiss it: As can be seen from Figure 4, its main effect is to enforce a somewhat tight relationship between the parameters α and β, but it retains the tendency to prefer large values of γ. When used in conjunction with additional criteria to be discussed below, this will become an essential ingredient in finding a definition of habitability that renders all three probabilities very likely.

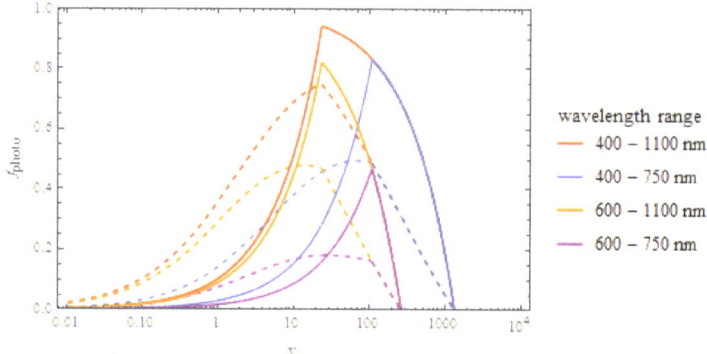

Figure 3. The fraction of stars which are capable of supporting photosynthesis as a function of the composite parameter Y defined in the text. The different curves correspond to taking the minimal wavelength to be both 400 nm and 600 nm, and the maximal to be 750 and 1100 nm. The solid curves use the estimate in Equation (8), and the dashed curves use the more refined initial mass function (IMF) of Equation (26).

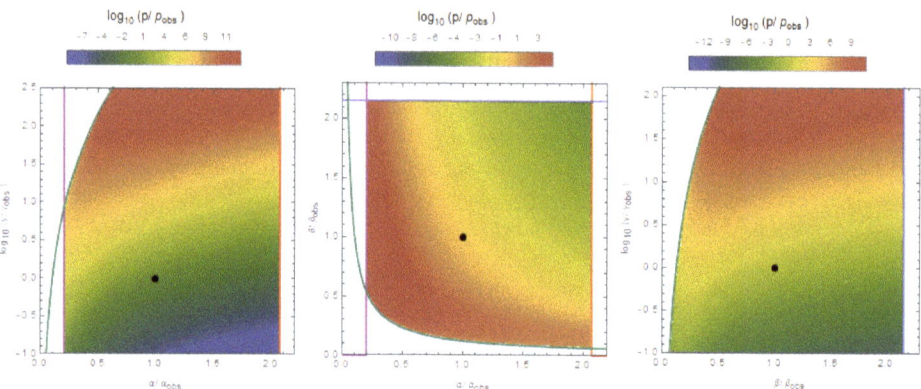

Figure 4. Distribution of observers from imposing the photosynthesis condition. A strong preference for the parameters to be restricted to the photosynthetic range is introduced, but there is still a preference for large γ.

Before moving on, there is also a lower limit on flux needed to support photosynthesis: This is commonly taken to be 1% of the average surface flux on Earth. (Though an organism has been found that subsists on 10^{-5} of this level, it would be incapable of sustaining the biosphere as we know it) [62]. The origin of this lower bound is that at least one photon should be incident on the photosystem molecules per cycling time, which from the appendix gives $\Phi_{min} \sim T_{mol}^3 = 5 \times 10^{-5} \alpha^6 m_e^{9/2}/m_p^{3/2}$, in excellent accord with the actual value. Comparing this to the flux from sunlike stars gives the following bound on the physical constants: $\alpha^{3/2}\beta\gamma^{-1/4} > 1.4 \times 10^{-4}$. This is a rather mild criterion, and will not be anthropically relevant. This would be useful in a more sophisticated analysis that determines the potentially photosynthetic stars by considering when the useful part of their spectrum falls below this threshold along the lines of [63], rather than our somewhat simplified prescription of scaling based off the surface temperature. We do not expect this more elaborate treatment to substantially alter our conclusions.

4.2. Is Photosynthesis Possible around Red Dwarfs?

Without understanding the extent of the range of wavelengths capable of giving rise to photosynthesis, there is some ambiguity to the amount of tuning that is required for it to hold. Additionally, we may wonder how our assumptions about the minimum or maximum wavelength thresholds affect the probabilities of our observed quantities, and whether these considerations can be used to inform our expectations of where photosynthesis can arise. We address these points here.

To split a hydrogen off a water molecule, it takes 1.23 eV [64]. In order to perform this splitting, it is necessary for life to utilize two of the sun's photons, at 680 nm and 700 nm (1.78 eV and 1.82 eV) [54], to perform a concatenated cascade of excitations known as the Z process.

If we assume that the maximal efficiency of oxygenic photosynthesis is $\epsilon = E_{H_2O}/E_{tot}$, which is equal to $\epsilon = 0.33$ on Earth, and also that photosynthesis can occur by the collection of n photons per molecular bond (2 on Earth, as per our counting), then the longest wavelength possible is $\lambda_{fizzle} = \epsilon\, n\, 1008$ nm. The complicated nature of this two stage process likely delayed the evolution of oxygenic photosynthesis considerably, despite its great advantages [65].

On this basis, it was argued in [53] that photosynthesis could in principle take place around red dwarf stars, though it would take three photons per water splitting, and so may be proportionately harder to evolve. Additionally, it was found in [66] that it may be less productive, depending on the wavelength of light harvested. Ref. [58] argue that photosynthesis may take place around potentially F, G, K, and M star types, and [67] argue that even brown dwarfs and black smokers may support photosynthesis of some type.

In the above above analysis, we effectively assumed that photosynthesis will evolve in the entire physically allowed wavelength range. We may consider how our analysis changes with different assumptions, however. A few representative values are considered in Table 1, to illustrate the change in the probabilities that is effected. We can see that there is not a large difference introduced, so that the multiverse has little to say about whether we should expect photosynthesis around red dwarf stars in this regard. However, it will be useful to keep such effects in mind for future purposes, when we consider our location within this universe as well.

Table 1. Dependence of the probabilities of our observed quantities on the upper and lower limits of the photosynthetic range.

Wavelength Range	$\mathbb{P}(\alpha_{obs})$	$\mathbb{P}(\beta_{obs})$	$\mathbb{P}(\gamma_{obs})$
400–1100 nm	0.318	0.231	5.18e-07
400–750 nm	0.242	0.263	3.85e-07
600–1100 nm	0.444	0.175	7.35e-07
600–750 nm	0.334	0.221	5.40e-07

4.3. Is There a Minimum Timescale for Developing Intelligence?

Until this point, we have not specified any bound that places an upper limit on the strength of gravity, and so the probabilities we have reported can be viewed as optimistic(!) estimates. However, our treatment has disregarded any mention of the actual lifetimes of the stars considered, which should surely influence the habitability properties of their surrounding planets. Here, we rectify this. The purpose is not to fully investigate the influence of stellar lifetime on habitability, which will be treated more fully in [22], but rather to investigate the potential importance of this restriction. As a byproduct, we will find a maximal value of γ for use throughout our calculations.

This can be done by noting that there is a maximum allowable mass for a habitable star, based on the criterion that it last long enough for life to take hold. Here, we make the crude approximation that all stars above this mass are inhospitable, and all below are equally habitable. How to define the timescale necessary for life is very uncertain—here we simply take it to be $t_{bio} = N_{bio} t_{mol}$, with $N_{bio} \sim 10^{30}$ and, t_{mol} is the molecular timescale given in the appendix. For our universe, this should be on the gigayear timescale, an estimate suggested in [68]. This subscribes to the notion that this amount of time is both necessary and sufficient for a biosphere to achieve the complexity we observe, as advocated for example in such papers as [69]. Alternative viewpoints are certainly taken on this matter, which makes the adoption of this criterion that of a personal preference at the moment. These alternatives will be explored fully in [22].

Then, using the formula for stellar lifetime in the appendix, the maximal mass is:

$$\lambda_{bio} = 1.8 \times 10^{-13} \alpha^{8/5} \beta^{-1/5} \gamma^{-4/5} \tag{12}$$

The normalization has been set to match that observed in our universe, $\lambda_{bio} = 1.2$ [70], corresponding to the largest stars that last 1 Gyr, around 2 solar masses.

Ensuring that this maximum mass is larger than the minimum stellar mass yields an upper bound for γ and unimportant lower bounds for α and β. The global upper bound on γ is found to be $\gamma_{max} = 134$, the value implicitly used in all calculations above.

The habitability can then be expressed as:

$$\mathbb{H}_{bio} = N_\star f_{bio} \propto \alpha^{-3/2} \beta^{3/4} \gamma^3 \left(1 - \min\left\{1, 1.48 \times 10^{15} \alpha^{-0.14} \beta^{-0.74} \gamma^{1.08}\right\}\right) \tag{13}$$

This is displayed in Figure 5. The most interesting feature that this criterion entails is the fact that the maximum value of γ is now dependent on α to some extent, and especially β. This gives rise to a preference for larger values of the latter quantity by virtue of there being more observers for larger γ. This effect serves to make the probability of our observed value of β less likely, as follows:

$$\mathbb{P}(\alpha_{obs}) = 0.28, \quad \mathbb{P}(\beta_{obs}) = 0.12, \quad \mathbb{P}(\gamma_{obs}) = 9.5 \times 10^{-6} \tag{14}$$

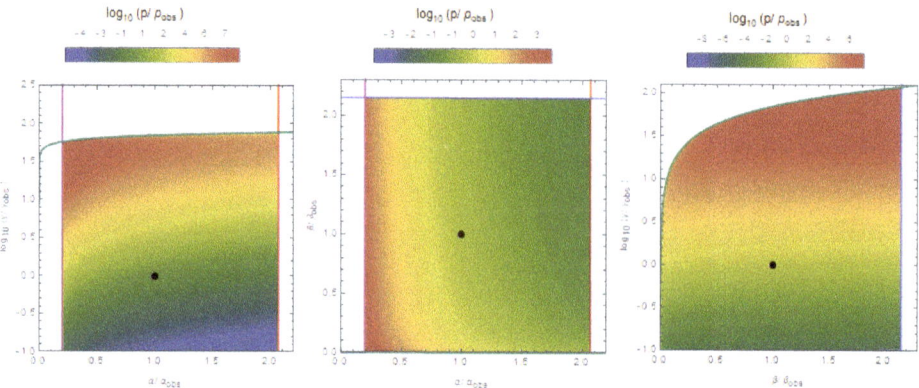

Figure 5. Distribution of observers from imposing the biological timescale condition. Of note is the secondary preference for large β the anthropic boundary induces.

4.4. Are Tidally Locked Planets Habitable?

Traditionally, it has been argued that life would have a very hard time evolving on tidally locked planets (for a review see [71]): One side is perpetually scorched by its host star, and the other eternally shrouded in frigid darkness. Winds between the two hemispheres would tear over the surface, and rotation would be too slow to generate a magnetic shield. However, some researchers feel this dismissiveness may be unjustified: Recent climate modeling that suggests even a thin atmosphere would serve to taper these extreme conditions [72]. Even still, a recent paper [73] argues that stratospheric circulation may not be as clement on worlds like these after all. Clearly, it is premature to think that we know enough about all the complex processes on these worlds to be able to definitively conclude whether they may be potentially habitable or not, and new surprises are sure to abound. In this section we investigate how the estimates for the number of observers in a universe would change if we assume that tidally locked stars are not habitable.

If we adopt this, it will define a minimum allowable mass, as planets around smaller stars must orbit much closer to remain inside the temperate zone[3]. For this we use the standard formula for the time it takes a planet of mass M, radius R, and distance a away from its mass $M_\star$ star to spin down from initial angular frequency ω [74]:

$$t_{TL} = \frac{40}{3} \frac{\omega a^6 M}{G M_\star^2 R^3} \qquad (15)$$

The coefficient $40/3$ assumes the planet is rocky and roughly spherical. When planets are formed, they are nearly marginally bound: This sets the initial centrifugal force to be approximately equal to the force of gravity at the planet's surface, yielding $\omega \sim \sqrt{G\rho}$. For the Earth this gives a period of roughly 3 h, about twice as fast as our planet's initial rotation speed.

Using these expressions, the tidal locking time can be expressed as:

$$t_{TL} \sim 566 \frac{\lambda^{17/2} m_p^{17/2}}{\alpha^{51/2} m_e^{15/2} M_{pl}^2} \qquad (16)$$

Note however the high powers involved, indicating that tidal locking is extremely sensitive to stellar mass.

We now have to compare this to the total habitable lifetime as a function of mass. Because stars steadily increase in luminosity as they age, the habitable zone migrates outwards, eventually causing even ideally situated planets to boil over. The habitable time of a star can then be defined as the average amount of time its orbits stay within the habitable zone. This requires knowledge of how quickly the star's luminosity changes, but the end result is just an order one factor of the total stellar lifetime, $t_{hab} = 0.4 t_\star$, independent of mass, and, more importantly, independent (or only very weakly dependent) on the physical parameters. The condition that $t_{TL} > t_\star$ gives $\lambda > \lambda_{TL}$, where:

$$\lambda_{TL} = 0.89 \, \alpha^{5/2} \, \beta^{1/2} \, \gamma^{-4/11} \qquad (17)$$

---

[3] We entertain adopting a different stance as to whether planets must be in the temperate zone to be habitable in [29].

Here, the coefficient is set to agree with the estimate $\lambda > 0.47$ $(0.85 M_\odot)$ [75] which was found to be the threshold mass in our universe[4]. This can then be used in Equation (4) to yield:

$$\mathbb{H}_{TL} = N_\star \, f_{TL} \propto \alpha^{-3/2} \beta^{3/4} \gamma^3 \min \left\{ 1, 0.154 \, \alpha^{-1.35} \beta^{-1.69} \gamma^{0.49} \right\} \tag{18}$$

This gives probabilities of our observed values as:

$$\mathbb{P}(\alpha_{obs}) = 0.12, \quad \mathbb{P}(\beta_{obs}) = 0.30, \quad \mathbb{P}(\gamma_{obs}) = 1.8 \times 10^{-7}, \tag{19}$$

The distributions are visualized in Figure 6. Adding this consideration does not appreciably alter these probabilities, amounting to a factor of 1/6 amongst all three. However, the distribution looks markedly different: There is a much steeper dependence on α and β, favoring smaller values for each.

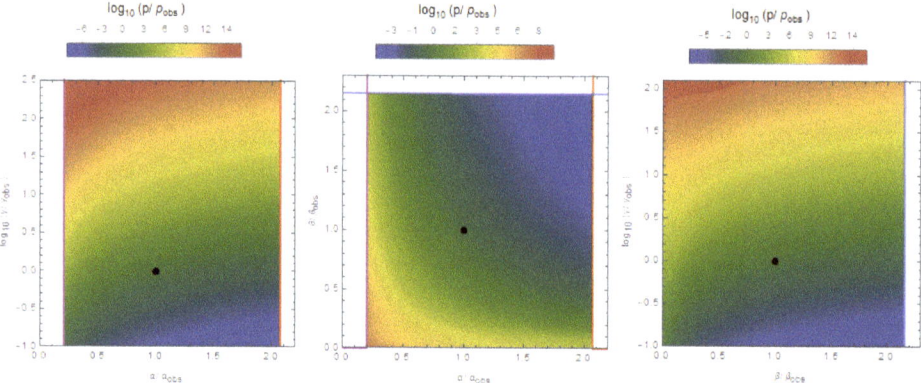

Figure 6. Distribution of observers from imposing the tidal locking condition.

4.5. Are Convective Stars Habitable?

In our universe, stars are divided into two broad classes, in accordance with the dominant mode of energy transfer between layers [76]. Stars below a certain size $0.35 M_\odot$ are convective, in that stellar material physically moves in order to attain local thermal equilibrium. Stars above $1.5 M_\odot$ are radiative, in that heat is transferred with relatively little radial mixing. Evidently, our sun lies between these two regimes, and accordingly it transfers heat using both methods. These two types of stars have a markedly different behavior, which may affect the overall habitability of the system.

Given the paucity of direct evidence, it is currently unknown how the heat transport of a star affects its habitability. Several reasons have been given to suspect that convective stars are indeed uninhabitable, all stemming from their pronounced churning. This induces strong XUV flux [77], flares [78] and space weather [79], all of which are capable of contributing to a severe level of atmospheric erosion of a planet orbiting within the habitable zone. As a counterpoint, these flares have been suggested to result in periods of potentially increased productivity in [80]. While there is a considerable effort currently devoted to determining whether these convective processes preclude life [81,82], here we may examine the consequences of adopting the viewpoint that convective stars are uninhabitable, and determine whether this is consistent with the multiverse framework.

---

[4] The reader may object that even if a star's planets become tidally locked before it expires, they may remain tidally unlocked for sufficient time for complex life to develop. We have adopted this more rough criterion to make this section self contained, but one may instead compare it to the biological time discussed in the previous subsection, for example. If this is done, they would find $\lambda_{TL} = 3556 \alpha^{47/17} \beta^{12/17} \gamma^{-4/17}$, which does not alter the conclusions by much.

We take the threshold for stellar convection from [57], based off the dominance of Thomson scattering, and updated to incorporate our radial and luminosity dependence:

$$\lambda_{\text{conv}} = 0.57\, \alpha^3\, \beta\, \gamma^{-1/2} \tag{20}$$

Note the resemblance of this quantity to that of the photosynthesis condition from Equation (7). This striking fact traces its origins back to the original interpretation of this condition in [56], where it was assumed that the existence of both types of star is essential to life. The apparent coincidence of these two conditions is not as mysterious as it may first appear, either: Convection occurs in stars when their surface temperature drops below the point where molecular bonds are able to form, and so stars that exhibit marginal convection will have their surface temperatures automatically set by molecular energies. Then, this dichotomy between the behavior of small and large stars is a generic feature in any universe where photosynthesis is possible.

Using this condition gives:

$$\mathbb{H}_{\text{conv}} = N_\star\, f_{\text{conv}} \propto \alpha^{-3/2}\, \beta^{3/4}\, \gamma^3\, \min\left\{1, .285\, \alpha^{-2.03}\, \beta^{-2.36}\, \gamma^{0.68}\right\} \tag{21}$$

The parameter dependence is very similar to the tidal locking case, with preference for small values of α and β tamed by eventually entering a regime of parameter space where no stars are purely convective. Due to the strong similarities between these two scenarios, we refrain from plotting the probability distributions for this criterion. This gives probabilities of our observed values as:

$$\mathbb{P}(\alpha_{\text{obs}}) = 0.16, \quad \mathbb{P}(\beta_{\text{obs}}) = 0.41, \quad \mathbb{P}(\gamma_{\text{obs}}) = 2.6 \times 10^{-7}, \tag{22}$$

From here we see that this criterion performs about the same, though overall slightly better, than the tidal locking criterion.

4.6. Is Habitability Dependent on Entropy Production?

Up until now, the various hypotheses we have considered have failed to account for the observed values of our physical constants. This indicates that treating all stars that meet some threshold criteria as equally habitable may be the wrong approach. Though extensions to this simplistic scheme fall under the purview of the other factors in the Drake equation, which will be dealt with in subsequent papers, we take this opportunity to report on a prescription which manages to bring all predicted values into accord with observation.

The successful habitability criteria is that the presence of life should be proportional to the total entropy processed by a planet over its lifetime. On Earth, this is predominantly given by the downconversion of sunlight to lower frequencies, which in the process generates biologically useful chemical energy. The amount available will then depend on the total number of photons generated by the host star over its lifetime, as well as the solid angle subtended by the planet that collects them. For this, we specify to terrestrial mass planets that orbit within the temperate zone, as outlined in the Appendix A.

The reason one might consider this to play an important factor is that entropy production play a key role in regulating biosphere size [83,84], as evidenced by the fact that Earth's biosphere operates close to the theoretical limit for how much information can be processed. There are a number of subtleties in this argument that we do not address here, but will be dealt with in full in [22]. It suffices for the present purposes to merely introduce this criterion and demonstrate that it yields the desired results. One point we do wish to make, however, is that it seems inextricably linked with the necessity of photosynthesis: If the star only produced photons that could not be used for chemical energy, they would all be instantly recycled as waste heat, rather than contributing to the biosphere. Therefore, we do not consider this criterion in isolation ever, but only in conjunction with the photosynthesis requirement.

To this end, we estimate the total amount of entropy incident on a planet situated a temperate distance from its host star as:

$$\Delta S_{\text{tot}}(\lambda) \sim \frac{L_\star}{T_\star} \frac{R_{\text{terr}}^2}{4 a_{\text{hab}}^2} t_\star \sim \frac{\alpha^{17/2} \beta^2}{\lambda^{119/40} \gamma^{17/4}} \sim 10^{54} \qquad (23)$$

In this case, habitability will not simply be proportional to the fraction of stars meeting some certain criteria, but instead each star must be weighted by its entropy production. This will yield:

$$\mathbb{H}_S \propto \frac{\alpha^{203/80} \beta^{797/160}}{\gamma^{5/4}} \left(\min\left\{1, 0.45 \frac{L_{\text{fizzle}}}{1100 \text{ nm}} Y^{1/4}\right\}^{9.11} - \min\left\{1, 0.16 \frac{L_{\text{fry}}}{400 \text{ nm}} Y^{1/4}\right\}^{9.11} \right) \qquad (24)$$

With Y defined as before in Equation (10). This greatly ameliorates the smallness of γ by virtue of the prefactor being very close to a scale invariant distribution. The distribution of observers is displayed in Figure 7.

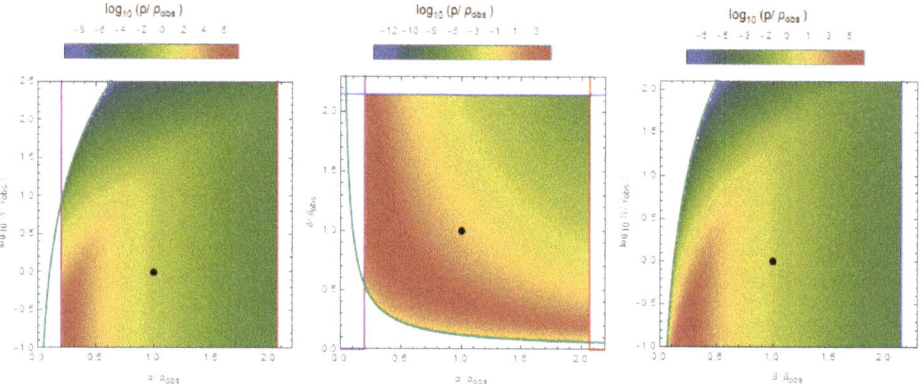

Figure 7. Distribution of observers from imposing the entropy condition.

Using the optimistic estimates for the photosynthetic range, the corresponding probabilities are:

$$\mathbb{P}(\alpha_{obs}) = 0.19, \quad \mathbb{P}(\beta_{obs}) = 0.45, \quad \mathbb{P}(\gamma_{obs}) = 0.32 \qquad (25)$$

This is the first fully satisfactory synthesis of habitability criteria that is consistent with the multiverse hypothesis. It has implications for the distribution of observers that may be eventually tested: We should expect to find complex life in those locales with the most amount of entropy production. While fully determining the places this distinguishes will rely on an in-depth analysis, this would include planets that orbit more active stars, for longer, and able to collect more incident radiation.

5. Discussion: Comparing 40 Hypotheses

Until this point, we have considered the number of observers throughout universes with different microphysical constants and, weighing against the expected relative frequencies of such universes in a generic multiverse context, have determined the probability of measuring the three values of our constants as they are. Our findings show that these probabilities depend sensitively on the precise requirements for habitability that are assumed, as we have demonstrated by separately considering the expectations that complex life is proportional to the number of stars, that it is dependent on photosynthesis, the absence of tidal locking, that it can only arise around tame stars, that it requires a certain length of time to develop, and that its presence is proportional to the total amount of entropy processed by the system.

It is worth pausing to reflect on why this sensitivity should occur: Our estimates show that the number of observers in a universe is much more sensitive to the parameters than the underlying distributions we have taken. Due to this, our location in the multiverse will be dictated more by the actual requirements of life, rather than the availability of universes with those particular features. This may be contrasted with some of the cosmological parameters, where the underlying distribution can be exponentially sensitive, so that the expectation is simply to be in the most abundant locale, rather regardless of its habitability (recall that the volume fraction of our universe that is conventionally habitable is perhaps 10^{-40}). More poetically, for the microphysical parameters we expect to be in 'the best of all possible worlds', whereas for the cosmological parameters, we expect to be in 'the cheapest of all possible worlds'.

However, being independent criteria, these may all also be considered in conjunction, yielding at this stage 40 distinct potential criteria for habitability. Because the effect of each condition was to restrict the range of habitable stellar masses, when taken together some effects will be more dominant that others in certain regions of parameter space. The various stellar thresholds are displayed in Figure 8.

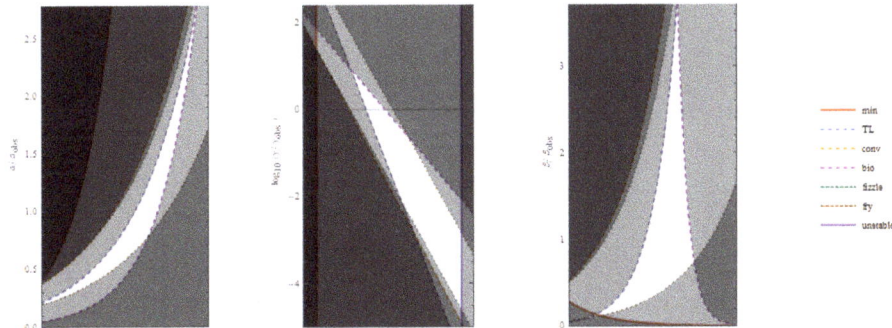

Figure 8. Habitable range of stellar masses for various choices of requirements. The shaded regions are treated as inhospitable for the various habitability assumptions.

We also take this opportunity to use a more realistic initial mass function, since including multiple criteria quickly make it extremely difficult to present results in an analytic form anyway. This takes into account the feedback that stars have on the collapsing protostellar dust cloud, which serves as a regulating mechanism guaranteeing that stars which are formed are within an order of magnitude or two of the scale $M_0 = (8\pi)^{3/2} M_{pl}^3/m_p^2$. The resultant initial mass function from this process resembles a broken power law with a transition scale of 0.2 $M_\odot$ set by the minimum stellar mass; an analytic expression for this scale was presented in [85]. Various parameterizations of this distribution appear in the literature, but here we use the one of [86], which can be simply implemented as:

$$p_{\text{Masch}}(\lambda) = p_{\text{IMF}}(\lambda) \frac{8.94}{\left(1 + \left(\frac{5.06\lambda_{\min}}{\lambda}\right)^{\beta_{\text{IMF}}-1}\right)^{1.4}} \quad (26)$$

Employing this usually alters the probabilities we find by $\mathcal{O}(10\%)$ and occasionally by a factor of $\mathcal{O}(2)$, which is not enough to change any of our conclusions qualitatively.

In Table 2 we enumerate the full list of probabilities without including the entropy condition. The most salient feature of this list is that the probability of measuring our strength of gravity is never more that 0.0114 (2.5σ), universes with larger values being by our considerations more favorable than ours. We remind the reader that at this stage, we have restricted our attention to only the first factor of the Drake equation, and so we may not find it surprising that we have not captured enough detail to yield a coherent account of our observations.

Other features may be noted if one examines the chart for long enough. There is a high degree of 'epistasis', to borrow a term from genetics: The inclusion of several requirements often does not influence the final result in a naively multiplicative manner. This can be seen by the inclusion of the biological timescale and tidal locking conditions, for instance: In isolation neither affect $\mathbb{P}(\alpha_{obs})$ to a strong degree, whereas in conjunction $\mathbb{P}(\alpha_{obs})$ is decreased by an order of magnitude. Including any additional habitability factors in isolation only decreases $\mathbb{P}(\beta_{obs})$, and yet when combined the effects are not nearly as pronounced. However, we caution against interpreting from this observation, since none of these criteria are fully satisfactory.

Table 2. Probabilities of observing our values of parameters for various habitability hypotheses. Here the shorthands are photo: Photosynthesis criterion, TL: Tidal locking, conv: Convective stars, and bio: The biological timescale criterion.

Criteria	$\mathbb{P}(\alpha_{obs})$	$\mathbb{P}(\beta_{obs})$	$\mathbb{P}(\gamma_{obs})$
number of stars	0.198	0.437	4.15×10^{-7}
bio	0.281	0.116	2.52×10^{-5}
conv	0.183	0.426	3.1×10^{-7}
conv bio	0.0564	0.159	2.59×10^{-5}
TL	0.152	0.37	2.34×10^{-7}
TL bio	0.0101	0.413	8.27×10^{-5}
TL conv	0.152	0.37	2.34×10^{-7}
TL conv bio	0.0101	0.413	8.27×10^{-5}
photo	0.439	0.183	8.16×10^{-7}
photo bio	0.0631	0.103	1.7×10^{-5}
photo conv	0.439	0.183	8.06×10^{-7}
photo conv bio	0.0637	0.104	1.7×10^{-5}
photo TL	0.48	0.232	6.88×10^{-7}
photo TL bio	0.0352	0.281	0.000139
photo TL conv	0.48	0.232	6.88×10^{-7}
photo TL conv bio	0.0352	0.281	0.000139
yellow	0.486	0.162	8.78×10^{-7}
yellow bio	0.0351	0.102	1.72×10^{-5}
yellow conv	0.486	0.162	8.78×10^{-7}
yellow conv bio	0.0351	0.102	1.72×10^{-5}
yellow TL	0.0303	0.0308	1.63×10^{-6}
yellow TL bio	0.324	0.335	0.0114
yellow TL conv	0.0303	0.0308	1.63×10^{-6}
yellow TL conv bio	0.324	0.335	0.0114

Additionally of note is that the inclusion of the photosynthesis condition renders the convective star condition almost completely superfluous, which makes sense in light of the fact that convective stars are always slightly lighter than the minimal photosynthetic star. This addresses an otherwise puzzling feature of our universe that arises if one believes convective stars are uninhabitable, which is why so many stars have this property, seemingly wasting the majority of opportunities for life to develop. If one simultaneously takes the viewpoint that photosynthesis is necessary for complex life, however, this puzzle is resolved because convective stars are a generic byproduct of the fact that photosynthetic light is just barely able to break chemical bonds.

Most importantly, however, is the observation that the criteria used change the probabilities by several orders of magnitude. The biological timescale condition, in particular, increases $\mathbb{P}(\gamma_{obs})$ by up to a factor of 353 by limiting the range of the strength of gravity. The spread in the probabilities of our observed values are 48, 14, and 49,000 for α, β and γ, respectively. The overall spread of the product of these values is 813,000. This gives the indication that the relative confidence that can be gained about certain habitability conditions will be of the same order of magnitude.

Table 3 incorporates the entropy condition as well. As can be seen, this brings all probabilities well into agreement with observations, irrespective of the inclusion of the various other hypotheses.

Interestingly, the spread of values once the entropy condition is included is much narrower than without, being 3.8, 1.3 and 1.5 for α, β, and γ respectively. The spread of the product of these values is now 3.6. This tempering is due to the fact that the entropy and photosynthesis conditions place such restrictive bounds on habitability, the effects of the other conditions (which primarily effect other regions of parameter space) play little role. As of now, this may seem somewhat disappointing, as the multiverse hypothesis has seemingly nothing to say about the role of tidal locking, stellar lifetime, convective habitability, or range of photosynthetic wavelengths, as long as one imposes the photosynthesis and entropy conditions as necessary for complex life. Indeed, one should not expect to come away with strong expectations for every proposed requirement based off these arguments. However, more information can in fact be gleaned about which hypotheses are viable based off a few additional considerations regarding our actual location within our own universe: This will be explored fully in [22].

Table 3. Probabilities of observing our values of parameters for various habitability hypotheses with entropy condition (denoted by S) included. The other shorthands are the same as above.

Criteria	$\mathbb{P}(\alpha_{obs})$	$\mathbb{P}(\beta_{obs})$	$\mathbb{P}(\gamma_{obs})$
photo S	0.24	0.386	0.376
photo bio S	0.178	0.414	0.426
photo conv S	0.256	0.401	0.368
photo conv bio S	0.191	0.433	0.421
photo TL S	0.394	0.446	0.356
photo TL bio S	0.278	0.465	0.453
photo TL conv S	0.394	0.446	0.356
photo TL conv bio S	0.278	0.465	0.453
yellow S	0.191	0.45	0.317
yellow bio S	0.125	0.486	0.38
yellow conv S	0.191	0.45	0.317
yellow conv bio S	0.125	0.486	0.38
yellow TL S	0.481	0.396	0.44
yellow TL bio S	0.343	0.476	0.31
yellow TL conv S	0.481	0.396	0.44
yellow TL conv bio S	0.343	0.476	0.31

Remember that, of the 40 possible conditions we started with, less than half have turned out to be compatible with the multiverse hypothesis. Considering additional habitability criteria will yield similarly strong predictions for multiple other leading schools of thought about what conditions are necessary for complex life. This demonstrates the power of this method of reasoning: We have utilized this method to uncover readily discoverable facts about the world that are currently unknown. The true conditions for habitability will eventually be found, and sooner than one might be prepared for: Upcoming experiments probing the solar system and galaxy promise to shed light on these issues, and inform our understanding of life's place in the universe, and, depending on their findings, multiverse.

Funding: This research received no external funding.

Acknowledgments: I would like to thank Fred Adams and Alex Vilenkin for useful discussions.

Conflicts of Interest: The author declares no conflict of interest.

Appendix A. Stellar Properties

Throughout, we have made use of how the main features of stars scale with mass. These are well known, as can be found in [23,42], for example, who take a particular emphasis on dependence on physical constants. We restrict our summary to main sequence stars.

One of the most important stellar characteristics is its luminosity, which is given by:

$$L_\star = 9.7 \times 10^{-4} \lambda^{q_L} \frac{m_e^2 \, M_{pl}}{\alpha^2 \, m_p} \tag{A1}$$

The dependence on mass is sometimes given as a broken power law, which reflects the fact that the opacity inside the star is set by different scattering processes depending on the temperature and pressure. Here, we neglect this subtlety, as it does not greatly affect our analysis other than making it far less amenable to analytic study, and take the value $q_L = 3.5$ from [76]. This most accurately characterizes smaller stars, which dominate the population, and so are most important to consider.

Equally important is the lifetime of a star, which can be found through the approximate scaling relation $t_\star \approx \epsilon_{nuc} M_\star / L_\star$, where ϵ_{nuc} is the energy yield per nucleon:

$$t_\star = 85.6 \frac{\alpha^2}{\lambda^{5/2}} \frac{M_{pl}^2}{m_p \, m_e^2} \tag{A2}$$

where we can see the characteristic scaling that massive stars live for a shorter duration than less massive stars.

The radius of a star is observed to scale as:

$$R_\star = 108.6 \, \lambda^{q_\xi} \frac{M_{pl}}{\alpha^2 \, m_p} \tag{A3}$$

Here, $q_\xi = 4/5$ [87]. Using this result, we can derive the star's surface temperature to be:

$$T_\star = 0.014 \, \lambda^{\frac{q_L - 2q_\xi}{4}} \frac{\alpha^{1/2} \, m_e^{1/2} \, m_p^{3/4}}{M_{pl}^{1/4}} \tag{A4}$$

For this, the expression for interior temperature of a star given in [23] was used. This estimates the interior temperature of the star by imposing the condition $\sqrt{T/m_p} \sim \alpha$ in order for thermal effects to balance out the energetic suppression that comes from the Coulomb barrier in reactions. As can be seen, the dependence of temperature on stellar mass is actually quite weak, $T \propto \lambda^{19/40}$, signifying that all stars emit light in approximately the same wavelength regime. This lends credence to the claim that a star's suitability for photosynthesis is largely independent of mass, but rather only contingent on the relation imposed on physical constants.

The stellar temperature can be compared to the temperature needed for life: This is commonly defined as the value for which liquid water is possible, but more generically, it can be identified as the temperature that matches the energy levels of typical molecular bonds, so that life is free to manipulate and store energy by subtly rearranging local chemical conditions. This requirement was identified in [23] to give:

$$T_{mol} = 0.037 \frac{\alpha^2 \, m_e^{3/2}}{m_p^{1/2}} \tag{A5}$$

This is a factor $(m_e/m_p)^{1/2}$ smaller than the Rydberg temperature that governs atomic ionization due to the lower energy vibrational modes of the molecules. Additionally, we include the factor $\epsilon_T \sim 0.037$, which encodes "the abhorrent details of chemistry that are omitted" from [23].

This defines the typical timescale for molecular processes as $t_{\text{mol}} = 1/T_{\text{mol}}$. This definition is based off the rotational frequencies of molecules—it can be viewed as the time it takes energy to redistribute throughout a molecule. This can be compared with the timescale set by the mean free path over the sound speed, L_{mfp}/c_s, the typical time between molecular interactions. For room temperature solutions, these scales differ only by a factor of $\beta^{-1/2}$, and so the distinction will be unimportant in this setting, but crucial for other purposes. These can also be compared with the atomic timescale used in [88] to set an upper bound on γ, which nevertheless produces similar results.

The traditional habitable orbital distance from the star can be found by demanding that the planet be at the habitable temperature. Though this depends on many planetary factors, these will be dealt with more properly in [29]. For now we content ourselves with a simple blackbody estimate, which yields $a = (T_\star/T)^2 R_\star/2$. Then:

$$a_{\text{temp}} = 7.6\, \lambda^{q_L/2}\, \frac{m_p^{1/2} M_{pl}^{1/2}}{\alpha^5 m_e^2} \tag{A6}$$

Finally, the radius of a terrestrial planet can be found by the condition that the escape velocity is of the same order as the thermal velocity for the molecular temperature, which yields:

$$R_{\text{terr}} = 3.6\, \frac{M_{pl}}{\alpha^{1/2} m_e^{3/4} m_p^{5/4}} \tag{A7}$$

References

1. Vilenkin, A. Predictions from Quantum Cosmology. *Phys. Rev. Lett.* **1995**, *74*, 846–849. [CrossRef] [PubMed]
2. Weinberg, S. Anthropic bound on the cosmological constant. *Phys. Rev. Lett.* **1987**, *59*, 2607–2610. [CrossRef] [PubMed]
3. Garriga, J.; Vilenkin, A. Anthropic Prediction for Λ and the Q Catastrophe. *Prog. Theor. Phys. Suppl.* **2006**, *163*, 245–257. [CrossRef]
4. Garriga, J.; Livio, M.; Vilenkin, A. Cosmological constant and the time of its dominance. *Phys. Rev. D* **2000**, *61*, 023503. [CrossRef]
5. Tegmark, M.; Rees, M.J. Why Is the Cosmic Microwave Background Fluctuation Level 10^{-5}? *Astrophys. J.* **1998**, *499*, 526–532. [CrossRef]
6. Hogan, C.J. Quarks, Electrons, and Atoms in Closely Related Universes. *arXiv* **2004**, arXiv:astro-ph/0407086.
7. Damour, T.; Donoghue, J.F. Constraints on the variability of quark masses from nuclear binding. *Phys. Rev. D* **2008**, *78*, 014014. [CrossRef]
8. Barnes, L.A. Binding the diproton in stars: Anthropic limits on the strength of gravity. *J. Cosmol. Astropart. Phys.* **2015**, *2015*, 050. [CrossRef]
9. Adams, F.C. The degree of fine-tuning in our universe—And others. *arXiv* **2019**, arXiv:1902.03928.
10. Borucki, W.J.; Koch, D.; Basri, G.; Batalha, N.; Brown, T.; Caldwell, D.; Caldwell, J.; Christensen-Dalsgaard, J.; Cochran, W.D.; DeVore, E.; et al. Kepler planet-detection mission: Introduction and first results. *Science* **2010**, *327*, 977–980. [CrossRef]
11. Ehrenfreund, P.; Charnley, S.B. Organic molecules in the interstellar medium, comets, and meteorites: A voyage from dark clouds to the early Earth. *Annu. Rev. Astron. Astrophys.* **2000**, *38*, 427–483. [CrossRef]
12. Matson, D.L.; Spilker, L.J.; Lebreton, J.P. The Cassini/Huygens mission to the Saturnian system. In *The Cassini-Huygens Mission*; Springer: Dordrecht, The Netherlands, 2003; pp. 1–58.
13. Stern, S.; Bagenal, F.; Ennico, K.; Gladstone, G.; Grundy, W.; McKinnon, W.; Moore, J.; Olkin, C.; Spencer, J.; Weaver, H.; et al. The Pluto system: Initial results from its exploration by New Horizons. *Science* **2015**, *350*, aad1815. [CrossRef] [PubMed]
14. Tsiaras, A.; Waldmann, I.; Zingales, T.; Rocchetto, M.; Morello, G.; Damiano, M.; Karpouzas, K.; Tinetti, G.; McKemmish, L.; Tennyson, J.; et al. A Population Study of Gaseous Exoplanets. *Astron. J.* **2018**, *155*, 156. [CrossRef]

15. Ricker, G.R.; Winn, J.N.; Vanderspek, R.; Latham, D.W.; Bakos, G.Á.; Bean, J.L.; Berta-Thompson, Z.K.; Brown, T.M.; Buchhave, L.; Butler, N.R.; et al. Transiting exoplanet survey satellite. *J. Astron. Telesc. Instrum. Syst.* **2014**, *1*, 014003. [CrossRef]
16. Broeg, C.; Fortier, A.; Ehrenreich, D.; Alibert, Y.; Baumjohann, W.; Benz, W.; Deleuil, M.; Gillon, M.; Ivanov, A.; Liseau, R.; et al. CHEOPS: A transit photometry mission for ESA's small mission programme. In Proceedings of the EPJ Web of Conferences, Geneva, Switzerland, 13–16 February 2012.
17. Gardner, J.P.; Mather, J.C.; Clampin, M.; Doyon, R.; Greenhouse, M.A.; Hammel, H.B.; Hutchings, J.B.; Jakobsen, P.; Lilly, S.J.; Long, K.S.; et al. The james webb space telescope. *Space Sci. Rev.* **2006**, *123*, 485–606. [CrossRef]
18. Schwieterman, E.W.; Kiang, N.Y.; Parenteau, M.N.; Harman, C.E.; DasSarma, S.; Fisher, T.M.; Arney, G.N.; Hartnett, H.E.; Reinhard, C.T.; Olson, S.L.; et al. Exoplanet Biosignatures: A Review of Remotely Detectable Signs of Life. *Astrobiology* **2018**, *18*, 663–708, [CrossRef] [PubMed]
19. Catling, D.C.; Krissansen-Totton, J.; Kiang, N.Y.; Crisp, D.; Robinson, T.D.; DasSarma, S.; Rushby, A.J.; Del Genio, A.; Bains, W.; Domagal-Goldman, S. Exoplanet Biosignatures: A Framework for Their Assessment. *Astrobiology* **2018**, *18*, 709–738, [CrossRef]
20. Fujii, Y.; Angerhausen, D.; Deitrick, R.; Domagal-Goldman, S.; Grenfell, J.L.; Hori, Y.; Kane, S.R.; Pallé, E.; Rauer, H.; Siegler, N.; et al. Exoplanet biosignatures: Observational prospects. *Astrobiology* **2018**, *18*, 739–778. [CrossRef]
21. Walker, S.I.; Bains, W.; Cronin, L.; DasSarma, S.; Danielache, S.; Domagal-Goldman, S.; Kacar, B.; Kiang, N.Y.; Lenardic, A.; Reinhard, C.T.; et al. Exoplanet Biosignatures: Future Directions. *Astrobiology* **2018**, *18*, 779–824, [CrossRef]
22. Sandora, M. Multiverse Predictions for Habitability III: Fraction of Planets That Develop Life. *arXiv* **2019**, arXiv:1903.06283.
23. Press, W.H.; Lightman, A.P. Dependence of macrophysical phenomena on the values of the fundamental constants. *Philos. Trans. R. Soc. Lond. Ser. A* **1983**, *310*, 323–334, [CrossRef]
24. Adams, F.C.; Grohs, E. On the habitability of universes without stable deuterium. *Astropart. Phys.* **2017**, *91*, 90–104, [CrossRef]
25. Barnes, L.A.; Lewis, G.F. Producing the deuteron in stars: anthropic limits on fundamental constants. *J. Cosmol. Astropart. Phys.* **2017**, *2017*, 036, [CrossRef]
26. Hogan, C.J. Nuclear astrophysics of worlds in the string landscape. *Phys. Rev. D* **2006**, *74*, 123514, [CrossRef]
27. Oberhummer, H.; Csótó, A.; Schlattl, H. Stellar Production Rates of Carbon and Its Abundance in the Universe. *Science* **2000**, *289*, 88–90, [CrossRef]
28. Schellekens, A.N. Life at the Interface of Particle Physics and String Theory. *Rev. Mod. Phys.* **2013**, *85*, 1491–1540, [CrossRef]
29. Sandora, M. Multiverse Predictions for Habitability II: Number of Habitable Planets. *arXiv* **2019**, arXiv:1902.06784.
30. Graesser, M.L.; Salem, M.P. The scale of gravity and the cosmological constant within a landscape. *Phys. Rev.* **2007**, *D76*, 043506, [CrossRef]
31. Frank, A.; Sullivan, W.T., III. A New Empirical Constraint on the Prevalence of Technological Species in the Universe. *Astrobiology* **2016**, *16*, 359–362, [CrossRef]
32. Gleiser, M. Drake equation for the multiverse: From the string landscape to complex life. *Int. J. Mod. Phys. D* **2010**, *19*, 1299–1308. [CrossRef]
33. Sandora, M. Multiverse Predictions for Habitability IV: Fraction of Life that Develops Intelligence. *arXiv* **2019**, arXiv:1904.11796.
34. Donoghue, J.F.; Dutta, K.; Ross, A. Quark and lepton masses and mixing in the landscape. *Phys. Rev. D* **2006**, *73*, 113002, [CrossRef]
35. Gibbons, G.W.; Turok, N. Measure problem in cosmology. *Phys. Rev. D* **2008**, *77*, 063516, . [CrossRef]
36. Freivogel, B. Making predictions in the multiverse. *Class. Quantum Gravity* **2011**, *28*, 204007, [CrossRef]
37. Bousso, R.; Freivogel, B.; Yang, I.S. Properties of the scale factor measure. *Phys. Rev.* **2009**, *D79*, 063513, [CrossRef]
38. Salem, M.P.; Vilenkin, A. Phenomenology of the CAH+ measure. *Phys. Rev.* **2011**, *D84*, 123520, [CrossRef]
39. De Simone, A.; Guth, A.H.; Salem, M.P.; Vilenkin, A. Predicting the cosmological constant with the scale-factor cutoff measure. *Phys. Rev. D* **2008**, *78*. [CrossRef]

40. Behroozi, P.S.; Wechsler, R.H.; Conroy, C. The Average Star Formation Histories of Galaxies in Dark Matter Halos from z = 0–8. *Astrophys. J.* **2013**, *770*, 57. [CrossRef]
41. Silk, J.; Mamon, G.A. The current status of galaxy formation. *Res. Astron. Astrophys.* **2012**, *12*, 917–946, [CrossRef]
42. Adams, F.C. Stars in other universes: Stellar structure with different fundamental constants. *J. Cosmol. Astropart. Phys.* **2008**, *2008*, 010. [CrossRef]
43. Salpeter, E.E. The Luminosity Function and Stellar Evolution. *Astrophys. J.* **1955**, *121*, 161. [CrossRef]
44. Krumholz, M.R. The big problems in star formation: The star formation rate, stellar clustering, and the initial mass function. *Phys. Rep.* **2014**, *539*, 49–134. [CrossRef]
45. Chabrier, G. Galactic Stellar and Substellar Initial Mass Function. *Publ. ASP* **2003**, *115*, 763–795. [CrossRef]
46. Loeb, A.; Batista, R.A.; Sloan, D. Relative likelihood for life as a function of cosmic time. *J. Cosmol. Astropart. Phys.* **2016**, *2016*, 040. [CrossRef]
47. Burrows, A.S.; Ostriker, J.P. Astronomical reach of fundamental physics. *Proc. Natl. Acad. Sci. USA* **2014**, *111*, 2409–2416. [CrossRef] [PubMed]
48. Rothschild, L.J. The evolution of photosynthesis . . . again? *Philos. Trans. R. Soc. Lond. Ser. B Biol. Sci.* **2008**, *363*, 2787–2801. [CrossRef] [PubMed]
49. des Marais, D.J. When Did Photosynthesis Emerge on Earth? *Science* **2000**, *289*, 1703–1705.
50. Blankenship, R.E. Early Evolution of Photosynthesis. *Plant Physiol.* **2010**, *154*, 434–438. . [CrossRef]
51. Schopf, J.W.; Kitajima, K.; Spicuzza, M.J.; Kudryavtsev, A.B.; Valley, J.W. SIMS analyses of the oldest known assemblage of microfossils document their taxon-correlated carbon isotope compositions. *Proc. Natl. Acad. Sci. USA* **2018**, *115*, 53–58. [CrossRef]
52. Buick, R. The Antiquity of Oxygenic Photosynthesis: Evidence from Stromatolites in Sulphate-Deficient Archaean Lakes. *Science* **1992**, *255*, 74–77. [CrossRef]
53. Wolstencroft, R.D.; Raven, J.A. Photosynthesis: Likelihood of Occurrence and Possibility of Detection on Earth-like Planets. *Icarus* **2002**, *157*, 535–548. [CrossRef]
54. Kiang, N.Y.; Siefert, J.; Govindjee; Blankenship, R.E. Spectral Signatures of Photosynthesis. I. Review of Earth Organisms. *Astrobiology* **2007**, *7*, 222–251. [CrossRef] [PubMed]
55. Catling, D.C.; Glein, C.R.; Zahnle, K.J.; McKay, C.P. Why O_2 Is Required by Complex Life on Habitable Planets and the Concept of Planetary "Oxygenation Time". *Astrobiology* **2005**, *5*, 415–438. [CrossRef] [PubMed]
56. Carter, B. Republication of: Large number coincidences and the anthropic principle in cosmology. *Gen. Relativ. Gravit.* **2011**, *43*, 3225–3233. [CrossRef]
57. Barrow, J.D.; Tipler, F.J. *The Anthropic Cosmological Principle*; Oxford University Press: Oxford, UK, 1986.
58. Kiang, N.Y.; Segura, A.; Tinetti, G.; Govindjee; Blankenship, R.E.; Cohen, M.; Siefert, J.; Crisp, D.; Meadows, V.S. Spectral Signatures of Photosynthesis. II. Coevolution with Other Stars And The Atmosphere on Extrasolar Worlds. *Astrobiology* **2007**, *7*, 252–274, [CrossRef] [PubMed]
59. O'Malley-James, J.T.; Kaltenegger, L. Biofluorescent Worlds: Biological fluorescence as a temporal biosignature for flare star worlds. *arXiv* **2016**, arXiv:1608.06930.
60. Krishtalik, L.I. Energetics of multielectron reactions. Photosynthetic oxygen evolution. *Biochim. Biophys. Acta* **1986**, *849*, 162–171. [CrossRef]
61. Nürnberg, D.J.; Morton, J.; Santabarbara, S.; Telfer, A.; Joliot, P.; Antonaru, L.A.; Ruban, A.V.; Cardona, T.; Krausz, E.; Boussac, A.; et al. Photochemistry Beyond the Red Limit in Chlorophyll f–containing Photosystems. *Science* **2018**, *360*, 1210–1213, [CrossRef]
62. Raven, J.; Kübler, J.; Beardall, J. Put out the light, and then put out the light. *J. Mar. Biol. Ass. UK* **2000**, *80*, 1–25. [CrossRef]
63. Lingam, M.; Loeb, A. Photosynthesis on habitable planets around low-mass stars. *arXiv* **2019**, arXiv:1901.01270.
64. Gale, J.; Wandel, A. The potential of planets orbiting red dwarf stars to support oxygenic photosynthesis and complex life. *Int. J. Astrobiol.* **2017**, *16*, 1–9. [CrossRef]
65. Lenton, T.; Watson, A. *Revolutions that Made the Earth*; Oxford University Press: Oxford, UK, 2011.
66. Shields, A.L.; Ballard, S.; Johnson, J.A. The habitability of planets orbiting M-dwarf stars. *Phys. Rep.* **2016**, *663*, 1–38. [CrossRef]

67. Raven, J.A.; Cockell, C. Influence on Photosynthesis of Starlight, Moonlight, Planetlight, and Light Pollution (Reflections on Photosynthetically Active Radiation in the Universe). *Astrobiology* **2006**, *6*, 668–675. [CrossRef] [PubMed]
68. Russell, D. Biodiversity and Time Scales for the Evolution of Extraterrestrial Intelligence. *Astron. Soc. Pac. Conf. Ser.* **1995**, *74*, 143–151.
69. Bains, W.; Schulze-Makuch, D. The Cosmic Zoo: The (Near) Inevitability of the Evolution of Complex, Macroscopic Life. *Life* **2016**, *6*, 25. [CrossRef] [PubMed]
70. Danchi, W.C.; Lopez, B. Effect of Metallicity on the Evolution of the Habitable Zone from the Pre-main Sequence to the Asymptotic Giant Branch and the Search for Life. *Astrophys. J.* **2013**, *769*, 27. [CrossRef]
71. Barnes, R. Tidal locking of habitable exoplanets. *Celest. Mech. Dyn. Astron.* **2017**, *129*, 509–536. [CrossRef]
72. Joshi, M.M.; Haberle, R.M.; Reynolds, R.T. Simulations of the Atmospheres of Synchronously Rotating Terrestrial Planets Orbiting M Dwarfs: Conditions for Atmospheric Collapse and the Implications for Habitability. *Icarus* **1997**, *129*, 450–465. [CrossRef]
73. Carone, L.; Keppens, R.; Decin, L.; Henning, T. Stratosphere circulation on tidally locked ExoEarths. *MNRAS* **2018**, *473*, 4672–4685. [CrossRef]
74. Gladman, B.; Quinn, D.D.; Nicholson, P.; Rand, R. Synchronous Locking of Tidally Evolving Satellites. *Icarus* **1996**, *122*, 166–192. [CrossRef]
75. Waltham, D. Star Masses and Star-Planet Distances for Earth-like Habitability. *Astrobiology* **2017**, *17*, 61–77. [CrossRef] [PubMed]
76. Hansen, C.J.; Kawaler, S.D.; Trimble, V. *Stellar Interiors: Physical Principles, Structure, and Evolution*; Springer Science & Business Media: New York, NY, USA, 2012.
77. Luger, R.; Barnes, R. Extreme Water Loss and Abiotic O2 Buildup on Planets Throughout the Habitable Zones of M Dwarfs. *Astrobiology* **2015**, *15*, 119–143. [CrossRef] [PubMed]
78. Airapetian, V.S.; Glocer, A.; Khazanov, G.V.; Loyd, R.O.P.; France, K.; Sojka, J.; Danchi, W.C.; Liemohn, M.W. How Hospitable Are Space Weather Affected Habitable Zones? The Role of Ion Escape. *Astrophys. J. Lett.* **2017**, *836*, L3. [CrossRef]
79. Garraffo, C.; Drake, J.J.; Cohen, O. The Space Weather of Proxima Centauri b. *Astrophys. J. Lett.* **2016**, *833*, L4. [CrossRef]
80. Mullan, D.J.; Bais, H.P. Photosynthesis on a Planet Orbiting an M Dwarf: Enhanced Effectiveness during Flares. *ApJ* **2018**, *865*, 101. [CrossRef]
81. Dong, C.; Lingam, M.; Ma, Y.; Cohen, O. Is Proxima Centauri b Habitable? A Study of Atmospheric Loss. *Astrophys. J. Lett.* **2017**, *837*, L26. [CrossRef]
82. Lingam, M.; Loeb, A. Physical constraints on the likelihood of life on exoplanets. *Int. J. Astrobiol.* **2018**, *17*, 116–126. [CrossRef]
83. Landenmark, H.K.; Forgan, D.H.; Cockell, C.S. An estimate of the total DNA in the biosphere. *PLoS Biol.* **2015**, *13*, e1002168. [CrossRef]
84. Kempes, C.; Wolpert, D.; Cohen, Z.; Pérez-Mercader, J. The thermodynamic efficiency of computations made in cells across the range of life. *Philos. Trans. Ser. A Math. Phys. Eng. Sci.* **2017**, *375*, 20160343. [CrossRef]
85. Krumholz, M.R. On the origin of stellar masses. *Astrophys. J.* **2011**, *743*, 110. [CrossRef]
86. Maschberger, T. On the function describing the stellar initial mass function. *Mon. Not. R. Astron. Soc.* **2012**, *429*, 1725–1733. [CrossRef]
87. Johnstone, C.; Güdel, M.; Brott, I.; Lüftinger, T. Stellar winds on the main-sequence-II. The evolution of rotation and winds. *Astron. Astrophys.* **2015**, *577*, A28. [CrossRef]
88. Adams, F.C. Constraints on Alternate Universes: Stars and habitable planets with different fundamental constants. *J. Cosmol. Astropart. Phys.* **2016**, *2016*, 042. [CrossRef]

 © 2019 by the author. Licensee MDPI, Basel, Switzerland. This article is an open access article distributed under the terms and conditions of the Creative Commons Attribution (CC BY) license (http://creativecommons.org/licenses/by/4.0/).

Article

Multiverse Predictions for Habitability: Number of Potentially Habitable Planets

McCullen Sandora [1,2]

1. Institute of Cosmology, Department of Physics and Astronomy, Tufts University, Medford, MA 02155, USA; mccullen.sandora@gmail.com
2. Center for Particle Cosmology, Department of Physics and Astronomy, University of Pennsylvania, Philadelphia, PA 19104, USA

Received: 14 May 2019; Accepted: 21 June 2019; Published: 25 June 2019

Abstract: How good is our universe at making habitable planets? The answer to this depends on which factors are important for life: Does a planet need to be Earth mass? Does it need to be inside the temperate zone? are systems with hot Jupiters habitable? Here, we adopt different stances on the importance of each of these criteria to determine their effects on the probabilities of measuring the observed values of several physical constants. We find that the presence of planets is a generic feature throughout the multiverse, and for the most part conditioning on their particular properties does not alter our conclusions much. We find conflict with multiverse expectations if planetary size is important and it is found to be uncorrelated with stellar mass, or the mass distribution is too steep. The existence of a temperate circumstellar zone places tight lower bounds on the fine structure constant and electron to proton mass ratio.

Keywords: multiverse; habitability; planets

1. Introduction

This paper is a continuation of [1], which aims to use our current understanding from a variety of disciplines to estimate the number of observers in a universe N_{obs}, and track how this depends on the most important microphysical quantities such as the fine structure constant $\alpha = e^2/(4\pi)$, the ratio of the electron to proton mass $\beta = m_e/m_p$, and the ratio of the proton mass to the Planck mass $\gamma = m_p/M_{pl}$. Determining these dependences as accurately as possible allows us to compare the measured values of these constants with the multiverse expectation that we are typical observers within the ensemble of allowable universes [2]. In doing so, there remain key uncertainties that reflect our ignorance of what precise conditions must be met in order for intelligent life to arise. Rather than treating this obstacle as a reason to delay this endeavor until we have reached a more mature understanding of all the complex processes involved, we instead view this as a golden opportunity: since the assumptions we make alter how habitability depends on parameters, sometimes drastically, several of the leading schools of thought for what is required for life are incompatible with the multiverse hypothesis. Generically, if we find that our universe is no good at a particular thing, then it should not be necessary for life because, if it were, we would most likely have been born in a universe which is better at that thing. Conversely, if our universe is preternaturally good at something, we expect it to play a role in the development of complex intelligent life, otherwise there would be no reason we would be in this universe. The requirements for habitability are in the process of being determined with much greater rigor through advances in astronomy, exoplanet research, climate modeling, and solar system exploration, so we expect that in the not too distant future our understanding of the requirements for intelligent life will be much more complete. At this stage of affairs, then, we are able to use the multiverse hypothesis to generate predictions for which of these habitability criteria will end up being true. These will either be vindicated, lending credence to the multiverse hypothesis, or not, thereby falsifying it.

In this work, we define the habitability of a universe $\mathbb{H}$ as the total number of observers it produces. In estimating the number of observers the universe contains, a great many factors must be taken into consideration. Thankfully, there has long been a useful way of organizing these factors: the Drake equation, which in a slightly modified form along the lines of [3] reads

$$\mathbb{H} = \int_{\lambda_{\min}}^{\infty} d\lambda \; p_{\text{IMF}}(\lambda) \times N_\star \times f_{\text{p}}(\lambda) \times n_{\text{e}}(\lambda) \times f_{\text{bio}}(\lambda) \times f_{\text{int}}(\lambda) \times N_{\text{obs}}(\lambda). \tag{1}$$

Here, $N_\star$ is the number of stars in the universe, f_p is the fraction of stars containing planets, n_e is the average number of habitable planets around planet-bearing stars, f_bio is the fraction of planets that develop life, f_int is the fraction of life bearing worlds that develop intelligence, and N_obs is the number of intelligent observers per civilization. We have included dependence of the size of the host star $\lambda = M_\star/((8\pi)^{3/2} M_{pl}^3/m_p^2)$, in order to more accurately reflect the fact that these quantities may depend on this. We then integrate over the stellar initial mass function given in [4], which approximates a broken power law with turnover at $0.2 M_\odot$. This can be related to the probability of being in our universe by incorporating the relative occurrence rates of different universes as $P \propto p_\text{prior} \mathbb{H}$. It was argued in [1] that a reasonable choice of prior is given by $p_\text{prior} \propto 1/(\beta\gamma)$. However, this is ultimately set by physics at high energies, and so may in principle be something else.

Previously, the fact that the strength of gravity γ can be two orders of magnitude higher caused the biggest problems with finding a successful criterion, since most stars are in universes with stronger gravity. Though we had set out to focus solely on the properties of stars, it was only when we weighted the habitability of a system by the total entropy processed by its planets over its entire lifetime that we hit upon a fully satisfactory criterion. This was also reliant on the condition that starlight be in the photosynthetic range, colloquially referred to as 'yellow' light here: conservatively, this corresponds to the 600–750 nm range. Relaxing this range to be from 400–1100 nm will not qualitatively affect our results. The other factors we considered may be freely included at will without hindering this conclusion. For the majority of this paper, we take the entropy and yellow light conditions as our baseline minimal working model, and incorporate factors that influence the availability and properties of planets to determine how these alter our estimates for habitability. While our previous analysis was not heavily reliant on cutting edge results from the field of astronomy, our understanding of planets, from their population statistics to their formation pathways, has undergone rapid expansion in the past decade, and a state-of-the-art analysis needs to reflect that.

To this end, we begin by estimating the fraction of stars with planets in Section 2. Most notable is the recent determination of a threshold metallicity below which rocky planets are not found [5], as well as the understanding of the origin of this threshold. We also find the conditions for the lifetime of massive stars to be shorter than the star formation time and the fraction of galaxies able to retain metals to be sizable, but find that these only impose mild constraints. Using these allows us to determine the dependence of f_p on the underlying physical parameters. Additionally, we incorporate the fraction of systems that host hot Jupiters into our analysis, and find conditions for this process to not wreck all planetary systems.

In Section 3, we turn to the average number of habitable planets n_e. Two commonly used requirements for a planet to be habitable are that it needs to be both terrestrial and temperate, and so we optionally include both of these when estimating habitability. Recent results indicate that the distribution of rocky planets peaks at $1.3 R_\oplus$ [6–9], which is rather close to the terrestrial radius capable of supporting an Earthlike atmosphere. We track how this quantity changes in alternate universes, and what implications this effect has on the number of habitable planets in these universes. Additionally, we track the location and width of the temperate zone and compare this to the typical inter-planet spacing that results from the dynamical evolution of stellar systems, which provides a rough estimate for the probability that a planet will end up in the temperate zone. Finally, we discuss the importance of planet migration and how this changes in other universes.

An appendix is provided to collect the relevant formulas for the dependence on the physical constants on the variety of processes and quantities that are needed, in order to avoid distraction from the main text.

The overarching message we derive from this analysis is that the presence of planets is not that important in determining our location in this universe, a direct consequence of the fact that the presence of planets is a nearly universal phenomenon throughout the multiverse. Including these effects barely alters the probabilities we derived before. We find that most effects act as thresholds, serving to limit the allowed parameter range rather than alter the probability distribution of observing any particular value. We find several new anthropic bounds, including the most stringent lower bound on the electron to proton mass ratio in the literature. We find that these results are relatively insensitive to the exact models of planet formation and occurrence rate used, and so are robust to these current uncertainties.

This is not the first work to address the question of whether planets are still present for alternative values of the physical parameters. Limitations on the strength of gravity and electromagnetism imposed by the existence of habitable planets was investigated in [10], where it was found that long lived, temperate, terrestrial planets can exist over a wide range of parameter space. This was continued in [11] with the investigation of the influence of the density of galaxies on planetary stability, again finding a broad allowable parameter region. Our current work is novel in not just examining the possibility of the existence of planets with desirable properties, but also taking care to incorporate modern theories of planet formation into determining whether planets with these properties are indeed produced.

Taken together, we analyze 12 distinct possible criteria for habitability in this paper (not counting migration or the different views in the planetary size distribution we consider). Coupled to the 40 we considered in [1], this represents a total of 480 different hypotheses to compare. The quantities used to compute these are displayed in Table 1 for convenience.

Table 1. The quantities computed in this work. Here, Q is the amplitude of perturbations, κ parameterizes the density of galaxies, λ parameterizes stellar size, GI stands for giant impact formation mechanism, and iso stands for isolation production mechanism.

Quantity	Description	Expression
f_{gal}	fraction of stars in galaxies that retain supernova ejecta	$\text{erfc}(4.1 Q^{-1} \alpha^2 \beta^{5/3})$
$f_{2nd\ gen}$	fraction of stars born after supernova enrichment	$\exp(-0.24 \kappa^{3/2} \alpha^{-7/4} \beta^{-1/8} \gamma^{-1})$
f_Z	fraction of stars with high enough Z for planets	$\theta(1 - 0.038 \lambda^{3/4} \alpha^{-3} \beta^{-1/2} \gamma^{1/2})$
f_{hj}	fraction of stars without hot Jupiters	$1 - 2.3 \times 10^8 \kappa^2 \lambda^2 \alpha^{-13/8} \beta^{-3}$
n_p	average number of planets around a star	GI: Equation (25), iso: Equation (31)
f_{terr}	fraction of terrestrial planets	GI: Equation (27), iso: Equation (30)
f_{temp}	fraction of temperate planets	$0.0053 \kappa^{-1/2} \lambda^{-1.77} \alpha^{11/2} \beta^{7/4} \gamma^{-5/8}$

2. Fraction of Stars with Planets f_p

Two of the factors in the Drake equation regard the existence of planets, which will be considered in turn in this paper. The first quantity to determine is f_p, the fraction of stars that form planets. Here, we represent this as a product of factors:

$$f_p = f_{gal} \times f_{2nd\ gen} \times f_Z \times (f_{hj})^{p_{hj}}. \tag{2}$$

In succession, we have: f_{gal}, the fraction of stars in galaxies large enough to retain supernova ejecta, $f_{2nd\ gen}$, the fraction of stars born after supernova enrichment, f_Z, the fraction of stars born with high enough metallicity for planets to form, and an optional f_{hj}, the fraction of stars that do not produce hot Jupiters. Here, the exponent $p_{hj} \in \{0, 1\}$ is introduced as a choice of whether to include this last criterion or not. The other two are not treated as optional.

2.1. What Sets the Size of the Smallest Metal-Retaining Galaxy?

The requirement for a galaxy to be habitable is that it must retain its supernova ejecta in order to reprocess it into another round of metal-rich stars [12]. This sets a minimum galactic mass by the condition that the velocity of supernova ejecta is less than the escape velocity,

$$v_{SN}^2 \sim \frac{G\, M_{ret}}{R_{ret}}. \tag{3}$$

This is the asymptotic speed the supernova ejecta attains, and, to find this, a bit of the ejecta dynamics must be used. The initial speed can be found from energy balance [13]: this can be written in the form

$$v_{SN}^0 \sim \sqrt{\frac{T_{SN}}{A\, m_p}}, \tag{4}$$

where the temperature of the supernova is set by the Gamow energy, which is the amount required to overcome the repulsive nuclear barrier and force fusion, $T_{SN} \sim \alpha^2 m_p$. The mass of a typical particle in the ejecta is related to the atomic number $A \sim 50$, which cannot conceivably vary by much. We find that $v_{SN}^0 \sim 0.03$. However, as the ejecta moves through the intergalactic medium, it cools and slows until it merges completely. The asymptotic speed is that at which the temperature becomes equal to (about an order of magnitude less than) the hydrogen binding energy $T_{H_2} \sim \alpha^2 m_e/32 \sim 10^4$ K, below which Hydrogen becomes predominantly neutral and no more cooling takes place [14].

How far does the ejecta of a supernova spread before it completely merges with the interstellar medium, and, more importantly for our purposes, why? The observed value is around 100 pc [15], and this is after the blast has gone through several successive phases. The first is known as the blast wave phase, where the ejecta spread out at their initial velocity for roughly 100 yr, traveling a total of a few pc. After the amount of interstellar material encountered rivals the initial mass of the ejecta, which occurs at $d_{ST} \sim (M_{ej}/\rho_{gal})^{1/3}$, the blast enters the self-similar Sedov–Taylor phase, where the blast slows and expands considerably. According to standard theory [15], the self-similarity of the dynamics dictates that the temperature of the blast falls off as $T \propto d^{-3}$. During this phase, the velocity will decrease with distance as $v(d) = 2/5(d_{ST}/d)^{5/2}$ until the snowplow phase begins, at which point the speed essentially does not decrease any further. The snowplow phase occurs after the temperature reaches the molecular cooling threshold, where it expands by a factor of a few until the density of the material falls to that of the surrounding medium, at which point it is completely merged. Since the bulk of the expansion takes place in the Sedov–Taylor phase, the size of the blast will be dictated by the dynamics that take place there, and so the ultimate speed is given by

$$v_{SN} = v_{SN}^0 \left(\frac{T_{SN}}{T_{H_2}}\right)^{5/6}. \tag{5}$$

Using the above relations, the asymptotic speed of supernova ejecta is found to be

$$v_{SN} = 2.4\, \alpha\, \beta^{5/6}. \tag{6}$$

Now, we can find an expression for the minimum mass by using $M_{\min} \sim \rho_{\text{gal}} R^3$. Using the density of galaxies given in the appendix, this gives

$$M_{\text{ret}} \sim \frac{M_{pl}^3 \, v_{\text{SN}}^3}{\rho_{\text{gal}}^{1/2}} = 90.1 \, \frac{\alpha^3 \, m_e^{5/2} \, M_{pl}^3}{\kappa^{3/2} \, m_p^{9/2}}, \tag{7}$$

where the coefficient in the last expression has been chosen to reproduce the observed minimal mass of $M_{\text{ret}} = 10^{9.5} M_\odot$ [16]. As explained in the appendix, the quantity κ determines the density of galaxies in terms of cosmological parameters.

While this critical mass is important conceptually, it is not as relevant to the retention of ejecta as the gravitational potential itself, which directly sets the escape velocity of the overdensity. The fraction of initial overdensities that exceed a potential of a given strength Φ is given by the Press–Schechter formalism as $f = \text{erfc}(\Phi/(\sqrt{2}Q))$ [17], where Q is the primordial amplitude of perturbations. Usually, this expression is immediately expressed in terms of mass and density, but this more primitive form will suffice for our purposes. With this, we can derive the fraction of matter that resides in potential wells deep enough to produce a second generation of stars as

$$f_{\text{gal}} = \text{erfc}\left(4.1 \, \frac{\alpha^2 \, \beta^{5/3}}{Q}\right). \tag{8}$$

Perhaps somewhat interestingly, this does not depend on the strength of gravity γ at all. This effectively acts as a step function, severely diminishing the habitability of universes where $M_{\text{ret}} > M_{\text{typ}}$. Note that this is a pessimistic estimate, as it ignores the potential for subsequent evolution that causes potential wells to deepen with time. Nevertheless, it is a very mild bound, and so a more thorough treatment is not called for.

2.2. Is Massive Star Lifetime Always Shorter Than Star Formation Time?

Since the formation of metal-rich systems is reliant on the evolution of the first stars through to their completion, if the lifetime of massive stars exceeds the duration of star formation, then no systems will form with any substantial metallicity. The second generation stars are not necessarily enriched enough to produce planets, but this serves as a sufficient condition for planets to be formed at all. Then, the fraction of stars with planets can be estimated as those that are born after a few massive stellar lifetimes have elapsed. It is worth considering how these two timescales compare for general values of the physical constants.

The star formation rate, averaged throughout the universe, is found to decline exponentially, as gas is depleted from the initial reservoir [18]. The most naive treatment one can perform is to relate the timescale of this depletion to the free fall time of a galaxy, $t_{\text{dep}} = 1/(\epsilon_{\text{SFR}}\sqrt{G\rho})$, where for simplicity the efficiency coefficient is taken to not vary with parameters. Then, the fraction of stars born after a time t is $f_\star(t) = e^{-t/t_{\text{dep}}}$.

This needs to be compared to the lifetime of massive stars, where massive here is taken to mean large enough to become a type II supernova. This threshold is eight solar masses in our universe, and is set by the inner core having a high enough temperature to undergo carbon fusion [19]. As such, this threshold is parametrically similar to the minimum stellar mass, which was shown in [20] to scale as $\lambda \propto \alpha^{3/2} \beta^{-3/4}$, which equated the Gamow energy of fusion with the internal temperature of the star. The resultant mass is about two orders of magnitude larger than the minimal mass, stemming from the larger repulsive barrier for large nuclei. Then, using the stellar lifetime from [1], we find

$$\frac{t_{\text{SN}}}{t_{\text{dep}}} = 0.24 \, \frac{\kappa^{3/2}}{\alpha^{7/4} \beta^{1/8} \gamma}. \tag{9}$$

The normalization has been set to the value of 0.01. This has been a bit loose on several counts, namely the assumption that the second stars are always metallic enough to form planets, and the neglect of even higher mass stars, which have correspondingly shorter lifetimes. However, in practice, these worries are of no consequence, precisely because the scales are separated by such a large amount. We find that, for all practical values of the parameters, the star formation timescale exceeds the lifetime of massive stars, so that $f_{\text{2nd gen}} \sim f_\star(t_{\text{SN}})$ is always close to 1.

2.3. What Is the Metallicity Needed to Form Planets?

Terrestrial planets must be formed out of heavy elements, and though even a minuscule amount would suffice in terms of actual planetary mass, the planet formation processes within the protoplanetary accretion disk can only occur after a high enough metallicity is reached. A recent analysis [5] has done a nice job characterizing the metallicity needed (as a function of stellar mass, and distance to the host star) in order for planets to form out of the initial protoplanetary disk.

The underlying physical picture is that there are two timescales, the lifetime of the disk t_{disk} and the dust settling time t_{dust}. If the second exceeds the first, the disk will dissipate before sufficient dust may settle into the midplane, and will disperse without forming planets. Both of these depend on metallicity monotonically, and so only above some critical value will the conditions for planetary formation hold. We detail the scaling of each in turn.

The physics of accretion disks was first laid out in [21]. An initially uncollapsed cloud first condenses, and then through angular momentum transfer begins to form a disk. The disk remains around the star until either UV light from the star photoevaporates the gas or it escapes thermally, and the disk evaporates. This phase of evolution occurs on the viscous timescale of the disk, which is much smaller than the total disk lifetime [22].

Though the precise time at which this crossover occurs depends on the exact mechanism of photoevaporation (X-ray versus extreme ultraviolet, other stellar sources in clusters, presence of high activity tau phase) the drop-off of accretion is set by the disk's secular evolution time [23], with crossover occurring on the order of this time. Therefore, the only relevant physics that needs to be kept track of is the free fall time. Since this is given by a cloud that has approximately virialized after Jeans collapse, it is simply set by the condition that the gravitational energy is equal to the thermal energy. Then, the accretion rate is given by $\dot{M} = c_s^3/G \sim 10^{-6} M_\odot/\text{yr}$, and the timescale is given by $t_{\text{disk}} = M_\star/\dot{M} \sim$ Myr. The sound speed and the temperature are given by bremsstrahlung from molecular line cooling, as found in the appendix. Then,

$$t_{\text{disk}} = 5.5 \times 10^5 \frac{\lambda \, m_p^{1/4} M_{pl}}{\alpha^3 \, m_e^{9/4}} \left(\frac{Z}{Z_\odot}\right)^p. \tag{10}$$

In [24], the metallicity dependence was studied, where it was found that cooling quickens with metallicity Z. The exact process was open to interpretation, with a handful of viable candidate mechanisms, but the overarching explanation is that the less shielding there is, the faster the disk will dissipate. Observationally and theoretically, it was found that the scaling with metallicity is consistent with $p = 1/2$, and so we will adopt this for our analysis.

The dust settling timescale is dictated by the rate at which dust grains sink into the midplane from their initial positions in the protoplanetary disk. It is given in terms of the Keplerian timescale of the disk, $t_{\text{dust}} = 0.72 t_{\text{Kepler}}/Z$, as derived in [5]. There it was related to the growth rate of grains, and found to depend inversely on the metallicity, since only dust can participate in the accretion process. This is a function of orbital distance, but specifying to planets that form within what will become a temperate orbit for simplicity, we have

$$t_{\text{dust}} = 3.0 \times 10^5 \frac{\lambda^{17/8} \, m_p^{7/4} M_{pl}^{1/4}}{\alpha^{15/2} \, m_e^3} \frac{Z_\odot}{Z}. \tag{11}$$

Here, we are explicitly assuming that significant migration does not occur, which would alter the metallicity needed for planet formation, given its dependence on orbit.

Equating these two timescales yields the critical metallicity

$$Z_{\min} = 4.2 \times 10^{-4} \lambda^{3/4} \frac{\gamma^{1/2}}{\alpha^3 \beta^{1/2}}. \qquad (12)$$

The critical value is found to be $Z_{\min} = 6.3 \times 10^{-4}$. As compared to the solar value $Z_\odot = 0.02$, we find $Z_{\min} = 0.03 Z_\odot$ [5].

This may be used to determine the fraction of stars that host planets by considering the amount of stars that are formed above this metallicity. This requires a model for how metallicity builds up inside a galaxy (of a given mass), as well as the distribution of galaxy masses). A full calculation of this sort would take us too far afield here so we will return to it in a later publication. However, a very reasonable approximation is to compare this threshold metallicity to the asymptotic metallicity a galaxy attains—this usually affords percent-level accuracy. In our universe, this is found to be $Z_\infty = 0.011$ [16,25], which is actually a factor of two below solar due to the intrinsic scatter in metallicities. This asymptotic value is set by the fraction of stellar mass that is transformed into heavy elements during stellar fusion, and is not expected to depend sensitively on the fundamental constants over the ranges considered.

If attention is restricted to solar mass stars, the requirement that the threshold metallicity be lower than the asymptotic value equates to $\alpha^3 \beta^{1/2} \gamma^{-1/2} > 0.80$, which is most sensitive to the fine structure constant. Leaving the other parameters fixed, this gives $\alpha > 1/361$, which is stronger than the bound $\alpha > 1/685$ based on galactic cooling found in [26,27]. A more forgiving bound is found using the smallest stellar mass, which is $\alpha^{5/2} \beta^{17/12} \gamma^{-2/3} > 0.32$. This latter boundary is displayed in Figure 1, where the distribution of observers is plotted as a function of the three variables α, β and γ.

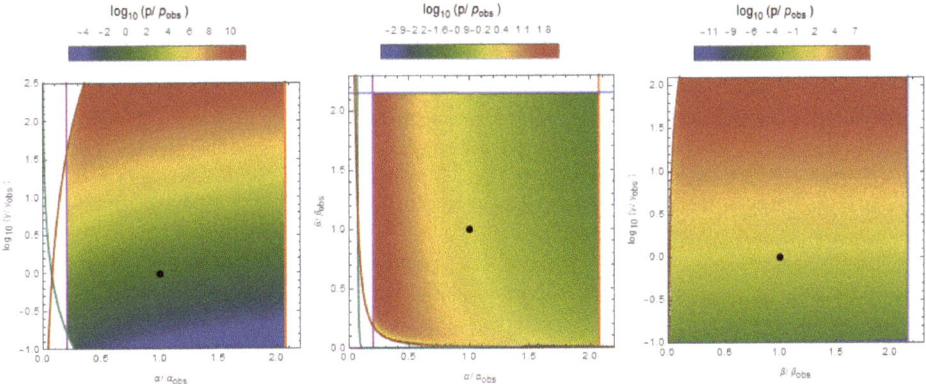

Figure 1. The distribution of observers when taking metallicity effects into account, in the α–γ, α–β, and β–γ subplanes. What is plotted is the logarithm of the probability of measuring any value of the coupling constants, with red being more probable than blue, and the black dot corresponding to our observed values. Included thresholds are an upper bound on α for hydrogen stability, as well as a lower bound for galactic cooling, and an upper bound on β for proton–proton fusion. The teal and brown curves correspond to the metallicity and supernova lifetime thresholds, respectively, and the supernova retention threshold is not relevant in this range.

Incorporating the above effects determining the fraction of stars that host planets, and again using the entropy and yellow light conditions, we can find that the probability of observing our particular values of each constant. These are defined as $\mathbb{P}(x_{\text{obs}}) = \min\{P(x > x_{\text{obs}}), P(x < x_{\text{obs}})\}$ for any observable quantity x, the others being integrated over, and the probability of measuring any value given by Equation (1). With this, we have

$$\mathbb{P}(\alpha_{\text{obs}}) = 0.19, \quad \mathbb{P}(\beta_{\text{obs}}) = 0.45, \quad \mathbb{P}(\gamma_{\text{obs}}) = 0.32. \tag{13}$$

These values are indistinguishable from those that were found in [1] without including these effects. We conclude that the presence of planets is a generic feature throughout most of the multiverse, and it does not alter where we expect to be situated in the slightest[1].

2.4. Are Hot Jupiter Systems Habitable?

The constraints above were all quite mild, indicating that the presence of planets is a fairly generic feature of the multiverse. However, we can make a further refinement by incorporating one of the most famous statistical correlations in the field of exoplanets, the hot Jupiter–metallicity correlation [28]. This finds that the fraction of stars that possess hot Jupiters, that is, Jupiter sized planets on orbits extremely close to the star increase with metallicity as Z^2. The general (though not universal— [29]) consensus is that these planets must have formed in the outer system before moving inward, to avoid the necessity of a disk that would be so massive as to be unstable [30]. In this scenario, the migration of the planet through the inner solar system would have certainly ejected any preexisting planets from their orbits (or worse), precluding them from sustaining life. However, the authors in [31] hypothesize that this entire process could happen early enough that the main stage of planet formation could occur after this migration had already taken place. If this turns out to be the case, then there may in fact be no correlation between hot Jupiters and habitability (barring other factors that may impact habitability [32]). If one wishes to include this effect, however, then the fraction of stars that host Earths and not Hot Jupiters is given by

$$f_{\text{hj}} = 1 - \frac{\bar{Z}^2}{Z_{\text{max}}^2}. \tag{14}$$

Here, we have used the mean metallicity rather than averaging this fraction over the metallicity distribution, but this approximation is sufficient for a first analysis. The normalization Z_{max} is the threshold above which the stellar system is almost assured to possess a hot Jupiter. It is set to reproduce the observed abundance of hot Jupiter systems of 3% as $Z_{\text{max}} = 5.77 Z_\odot$. In a multiverse setting, we would expect to inhabit a universe where Z_{max} is safely above the average metallicity, beyond which any further increase would result in little increase in habitability. Somewhat in line with this expectation, then, is the fact that in our universe hot Jupiters exist in only a few percent of systems, and mainly in those that are highly metal-rich.

To investigate whether this is the result of some selection effect, we must know what determines this metallicity. The functional dependence is a clue: as the effect becomes more pronounced with the square of the nongaseous material present, this is indicative of an interaction process. What remains, however, is the question of whether this migration is a result of planet–planet or planet–disk interactions. In fact, both explanations have been considered in the literature: references may be found in [33] [2]. The planet–planet hypothesis is supported by the fact that the eccentricities of observed hot Jupiters are correlated with metallicity as well, indicating a more chaotic, violent origin, rather than the steady, deterministic process indicative of planet–disk interaction. Additionally, [35] note a substantial misalignment between the orbits of known hot Jupiters and the spins of their host stars, which is most easily explained through a violent migration scenario. However, systems like WASP-47 [36], which possess both a hot Jupiter and smaller companions, demonstrate that more

---

[1] The code to compute all probabilities discussed in the text is made available at https://github.com/mccsandora/Multiverse-Habitability-Handler.
[2] A third explanation was additionally given in [34] that the disk dispersal timescale increases with metallicity, allowing a longer period of accretion onto seed cores.

dynamically quiet migration pathways are possible, if not necessarily the norm. Here, we only expound upon the planet–planet scenario, though the others could just as readily be incorporated into our analysis.

We start by determining the value of Z_{max} due to planet–planet interactions, as proposed in [37]. As noted in [38], the coexistence of hot Jupiters and low mass planets is impossible in this paradigm, as migration occurs after the disk has dissipated. From [33], this is set by the expected number of Jupiter mass planets initially formed in the outer system. They provide a framework for estimating this by determining the probability that a core will attain Jupiter mass as a result of planetesimal accretion as $p \propto \Delta M/M_{crit}$, where $M_{crit} \sim 10 M_\oplus$ is the mass above which runaway gas accretion is possible and ΔM is the typical total accreted mass. For this, we use the analytic expression for the accretion rate from [39],

$$\dot{M} \approx (18\pi)^{1/3} \frac{G M_{crit} M_\star^{1/3} \Sigma_{planetesimals}}{\rho^{1/3} a}. \tag{15}$$

This employs the strong gravitational focusing limit, and treats the typical relative velocities of planetesimals as roughly given by the Hill velocity. This can be used to determine the total mass accreted by simply multiplying by the disk lifetime given in Equation (10) (making use of the simplifications that the nonlinear oligarchic regime is not quite reached, and the initial isolation timescale is small compared to the disk lifetime). The probability that there will be at least two gas giants to trigger the instability will scale as $p(\geq 2) \sim N_{jup}^2 p^2$, where N_{jup} is the typical number of planets in the outer system, and we have assumed that p is small. The quantity N can be found by dividing the total mass of the planetary disk by the typical mass of a planet at the typical location of formation. Here, we use the initial seed being set by the isolation mass, have fixed the orbital radius to be given by the snow line, and have taken the disk temperature to be given by viscous accretion, all of which are discussed in the appendix. This gives

$$N_{jup} \sim \frac{M_{disk}}{M_{crit}} \sim 2.2 \times 10^{-5} \frac{\lambda}{\alpha^{3/2} \beta^{3/4}}. \tag{16}$$

This scales linearly with stellar mass, in agreement with the observations in [40]. The maximal value of metallicity is found to be

$$Z_{max} \sim \frac{1}{N_{jup}} \frac{M_{crit}}{\dot{M} t_{disk}} = 2.3 \times 10^{-8} \frac{\alpha^{13/6} \beta^{3/2}}{\kappa \lambda}. \tag{17}$$

This quantity is somewhat sensitive to both α and β, but not at all to γ. This also defines a stellar mass $\lambda_{hj} = 2.3 \times 10^{-6} \alpha^{13/6} \beta^{3/2}/\kappa$: stars above this mass, equal to 11 $M_\odot$ for our values, will always host hot Jupiters. This will be below the smallest stellar mass if $789 \alpha^{8/27} \beta < 1$, which will occur when the electron to proton mass ratio is about 10 times smaller. However, this criteria does not alter the probabilities much:

$$\mathbb{P}(\alpha_{obs}) = 0.18, \quad \mathbb{P}(\beta_{obs}) = 0.44, \quad \mathbb{P}(\gamma_{obs}) = 0.31. \tag{18}$$

With this, notice that the probabilities are only changed from Equation (13) by a few percent. Thus, even when including the demand that hot Jupiter systems are uninhabitable, the fraction of systems with planets seems to be relatively insensitive to the physical constants.

We now turn to the next factor in the Drake equation, which deals with the characteristics of planets, rather than just their presence.

3. Number of Habitable Planets per Star n_e

Next, we focus on the number of habitable planets per star, n_e. The determination of habitability may depend on many factors, such as amount of water, eccentricity, presence of any moons, magnetic field, distance from its star, atmosphere, composition, etc. Here, we focus on two: temperature and size, and determine the fraction of stars that have planets with each of these characteristics.

As usual, it is possible that habitability is completely independent from these properties; which viewpoint one adopts depends on how habitable one expects environments without liquid surface water and thin atmospheres can be. In this work, we remain agnostic to either expectation and report the number of observers for all combinations of choices, where a planet will only be habitable if it is approximately Earthlike, and where the size and/or temperature of the planet have no effect on its habitability.

It should be noted that we are restricting our attention here to surface dwelling life on planets orbiting their star. Thus, life in subsurface oceans and/or on icy moons like Enceladus [41], or even more exotic types of life (e.g., [42,43]), are not considered. It is our plan to consider these alternative environments in future work.

The number of habitable planets can be broken down as

$$n_e = n_p \times (f_{terr})^{p_{terr}} \times (f_{temp})^{p_{temp}}. \tag{19}$$

Here, the average total number of planets around a star is n_p. The fraction of terrestrial mass planets is denoted f_{terr}, and f_{temp} is the fraction of planets that reside within the temperate zone. The exponents $p_i \in \{0,1\}$ parameterize the choice of whether to include these conditions in the definition of habitability or not. These quantities all depend on stellar mass, giving preference to large stars because they make larger planets, and small stars in that their temperate zone is wider compared to the interplanetary spacing.

Estimating these quantities is somewhat muddled by the current uncertainties in planet formation theory. Not only is the distribution of planet masses contested in the literature, but the exact formation pathways, as well as the physics that dictates the results, is not completely settled. Where we come across disagreement, we separately try each proposal, in order to understand the sensitivity of our analysis to present uncertainties. While the results for the overall probabilities can vary by a factor of 2, the upshot is that our estimates are relatively robustx.

3.1. Why Does Our Universe Naturally Make Terrestrial Planets?

The size of a planet is of crucial importance because it dictates what kind of atmosphere it can retain. If it is too small, all atmospheric gases will eventually escape, whereas if it is too large, it will retain a thick hydrogen and helium envelope, leading to a runaway growth process. Terrestrial planets must have a very specific size in order that the escape velocity exceeds the thermal velocity for heavy gases such as H_2O, CO_2 and N_2, but not that of the lightest gas H and He. In our universe, and for temperatures within the range where liquid surface water is possible, this restricts the range of planetary radii to be within 0.7 and 1.6 that of Earth's [44]. This is a narrow sliver compared to the eight orders of magnitude mass range of spherical, non-fusing bodies, ranging from the potato radius of 200 km to 10 Jupiter masses. In terms of fundamental parameters, this requirement gives the mass to be

$$M_{terr} = 92 \frac{\alpha^{3/2} M_{pl}^3 m_e^{3/4}}{m_p^{11/4}}. \tag{20}$$

The coefficient has been set to reproduce Earth's mass, but the allowed spread in masses is taken to be between 0.3–4 $M_\oplus$.

There are compelling arguments that complex may only be possible on terrestrial planets, with atmospheres composed of only heavy gases [45]. Any smaller, and the planet would be Marslike, an apparently barren wasteland incapable of sustaining any appreciable liquid ar atmosphere. The other extreme would be Neptunelike, with its hellish surface temperatures, pressures and wind speeds. Of course, these arguments may be misguided, but here we explore the consequences of adopting them for the multiverse computations we perform.

It is important to note that the conditions that set the presence of atmospheres are completely separate from the physics that dictates the size of planets, which is set by the clumping of the initial circumstellar disk[3]. Nonetheless, the observed population of rocky planets is thought to peak at only slightly super-Earth mass, making the production of terrestrial planets the norm for stars throughout the universe. To be fair, the current exoplanet samples are biased towards large mass planets and become very incomplete below Earth mass [48], but a number of different groups have concluded that a detectable turnover is present near Earth masses: the authors of Ref. [9] find a good fit to a log-normal distribution that peaks at $1.3R_\oplus$. In Ref. [7] a Rayleigh distribution with width $3M_\oplus$ is used. The authors of Ref. [8] advocate for a broken power law with turnover at $5M_\oplus$. It was noted in Ref. [6] that the distribution appears to be flat below $2.8R_\oplus$. Use of these differing proposed distributions make very little difference to our final outcome.

However, not everyone is convinced that the mass distribution exhibits a peak, and even if there is one, it is just as reasonable to assume that there are many more smaller mass planets for every planet of Earth size, as the plethora of small asteroids and comets in our system indicates. Because of the incompleteness of current exoplanet surveys for small mass planets, there is room for such disagreement at the current moment. Additionally, even if the mass peak is real, it is only observed for close in exoplanets, and so requires an extrapolation to Earthlike orbits, where different dynamics may be at play. One possibility is that the peak at super-Earth mass may be due to their enhanced migration capability [49]. We will consider each scenario in turn.

3.1.1. What Sets the Size of Planets?

Why is the turnover so nearly equal to Earth mass planets, out of the potentially eight orders of magnitude that could have been selected instead? Simulations provide a means to address this question: it was found in Ref. [50] that the mass of planets is directly proportional to the amount of initial material present in the disk, so that increasing disk mass makes larger, rather than more, planets. In this scenario, nearly all the material initially present in the disk eventually gets constituted into planets, with negligible (perhaps a factor of two, but not an order of magnitude) losses throughout the evolution of the system. Determining the final planet mass in this setup requires knowledge of the initial disk mass, as well as the fraction of material within the inner solar system. This boundary is set by the snow line, the difference in composition interior and exterior to which dictates the formation of rocky versus icy planets. In the following, we adopt the conventional view that little migration takes place during planet formation, and comment further on alternative pathways when migration is discussed.

Current observations indicate that the initial mass of heavy elements in protoplanetary disks is roughly proportional to disk mass, $M_{\rm disk} = 0.01 M_\star$ [51]. In Ref. [52], a stronger dependence was found, but the scatter of 0.5 dex is larger than the trend, and the observed trend was suggested to possibly be due to a selection effect arising from processing into undetectably large grains, so we omit this stronger scaling for now. For solar mass stars, this works out to be roughly $10 M_{\rm Jupiter}$, or $3300 M_\oplus$. This mass is distributed out to a radius of ~100 AU, which is set by the conservation of angular momentum, and the initial size of the collapsing cloud. From the appendix, we arrive at the following expression for disk size:

$$r_{\rm disk} = 3.6 \times 10^{-6} \frac{\lambda^{1/3}}{\kappa} \frac{M_{pl}}{m_p^2}. \qquad (21)$$

---

[3] This is not entirely true: there is known to be some feedback between the irradiation of initial atmospheres from the host star that favors both small and large atmospheres [46]. While this leads to the interesting bimodal distribution of observed radii [47], this is driven by atmospheric size, and is certainly not enough to affect the terrestrial core.

This may be significantly altered if the young star is in a dense environment [53], but this scaling will suffice for our purposes.

To determine the amount of material present within the snow line, one must take the surface density profile into account. For this, we use the expression for $\Sigma(a)$ found in the appendix, which is inversely proportional to a. Using these, the typical mass of an interior planet becomes

$$M_{\text{inner}} \sim \eta_< \frac{a_{\text{snow}}}{R_{\text{disk}}} M_{\text{disk}}, \qquad (22)$$

where we have included the quantity $\eta_< \sim 0.25$ to account for the difference in composition interior to the snow line [54].

The location of the snow line is the point in the disk beyond which water condenses, equal to 2.7 AU in our solar system [55]. In order to determine its location, the temperature as a function of radius must be used, which will depend on the dominant source of heating. For most of the present paper, we use the temperature given by viscous accretion, and find

$$a_{\text{snow}} = 30.4 \frac{\lambda M_{pl}}{\alpha^{5/3} m_e^{5/4} m_p^{3/4}}. \qquad (23)$$

Note that, perhaps unsurprisingly, the snow line is situated outside the temperate zone for all relevant parameter values. The dependence on stellar mass was found taking $\dot{M} \propto \lambda^2$, but is generically found to scale as $a_{\text{snow}} \propto \lambda^{2/3-2}$ [56]. This is now enough to determine the parameter dependence of M_{inner}, which will ultimately be used to derive the expected number of planets per star as a function of these parameters.

Though we tend to favor accretion dominated disks throughout this work, irradiation from the central star can actually play a significant role as well [22]. If this is the dominant mode of heating, then the snow line will instead be given by the expression $a_{\text{snow}} = 297 \lambda^{43/30} \alpha^{-4} m_e^{-4/3} m_p^{-1/3} M_{pl}^{2/3}$. Though this is functionally quite similar to the accretion dominated case from above, in Table 2, we investigate the effect of assuming this form instead. Actually, both are almost equally relevant in determining the position of the snow line, which helps to greatly complicate the disk structure (as well as enhance the variability between different star systems [57]). Additionally, irradiation from other neighboring stars may be important as well, especially in clusters [58], but we do not consider this contribution in this work.

Table 2. Display of the insensitivity of the probabilities of our observables to the choices made regarding planet formation. In addition to the options displayed, they may have been combined, but this would belabor the point.

Choices	$\mathbb{P}(\alpha_{\text{obs}})$	$\mathbb{P}(\beta_{\text{obs}})$	$\mathbb{P}(\gamma_{\text{obs}})$
standard	0.229	0.260	0.409
Rayleigh distribution	0.229	0.251	0.411
irradiation	0.255	0.320	0.426
with no λ dependence	0.380	0.254	0.007
shot noise	0.232	0.260	0.306

Determining the average planet mass is somewhat involved, given the many distinct stages of growth that occur as microscopic dust grains agglomerate to the size of planets. For reviews of this multi-stage process, see [59–61]. In brief, planetesimals form characteristic masses as the final outcome of pebble (or dust) accretion. This arrangement is unstable, and ultimately leads to a phase of giant impacts, wherein planetesimals collide together to form full planets.

An estimate for the maximum mass a planet can attain after a phase of growth through chaotic giant impacts was found in [62]. There, they assumed no migration, small eccentricity, and determined the width of the 'feeding zone' to be $\Delta a = 2v_{esc}/\Omega$ by noting that within this region planet–planet interactions result in collisions rather than velocity exchange. This yields

$$M_{planet} \sim \left(\frac{4\pi \Sigma a^{5/2} \rho^{1/6}}{M_\star^{1/2}} \right)^{3/2}, \qquad (24)$$

where ρ is the average density of the planet and Σ is the disk density. With this, we can use the appendix to reinstate parameter dependence into the expressions for the average number of planets (normalized to 3 for the solar system) as well as the typical planet mass:

$$n_p = 0.0061 \frac{\alpha^{4/3} \beta^{13/16}}{\kappa^{1/2} \lambda^{5/6}}, \qquad (25)$$

$$\frac{M_{planet}}{M_{terr}} = 1.6 \times 10^6 \frac{\kappa^{3/2} \lambda^{5/2}}{\alpha^{9/2} \beta^{45/16}}. \qquad (26)$$

The latter has quite a steep dependence on stellar size. This is expected from the simulations [50], with a dependence closer to linear, but is not particularly observed in exoplanet catalogs due to large scatter [63]. In our calculations, we explore the effect of ignoring this dependence altogether, displayed in Table 2. A full treatment would take the dependence on semimajor axis into account, rather than simply evaluating at the snow line: in fact, this would be somewhat unnecessary, as the scatter observed in exoplanet surveys, simulations, and indeed within the solar system masks any dependence that may exist.

For the fraction of planets that are terrestrial, we use the log-normal distribution of [9] since this is what is expected of the core accretion process, though we check that the actual distribution used does not alter the outcome much. Under these assumptions, we can find the probability that a planet will be terrestrial as

$$f_{terr} = \frac{1}{2} \mathrm{erf} \left(\frac{\log\left(\frac{4 M_{terr}}{M_{planet}} \right) + \frac{1}{2}\sigma_M^2}{\sqrt{2}\sigma_M} \right) - \frac{1}{2} \mathrm{erf} \left(\frac{\log\left(\frac{.3 M_{terr}}{M_{planet}} \right) + \frac{1}{2}\sigma_M^2}{\sqrt{2}\sigma_M} \right). \qquad (27)$$

This function peaks at $M_{planet} \sim M_{terr}$, and approaches are when M_{planet} is very different from M_{terr}. Being a two-parameter distribution, this requires not just the mean, but also the variance. As it is not currently known what sets this quantity, here we explore two options: the first uses maximization of entropy production to set $\sigma_M = 1/\sqrt{6}$, which is fairly widely observed in natural processes [64]. This estimate should occur for large systems, but for small systems one would expect the variance to be set by shot noise instead, $\sigma \sim \sqrt{M_{iso}/M_{inner}}$, making use of the isolation mass defined in the next section. The dependence on the various parameters is displayed in Figure 2.

The probabilities of observing the observed values of our constants are computed for the various choices we made in Table 2. These can be compared with Equation (13) that only considered the fraction of stars with planets. The most significant change for our most favored prescription is the probability of observing our electron to proton mass ratio, which is decreased by less than a factor of 2. Such insensitivity hardly constitutes any evidence for whether life should only appear on terrestrial planets. More interestingly, if the dependence on stellar mass is neglected, our strength of gravity becomes quite uncommon to observe. This gives us strong reason to suppose that, if life requires terrestrial planets, we will begin to see a correlation between the two soon, and, if we don't, then life requiring Earth mass planets is incompatible with the multiverse at the 2.7σ level.

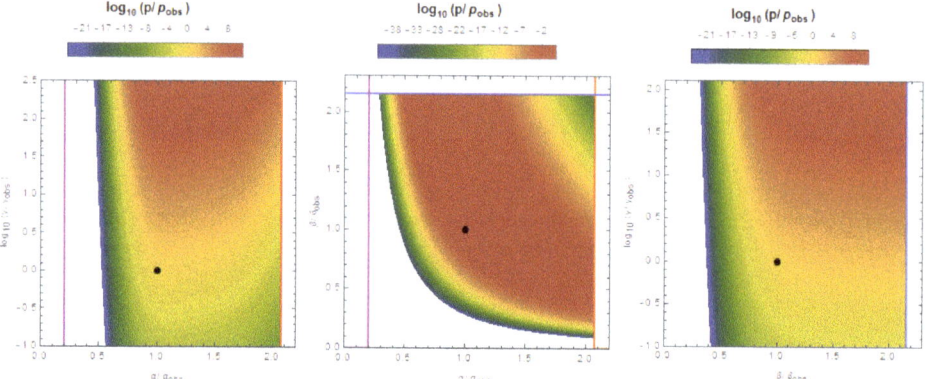

Figure 2. The distribution of observers if life can only arise on a terrestrial planet in the α–γ, α–β, and β–γ subplanes, size being dictated by a giant impact phase. Here, we have used a log-normal distribution with $\sigma = 1/\sqrt{6}$, excluded λ dependence on the average planet mass, and assumed accretion dominated disks.

Aside from this, though, the choices we made to come up with these estimates do not affect the outcome very much at all. On the one hand, this is disappointing, as the stronger the dependence these probabilities have on the assumptions of planet formation, the stronger our predictions can be about which to expect to be dominant. However, this is also heartening: because the current uncertainties about planet formation do not affect the outcome all that much, we are able to trust the broad conclusions we have reached a bit better.

3.1.2. Is Life Possible on Planetesimals?

The mass discussed above really refers to the maximal planet mass of a system, which form as a result of the secondary stage of collisions after the isolated planetesimals form. However, this agglomeration will likely not completely deplete the system of its primordial planetesimals, and so there are also expected to be numerous smaller planets accompanying each large one, as is the case in and around the solar system's asteroid belt. If these smaller bodies are considered as potential abodes for life as well, the distribution continues past Earth masses, rather than having a peak there.

As a planet is condensing out of the protoplanetary disk, it eventually reaches what is termed as the pebble isolation mass, which from the appendix is given by

$$M_{\text{iso}} = 2.0 \times 10^8 \, \frac{\kappa^{3/2} \lambda^2 M_{pl}^3}{\alpha^{5/2} m_e^{15/8} m_p^{1/8}}. \tag{28}$$

The isolation mass is a function of the semi-major axis, but if we evaluate it at the edge of the inner system given by Equation (23), we find, using the value for the disk surface density from the Appendix A,

$$\frac{M_{\text{iso}}}{M_{\text{terr}}} = 2.2 \times 10^6 \, \frac{\kappa^{3/2} \lambda^2}{\alpha^4 \beta^{21/8}}. \tag{29}$$

Here, the density of the galaxy comes into play in setting the outer edge of the disk. The dependence on stellar mass in this expression is quite close to that found in [55], $M_{\text{iso}} \propto \lambda^{7/4}$.

It remains to specify the distribution of planetary masses in order to find the fraction that are terrestrial in this picture. It is generically expected to take a power law form that continues to the small mass cutoff, so that $N(M) = (M_{\text{iso}}/M)^q$. However, different authors prefer different values for the slope: Ref. [65] find $q = 0.31 \pm 0.2$. Ref. [66] find a nearly scale invariant distribution for the radius, which translates into $q = 0.30 \pm 0.03$ if we use $M \propto r^3$. Refs. [39,67] find $q = 0.6$ for simulations of planetesimals, and Ref. [54] extrapolate from the known asteroid population to find $q = 1$. Ref. [68] favors a value of $q = 2$ from population synthesis methods. Here, we report with various values of q to investigate its influence on the probabilities. The fraction of terrestrial planets is then

$$f_{\text{terr}} = \min\left\{1, \left(\frac{M_{\text{iso}}}{0.3\, M_{\text{terr}}}\right)^q\right\} - \min\left\{1, \left(\frac{M_{\text{iso}}}{4\, M_{\text{terr}}}\right)^q\right\}. \tag{30}$$

This is an increasing function of stellar mass until $M_{\text{iso}} > 0.3 M_{\text{terr}}$, reflecting the expectation [69] that earthlike planets should be rare among low mass stars. With this prescription, stars above a certain mass will not produce any earthlike planets because the isolation mass will exceed the largest terrestrial planet size. This defines the largest viable stellar mass $\lambda_{\text{iso}} = 0.0013 \kappa^{-3/4} \alpha^2 \beta^{21/16}$, which corresponds to 6.3 $M_\odot$ in our universe. This will only exceed the minimal stellar mass if $\alpha^{1/2} \beta^{33/16} \kappa^{-3/4} > 169$. This condition is most sensitive to the electron to proton mass ratio and, holding the other constants fixed, will be violated if it drops to about 11% of its observed value.

For $q < 1$, the average mass for this distribution is formally infinite, which presents a problem for using our expression for the expected number of planets in a system as given by $n_p = M_{\text{inner}}/\langle M_p \rangle$. However, the total mass in the inner disk introduces a large mass cutoff, for which we have

$$n_p = \frac{q-1}{q} \frac{\left(\frac{M_{\text{inner}}}{M_{\text{iso}}}\right) - \left(\frac{M_{\text{inner}}}{M_{\text{iso}}}\right)^{1-q}}{1 - \left(\frac{M_{\text{inner}}}{M_{\text{iso}}}\right)^{1-q}}. \tag{31}$$

This interpolates between $(q-1)/q M_{\text{inner}}/M_{\text{iso}}$ for $q > 1$ and $(q-1)/q$ for $q < 0$. The ratio of masses is

$$\frac{M_{\text{inner}}}{M_{\text{iso}}} = 0.0017 \frac{\alpha^{5/6} \beta^{5/8}}{\kappa^{1/2} \lambda^{1/3}}, \tag{32}$$

thus that this equates to 30 with our constants. With this viewpoint, the distribution of observers is plotted in Figure 3.

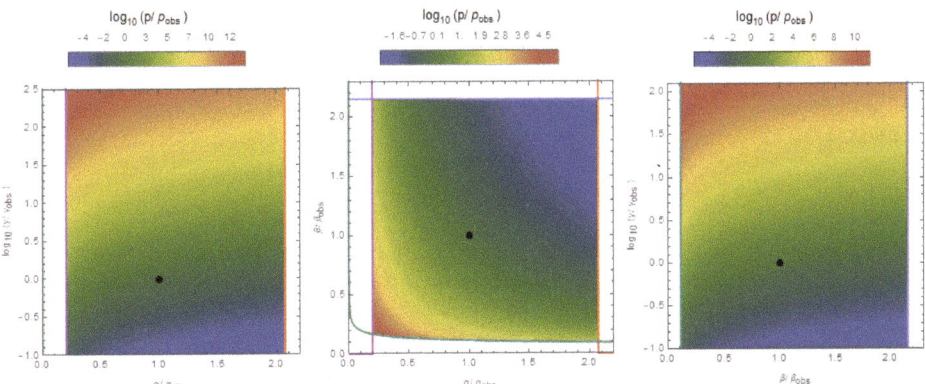

Figure 3. The distribution of terrestrial planetesimals that result from isolated accretion in the α–γ, α–β, and β–γ subplanes, without the subsequent phase of giant impacts. The slope here is $q = 1/3$. The teal line represents the region where the largest star capable of hosting terrestrial planets is smaller than the smallest possible star.

The overall probabilities are calculated for different representative values of the power law in Table 3. It can be seen that, for increasing q, the probability for α increases, while the other two decrease.

In particular, for $q = 2$, the probability of observing our value of the electron to proton mass ratio is disfavored by 2σ.

Table 3. Display of the probabilities of our observables for different values of the power law slope.

Exponent	$\mathbb{P}(\alpha_{obs})$	$\mathbb{P}(\beta_{obs})$	$\mathbb{P}(\gamma_{obs})$
$q = 1/3$	0.234	0.419	0.436
$q = 2/3$	0.302	0.260	0.424
$q = 1$	0.362	0.157	0.333
$q = 2$	0.469	0.044	0.228

3.2. Why Is the Interplanet Spacing Equal to the Width of the Temperate Zone?

Perhaps the most commonly employed habitability criteria is that a planet must be positioned a suitable distance away from its host star to maintain liquid water on its surface. This assumption is so pervasive that this region is usually referred to as the circumstellar habitable zone. In the spirit of remaining agnostic toward the conditions required for life, we will adhere to the recently proposed renaming as the 'temperate zone' [70]. If this is indeed essential for life, then the expected number of habitable planets orbiting a star will depend both on the interplanetary spacing, as well as the width and location of the temperate zone. A rather clement feature of our universe is that the width of the temperate zone is comparable to the interplanetary spacing (for sunlike stars). Because of this, it is relatively common that one of the planets in any stellar system is situated inside the temperate zone, no matter its particular arrangement. This could be contrasted to the hypothetical case where the temperate zone were much narrower than the interplanetary spacing, in which case the odds of a planet being situated inside it would be quite low. However, this coincidence of distance scales is not automatic: both these quantities are dependent on the underlying physical parameters, and so in universes with different parameter values the expected number of potentially habitable planets per star will be altered. We go through these length scales in turn, and then fold them into our estimate for the overall habitability of the universe.

The boundaries of the temperate zone depend on the characteristics of the planet in question, such as the atmospheric mass and composition [71], its orbital period [72], etc. However, these details only alter the location of the temperate zone to subleading order, and do not affect the scaling with fundamental parameters we are interested in. If the planet is assumed to be a simple blackbody, then the temperature is set solely by the amount of incident flux. In this case, the location of the temperate zone will be

$$a_{\text{temp}} = \frac{1}{2} \frac{T_\star^2}{E_{H_2O}^2} R_\star = 7.6 \frac{\lambda^{7/4} m_p^{1/2} M_{pl}^{1/2}}{\alpha^5 m_e^2}. \tag{33}$$

Albedo and greenhouse effects will change the coefficient, but not the overall scaling. Determining the width of the temperate zone entails finding the temperatures at which runaway climate processes occur, but we will now argue that both the inner and outer edge are dictated by (broadly speaking) the same underlying physical process of phase change, and so the width of the temperate zone will scale in the same way as its location.

The inner edge of the temperate zone is set by the runaway greenhouse effect: this occurs when a temperature threshold is crossed that allows an appreciable amount of water vapor to be sustained in the atmosphere. Since this serves to trap infrared light from escaping to space, this will increase the temperature further, in turn driving a further increase in atmospheric water vapor. Once the atmosphere is comprised primarily of water, it will be photodissociated and/or escape to space, leaving the Earth in a dry and Venus-like state [73]. Since there is always some level of outward flux, the ocean will escape into space eventually if a long enough time has elapsed. Therefore, the exact threshold for this process is defined as when the timescale for this process is of the order of one billion years, which in turn will depend on the mass of the planet's ocean. However, key to our discussion is that the change in the atmospheric water fraction occurs very abruptly when the temperature crosses the latent heat of vaporization, going from equilibrium values of $10^{-5} - 1$ in the span of 250–420 K. Because of this, the exact mass of the planet, ocean or atmosphere will only play a subleading role in determining this threshold, which instead is dictated solely by molecular processes. Since this transition is set by what is essentially the intermolecular binding energy, this scales identically with the condition for liquid water. For Earth, this occurs at a temperature of 330 K, which corresponds to 0.95 AU [74].

Similarly, the outer edge of the temperate zone is set by the runaway icehouse process. The temperature of a planet is set not only by the incident flux, but also by the carbon dioxide content content of the atmosphere and albedo. Carbonate weathering is an important regulatory feedback mechanism that serves to stabilize the temperature of a planet to remain within the temperate range by adjusting atmospheric carbon dioxide content to compensate for a change in stellar flux [75] (it does this because liquid water is necessary for efficient weathering to occur). However, this only works to an extent, since beyond a critical concentration atmospheric CO_2 increases the albedo of a planet, leading to a runaway icehouse effect [73]. Because the amount of atmospheric carbon dioxide is also set by reactions due to intermolecular forces, it scales the same as above. Therefore, both the inner and outer edges of the temperate zone are set by molecular binding energies, and so the width will always be of the same order as the mean[4].

The exact delineations are subject to the uncertainties of the atmospheric model used, but, for definiteness, we take $\Delta R_{HZ} = 0.73$ AU, from [71].

The next task is to determine how far apart planets typically reside from each other. It was hypothesized in [77] that stellar systems are dynamically packed, in that they are filled to capacity, and the insertion of any additional planet would render the system unstable. Though this may not strictly hold all the time [78], this spacing is roughly observed in our solar system [79], in Kepler data on multiplanetary systems [80], and found to occur naturally in simulations in [81]. The essential idea is that planetary scattering either ejects or collides excess planets from an initially overpacked system until the remainder are far enough apart to be stable on the timescale of the system. The stability condition is that planets are further than a certain multiple of Hill radii away, nominally around 10, and it was found in [80] that the distribution of separations was a shifted Rayleigh distribution with usual separation $21.7 \pm 9.5 R_{Hill}$.

As mentioned in [62], the typical multiple of mutual Hill radii is not universal, but can be shown to depend on the width of a planet's 'feeding zone' to be given by $(\rho a^3 / M_\star)^{1/4}$, where ρ is the density of matter and a is the semimajor axis. Though above we were interested in the typical size of a planet during this process, here we may use this condition to find the typical spacing as

$$a_{\text{spacing}} \sim \left(\frac{16\pi \Sigma a^{11/2} \rho^{1/2}}{M_\star^{3/2}} \right)^{1/2} = 1430 \frac{\kappa^{1/2} \lambda^{169/48} \gamma^{1/8}}{\alpha^{21/2} \beta^{15/4}} \qquad (34)$$

This was evaluated at the center of the temperate zone to give the ratio of these two length scales in terms of fundamental constants. Since this ratio loosely sets the fraction of planets within the temperate zone, we arrive at

$$f_{\text{temp}} \sim \frac{\Delta a_{\text{temp}}}{a_{\text{spacing}}} = 0.0053 \frac{\alpha^{11/2} \beta^{7/4}}{\kappa^{1/2} \lambda^{85/48} \gamma^{5/8}}. \qquad (35)$$

---

[4] This neglects other potential thresholds, such as the inner boundary set by the photosynthetic threshold of carbon dioxide abundance, which is remarkably close to the inner boundary discussed here [76].

Set this to the value of 1.6 in our solar system. Notice that, for α or β much smaller or γ much larger, the fraction of temperate planets will be diminished. Additionally, this quantity is larger for smaller mass stars, reflecting the comparatively broad temperate zones there.

This distribution is illustrated in Figure 4. The corresponding probabilities for observing our values of the constants are

$$\mathbb{P}(\alpha_{obs}) = 0.24, \quad \mathbb{P}(\beta_{obs}) = 0.37, \quad \mathbb{P}(\gamma_{obs}) = 0.15. \tag{36}$$

Though the probabilities shift around by a factor of a few, the inclusion of this criterion does not actually affect the results very much. This is a generic conclusion, even when this is included with other criteria. Whether or not this condition is essential for life, the multiverse hypothesis is relatively insensitive to it.

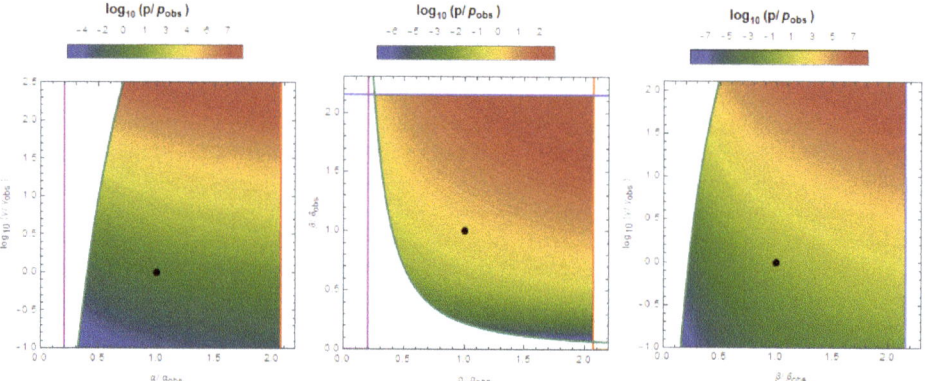

Figure 4. Distribution of observers if life may only exist within the temperate zone, in the α–γ, α–β, and β–γ subplanes. The teal line is the boundary for which planetary disks are smaller than the temperate zone.

Let us also take this opportunity to determine what values of parameters would render disks smaller than the habitable orbit, thereby precluding temperate planets from ever forming. Based off the expressions above and in the appendix, we have

$$\frac{a_{temp}}{R_{disk}} = 4.0 \times 10^5 \, \kappa \, \lambda^{17/12} \, \frac{\gamma^{1/2}}{\alpha^5 \beta^2}. \tag{37}$$

Aside from the obvious condition that, if the mean free path of the galaxy were 100 times smaller, the protoplanetary disk would be shrunk by the same amount, this also places restrictions on the fundamental parameters. Though these introduce lower bounds for α and β, this is the region with very few temperate observers anyway, and so this does not change the habitability estimate appreciably, especially since these scales would have to shift by two orders of magnitude before the threshold is reached. A more stringent boundary would require that disks be larger than the orbits of the gas giants. If not, they would be precluded from forming in the first place, and, if these were essential for life on Earth, universes like this would be altogether barren.

3.3. Planet Migration

Traditional models of planet formation assume that little orbital migration takes place during planet formation. However, the large number of gas and ice giants found situated extremely close to their host stars [82], the presence of orbital resonances found in some exoplanet systems [83], and icy composition of planets within the snow line [84] all point to the presence of migration. Migration may strongly affect the habitability of planetary systems, as if it proceeds for too long, all inner planets will drift toward the inner edge of the disk at around 0.014 AU [85], and giant planets migrating across the temperate zone would destabilize the orbits of any existing planets. The vast diversity of systems found, as well as analytic and numeric simulations of protoplanetary disks [86], point to an exquisitely complex array of migration scenarios, which will be selectively operational depending on the characteristics of the initial system such as disk mass, viscosity, and the size and locations of the planets. Nevertheless, it is possible to perform rough estimates for when migration will be present in a given system. In this section, we do not attempt to characterize all the complex features of planet migration in universes with alternate physical constants, but rather wish to provide a useful diagnostic for when migration will be important. We will find conditions that the physical constants must satisfy so that all planets do not migrate into their host stars at an early stage of evolution, our observed values being intermediate such that this scenario only afflicts some percentage of planetary systems with unlucky characteristics. It should be noted that some amount of migration appears to have occurred even in the outer solar system [87]. However, we evidently ended up with at least one habitable location. In our particular instance, the particular positions of the giant planets ultimately halted migration, preventing the obliteration of the inner solar system [88].

Migration occurs when a planet's influence on the surrounding disk produces a net torque on the planet itself, usually driving it inward. As such, migration halts after the time t_{disk} when the disk has cleared out. This may be compared to the migration timescale

$$t_{\text{mig}} \sim \frac{a}{\dot{a}} \sim \frac{L}{\Gamma} \sim \frac{\Omega a^2 M}{\Gamma}, \tag{38}$$

where M is the planet's mass, L its angular momentum and Γ the torque it experiences. A heuristic condition for when migration will be significant was found in [89] as $t_{\text{mig}} \lesssim 10\, t_{\text{disk}}$. There are various contributions to the torque, and which is dominant organizes migration into several different types, which depend on the circumstances of the case at hand. These are classified into type I, which arises when a trailing overdensity of dust behind a planet (and preceding in front) exerts a torque, and type II, in which the planet is capable of opening up a gap in the disk, resulting in a torque imbalance from the absence of material (for a recent review see [86]). The latter is more relevant for larger planets capable of significantly altering the disk structure, and the former more relevant for smaller planets, which are not. They will each be considered in turn, resulting in conditions on the fundamental constants that must be satisfied in order for a system with given characteristics to retain its initially habitable planets.

For type I migration, the timescale is derived in [90]:

$$t_{\text{I}} = \frac{h^2 M_\star^2}{\Sigma \Omega a^4 M}, \tag{39}$$

where h is the disk height, set by the sound speed. Then, when using the expressions from the Appendix A and Equation (10), and specifying to terrestrial planets situated in the temperate zone, we find

$$\frac{t_{\text{I}}}{t_{\text{disk}}} \propto \frac{\lambda^{43/48} \beta^{9/16} \gamma^{3/8}}{\kappa \alpha^{3/2}}. \tag{40}$$

If this quantity becomes too large, then Earthlike planets in all systems will migrate into their stars, and the universe would be uninhabitable in the traditional sense. Note the dependence $\lambda^{0.9}$, indicating that this type of migration is more important for low mass stars.

This estimate would also be relevant for the production mechanism for Trappist-1 type planets proposed in [91], whereby earthlike planets form behind the snow line before migrating inwards. Such an unconventional pathway is needed to explain this system [92], which has multiple Earth sized planets orbiting the temperate zone of a 0.08 solar mass star with planets near mean motion resonances [93] and possessing icy compositions [84]. This will not be undertaken here, however.

Even if we restrict our attention to parameter values where Earthlike planets do not undergo substantial migration, we may still run into trouble if Jupiter-like planets routinely barrel through their planetary systems. As this is governed by the physics of type II migration, the conditions that must be satisfied for (the majority of) these planets to stay put are somewhat different. Here, we restrict our attention to the case where a full gap is opened in the disk, and where the mass of the planet is substantially smaller than the mass contained in the disk. In this case, the timescale of this migration process is given by [86]:

$$t_{II} = \frac{2}{3} \frac{a^2}{\nu},\qquad(41)$$

where ν is the viscosity. Strictly speaking, these quantities are supposed to be evaluated not at the position of the planet, but rather the place of maximum angular momentum deposition, which can be approximated as $a + 2.5 R_{Hill}$. For our purposes, however, this correction is negligible for the scaling arguments. Then, the ratio of this timescale to the disk lifetime is

$$\frac{t_{II}}{t_{disk}} \propto \frac{\lambda^{15/16} \gamma^{5/8}}{\alpha_{disk} \alpha^4 \beta^{13/16}}.\qquad(42)$$

Here, α_{disk} is the standard parameterization of disk viscosity, discussed further in the appendix. Since this is inversely proportional to disk mass, type II migration is more important for smaller disks, opposite to the type I case. Additionally, since only the scaling with β is flipped from before, the conjunction of these two migration scenarios is capable of putting an upper bound on α and a lower bound on γ, if the other quantity is fixed. The thresholds for both types of migration are displayed in Figure 5. However, the absolute normalization of each of these timescales is uncertain, so we do not derive how the probabilities are altered due to these effects.

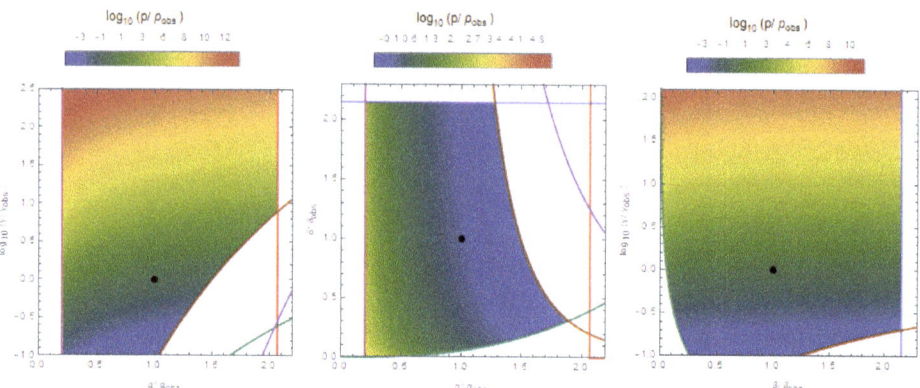

Figure 5. Display of when runaway migration takes hold, in the α–γ, α–β, and β–γ subplanes. The teal line is for type I migration, the orange line is for type II, and the purple line is when the mass of the fastest migrating body is equal to the terrestrial mass. All curves have been normalized so the timescales in our universe are five times larger than the critical value.

4. Discussion: Comparing 480 Hypotheses

Having spent the last two sections detailing the physics behind a multitude of processes that may influence the habitability of a system, we now synthesize these into an estimate of the probability of observing our measured parameter values, for each combination of individual habitability hypotheses. To summarize our results so far, we have firstly included several conditions necessary for planet formation, namely that the majority of galaxies should be larger than the minimal retentive mass that massive stars should have shorter lifetimes than the star formation timescale, and that the minimum metallicity needed to form planets should be smaller than the asymptotic value. Of these three, the third was most constraining. These are also presumably not optional, as opposed to the rest of the criteria that were considered: these were the absence of hot Jupiters, the production of terrestrial planets both through giant impact and isolation formation pathways, and the fraction of planets that end up in the temperate zone. Whether these are necessary for life are still intensely debated, and so each provides a prime opportunity to determine its compatibility with the multiverse hypothesis. Because they are all independent criteria, taken in conjunction they lead to a total of $2 \times 3 \times 2 = 12$ possibilities. This does not include the 16 different choices we made in terms of planet formation, as well as the potentially continuous parameter signifying the slope of the power law for smaller planets. We display the probabilities of observing our measured values in Table 4 for the criteria mentioned in this paper. Note that, though the spread in probabilities is around 2–3 for each, all choices are well within an acceptable range to explain our observations within the multiverse context.

Table 4. Probabilities of different combinations of the habitability hypotheses discussed in the text. Here, yellow stands for the photosynthesis criterion, S the entropy criterion (both explained below Equation (1)), temp for the temperate zone, GI and iso for the terrestrial planet criterion with the giant impact and isolation production mechanisms, and HJ the hot Jupiter condition.

Criteria	$\mathbb{P}(\alpha_{obs})$	$\mathbb{P}(\beta_{obs})$	$\mathbb{P}(\gamma_{obs})$	$\mathbb{L}$
yellow S	0.189	0.437	0.318	2.57
yellow temp S	0.241	0.363	0.151	1.54
yellow GI S	0.229	0.26	0.409	6.72
yellow GI temp S	0.377	0.196	0.42	6.65
yellow iso S	0.234	0.419	0.436	5.02
yellow iso temp S	0.313	0.48	0.245	3.54
yellow HJ S	0.181	0.428	0.308	2.59
yellow HJ temp S	0.237	0.359	0.147	1.54
yellow HJ GI S	0.23	0.261	0.41	6.72
yellow HJ GI temp S	0.378	0.197	0.419	6.64
yellow HJ iso S	0.231	0.424	0.431	5.05
yellow HJ iso temp S	0.311	0.483	0.241	3.54

When combined with the 40 additional combinations from [1] (including combinations of the tidal locking criterion, which posits that only planets that are not tidally locked are habitable, the biological timescale criterion, positing that only stars that last several billion years are habitable, the convective criterion, which states that only stars which are not purely convective are habitable, the photosynthesis and yellow criteria, which state that photosynthesis is necessary for complex life, defined with an optimistic and pessimistic wavelength range, respectively, and the entropy criterion, where habitability is proportional to the total amount of entropy processed by a system), there are a total of 480 separate habitability criteria that may be considered. Of these, only 43% of them give rise to probabilities of observing all three constants we consider of greater than 1%. The full suite of criteria is displayed in Table 5 at the end of this manuscript. For brevity, we omit the convective criteria of [1] because it only ever marginally changes the numerical values, and in all instances its inclusion does not affect the viability of the combination of other hypotheses one way or the other. Of the 190 habitability criteria which give probabilities of over 10%, all make use of the entropy condition. A further 16 which do not include the entropy condition have probabilities greater than 1%—all of them benefit from an interplay between the yellow, tidal locking and biological timescale criteria, which place both upper and lower bounds on the types of allowed stars. The rest of the habitability criteria can safely be regarded as incompatible with the multiverse hypothesis.

The inclusion of multiple hypotheses leads to nonlinear effects, as the interplay between the distribution of purported observers and the anthropic boundaries alter the overall probabilities in sometimes surprising ways. That being said, none of the criteria have that drastic of an effect on the probabilities, especially when including the entropy condition.

Some criteria, namely the terrestrial and temperate conditions, introduce lower bounds to some combination of α and β. In fact, lower bounds on these quantities are somewhat hard to come by in the anthropics literature, though it has always been clear that they should exist, as a world with massless electrons or no electromagnetism would certainly be very different from our own. The bounds we find are stronger than those that exist in the literature.

We also introduce a new measure of our universe's fitness, which we term the luxuriance. This is defined as the expected number of observers in our universe divided by the average number of observers per universe, restricting to universes that do have observers:

$$\mathbb{L} = \frac{P(\alpha_{obs}, \beta_{obs}, \gamma_{obs}) \int d\vec{\alpha}\, \theta(P(\alpha, \beta, \gamma))}{\int d\vec{\alpha}\, P(\alpha, \beta, \gamma)}. \tag{43}$$

Here, the integration is over all three constants, and $\theta(x)$ is the Heaviside step function: $\theta(0) = 0$, $\theta(x) = 1$ for $x > 0$. The rationale for including this is that, for most habitability criteria, the vast majority of universes will be sterile, obfuscating comparisons between different criteria. Restricting to universes that only contain life gives a better feeling for how good our universe is at satisfying the chosen criteria. If our universe is better than typical at making life, then this quantity will be greater than 1. While this is not actually the guiding principle for evaluating whether our observations are consistent with the multiverse, it is a somewhat interesting quantity to consider. It gives some indication of how strongly observers may cluster within the multiverse- and the strong dependence of the properties we discuss on physical constants leads us to expect that they will, so that the majority of observers do find themselves in overly productive universes. The luxuriance ranges by two orders of magnitude for the different possibilities we consider, but the maximum is $\mathbb{L} = 78.4$ for the condition that includes the yellow, tidal locking, biological timescale, hot Jupiter, isolation planet production mechanism, temperate, and entropy criteria.

In addition, recall that certain choices for the physical processes involved in determining the structure of planets greatly affected some of the probabilities: the absence of a dependence on the mass of a planet with stellar mass, and too large a slope for the isolation mass, are both disfavored from the multiverse perspective. These scenarios are not otherwise excluded, but it would count as strong evidence against the multiverse if either of these were verified to be the case.

The one conclusion that should be drawn from this study is that planets are a rather generic feature, and not especially atypical for our particular values of the fundamental constants. Metal buildup is generic, massive stars burn out quickly, and disks tend to clump faster than they dissipate, so the inclusion of these criteria barely influenced the numerical values of the probabilities at all, apart from adding some mild anthropic boundaries. Furthermore, even specifying the characteristics of the planets that are formed, or considering different scenarios of their formation, did not alter the probabilities by more than a factor of few for the most part. This points to a reassuring robustness of our predictions that they are not as highly sensitive to the vagaries of incompletely known planet formation processes as one may have feared.

While it is simple to imagine a universe where planets are almost never made (and indeed before the current plethora of exoplanets was discovered many wondered if our universe was of this character), the parameter values needed to realize this possibility are quite extreme. Additionally, there are several features of planets in our universe that are tantalizingly coincidental, seeming to beckon for an anthropic explanation: the tendency to produce Earth mass planets, the similarity of the interplanetary spacing around sunlike stars and the width of the habitable zone, and the small but nonzero fraction of stars with hot Jupiters. Nevertheless, this reasoning was not borne out when incorporating these criteria into our analysis. The multiverse hypothesis is consistent with the expectation that life may only arise on temperate, terrestrial planets in systems without hot Jupiters, but it is essentially as compatible with the complete converse. The existence of suitable planetary environments thus does not seem to be the most important factor in determining which of the potential universes we find ourselves situated in. The other factors of the Drake equation which we will explore in subsequent works [94,95] will uncover many additional predictions for the requirements of life.

5. Conclusions

We have demonstrated that there are plenty of habitability conditions that are completely incompatible with the multiverse: what this illustrates is that, if any of the ones we have uncovered so far are shown to be the correct condition for the emergence of intelligent life, then we will be able to conclude to a very high degree of confidence (up to 5.2σ) that the multiverse must be wrong. It should be stressed that there is a great deal more that these conditions omit: nothing at all is said about how habitability is affected by things like planetary eccentricity, elemental composition, water abundance, or a host of other potentially paramount aspects of a planetary system [96,97]. The de facto stance on all omissions is that they have no bearing on habitability, and it will only be through future work, including all possibly relevant aspects that a fully coherent list of predictions may be assembled. Further still, placing a priority on the relative availability of each type of universe based on reasonably generic arguments, the precise probabilities, and the conclusions that follow will tremendously benefit from a way of being able to derive this prior with absolute surety. Since only a single one of the myriad habitability criteria is ultimately true, and since we will eventually be able to determine which one that is once we have a large enough sample of life-bearing planets, this demonstrates that the multiverse is capable of generating strong experimentally testable predictions that are capable of being verified or falsified on a reasonable timescale, the hallmark of a sensible scientific theory.

Table 5. Probabilities of various hypotheses, including those from [1] (continued on following pages). In addition to the abbreviations from Table 4, the shorthand is: photo: photosynthesis (optimistic), yellow: photosynthesis (conservative), TL: tidal locking, bio: biological timescale, S: entropy, temp: temperate zone, GI and iso: terrestrial planet with giant impact and isolation production mechanisms, resp. and HJ: hot Jupiter.

Criteria	$\mathbb{P}(\alpha_{obs})$	$\mathbb{P}(\beta_{obs})$	$\mathbb{P}(\gamma_{obs})$	$\mathbb{L}$
number of stars	0.381	0.355	8.06×10^{-7}	0.000199
temp	0.175	0.0524	8.87×10^{-6}	0.000923
GI	0.424	0.281	1.16×10^{-6}	0.00021
GI temp	0.227	0.279	9.7×10^{-6}	0.00269
iso	0.422	0.493	8.69×10^{-7}	0.000239
iso temp	0.198	0.112	8.9×10^{-6}	0.00158
HJ	0.366	0.336	7.14×10^{-7}	0.000199
HJ temp	0.175	0.0523	8.87×10^{-6}	0.00092
HJ GI	0.425	0.282	1.15×10^{-6}	0.00021
HJ GI temp	0.227	0.279	9.7×10^{-6}	0.0027
HJ iso	0.419	0.499	8.53×10^{-7}	0.00024
HJ iso temp	0.198	0.112	8.9×10^{-6}	0.00158
bio	0.284	0.106	2.32×10^{-5}	0.00379
bio temp	0.145	0.0152	3.95×10^{-5}	0.00282
bio GI	0.211	0.0161	0.000405	0.0562
bio GI temp	0.282	0.0162	0.00153	0.324
bio iso	0.244	0.216	5.28×10^{-5}	0.00961
bio iso temp	0.167	0.0334	5.42×10^{-5}	0.00661
bio HJ	0.284	0.106	2.03×10^{-5}	0.00378
bio HJ temp	0.145	0.0152	3.95×10^{-5}	0.00281
bio HJ GI	0.211	0.0162	0.000405	0.0565
bio HJ GI temp	0.282	0.0163	0.00153	0.325
bio HJ iso	0.244	0.215	5.13×10^{-5}	0.00958
bio HJ iso temp	0.167	0.0334	5.41×10^{-5}	0.00661
TL	0.33	0.455	4.57×10^{-7}	3.39×10^{-5}
TL temp	0.455	0.257	1.93×10^{-7}	4.61×10^{-5}

Table 5. *Cont.*

Criteria	$\mathbb{P}(\alpha_{\text{obs}})$	$\mathbb{P}(\beta_{\text{obs}})$	$\mathbb{P}(\gamma_{\text{obs}})$	$\mathbb{L}$
TL GI	0.421	0.283	6.42×10^{-7}	3.09×10^{-5}
TL GI temp	0.237	0.29	8.16×10^{-7}	0.000146
TL iso	0.398	0.433	4.6×10^{-7}	5.6×10^{-5}
TL iso temp	0.454	0.356	3.63×10^{-7}	0.000107
TL HJ	0.306	0.433	3.11×10^{-7}	3.15×10^{-5}
TL HJ temp	0.454	0.256	1.8×10^{-7}	4.52×10^{-5}
TL HJ GI	0.422	0.284	6.38×10^{-7}	3.08×10^{-5}
TL HJ GI temp	0.236	0.291	8.12×10^{-7}	0.000146
TL HJ iso	0.393	0.439	4.39×10^{-7}	5.53×10^{-5}
TL HJ iso temp	0.455	0.355	3.55×10^{-7}	0.000106
TL bio	0.011	0.365	7.19×10^{-5}	0.00304
TL bio temp	0.0154	0.276	4.16×10^{-5}	0.00788
TL bio GI	0.0562	0.0184	0.000277	0.0109
TL bio GI temp	0.18	0.023	0.000429	0.0631
TL bio iso	0.0237	0.485	0.000107	0.0087
TL bio iso temp	0.0403	0.407	9.85×10^{-5}	0.0238
TL bio HJ	0.00968	0.362	4.84×10^{-5}	0.00301
TL bio HJ temp	0.0149	0.275	3.89×10^{-5}	0.00779
TL bio HJ GI	0.0561	0.0185	0.000276	0.0109
TL bio HJ GI temp	0.18	0.0231	0.000428	0.063
TL bio HJ iso	0.0228	0.488	0.000101	0.00861
TL bio HJ iso temp	0.0394	0.406	9.61×10^{-5}	0.0235
photo	0.439	0.183	8.43×10^{-7}	0.000127
photo S	0.241	0.382	0.381	8.95
photo temp	0.403	0.121	3.11×10^{-7}	7.34×10^{-5}
photo temp S	0.338	0.292	0.207	7.07
photo GI	0.016	0.296	8.09×10^{-6}	0.00169
photo GI S	0.335	0.142	0.33	34.8
photo GI temp	0.00942	0.276	8.96×10^{-6}	0.00513
photo GI temp S	0.497	0.113	0.424	45.7
photo iso	0.456	0.267	1.75×10^{-6}	0.000346
photo iso S	0.305	0.448	0.494	13.7
photo iso temp	0.329	0.18	6.74×10^{-7}	0.000207
photo iso temp S	0.403	0.458	0.321	12.7
photo HJ	0.439	0.183	8.22×10^{-7}	0.000126
photo HJ S	0.237	0.377	0.376	9.0
photo HJ temp	0.403	0.121	3.08×10^{-7}	7.31×10^{-5}
photo HJ temp S	0.337	0.291	0.205	7.08
photo HJ GI	0.016	0.297	8.04×10^{-6}	0.00168
photo HJ GI S	0.336	0.142	0.331	35.0
photo HJ GI temp	0.00942	0.276	8.92×10^{-6}	0.00513
photo HJ GI temp S	0.498	0.113	0.425	45.9
photo HJ iso	0.456	0.267	1.72×10^{-6}	0.000342
photo HJ iso S	0.302	0.452	0.497	13.8
photo HJ iso temp	0.329	0.18	6.68×10^{-7}	0.000206
photo HJ iso temp S	0.402	0.455	0.319	12.8
photo bio	0.0631	0.103	1.75×10^{-5}	0.00197
photo bio S	0.18	0.409	0.43	7.89
photo bio temp	0.123	0.0849	5.35×10^{-6}	0.000987
photo bio temp S	0.222	0.335	0.262	6.98
photo bio GI	0.353	0.0162	0.000924	0.166
photo bio GI S	0.316	0.131	0.308	30.9
photo bio GI temp	0.154	0.0141	0.000688	0.34
photo bio GI temp S	0.484	0.1	0.408	40.3
photo bio iso	0.104	0.188	5.14×10^{-5}	0.00757
photo bio iso S	0.262	0.42	0.448	11.6
photo bio iso temp	0.185	0.152	1.51×10^{-5}	0.00365
photo bio iso temp S	0.322	0.489	0.382	11.8

Table 5. Cont.

Criteria	$\mathbb{P}(\alpha_{obs})$	$\mathbb{P}(\beta_{obs})$	$\mathbb{P}(\gamma_{obs})$	$\mathbb{L}$
photo bio HJ	0.0631	0.103	1.71×10^{-5}	0.00195
photo bio HJ S	0.175	0.404	0.426	7.94
photo bio HJ temp	0.123	0.0848	5.31×10^{-6}	0.000983
photo bio HJ temp S	0.22	0.333	0.26	6.99
photo bio HJ GI	0.353	0.0162	0.000921	0.167
photo bio HJ GI S	0.317	0.132	0.309	31.0
photo bio HJ GI temp	0.154	0.0141	0.000687	0.34
photo bio HJ GI temp S	0.485	0.101	0.409	40.5
photo bio HJ iso	0.104	0.188	5.06×10^{-5}	0.0075
photo bio HJ iso S	0.259	0.424	0.451	11.7
photo bio HJ iso temp	0.185	0.151	1.5×10^{-5}	0.00363
photo bio HJ iso temp S	0.32	0.492	0.379	11.8
photo TL	0.478	0.232	7.36×10^{-7}	0.000102
photo TL S	0.376	0.423	0.382	10.8
photo TL temp	0.412	0.197	2.61×10^{-7}	6.01×10^{-5}
photo TL temp S	0.453	0.288	0.283	10.8
photo TL GI	0.016	0.312	4.34×10^{-6}	0.000271
photo TL GI S	0.312	0.431	0.447	13.9
photo TL GI temp	0.0096	0.306	1.93×10^{-6}	0.000332
photo TL GI temp S	0.485	0.317	0.345	16.7
photo TL iso	0.41	0.308	1.46×10^{-6}	0.000264
photo TL iso S	0.384	0.46	0.435	15.9
photo TL iso temp	0.323	0.26	5.48×10^{-7}	0.000154
photo TL iso temp S	0.447	0.331	0.365	15.5
photo TL HJ	0.478	0.232	6.94×10^{-7}	9.94×10^{-5}
photo TL HJ S	0.36	0.394	0.355	11.2
photo TL HJ temp	0.412	0.197	2.51×10^{-7}	5.92×10^{-5}
photo TL HJ temp S	0.458	0.276	0.272	10.9
photo TL HJ GI	0.016	0.312	4.31×10^{-6}	0.000269
photo TL HJ GI S	0.312	0.43	0.446	13.9
photo TL HJ GI temp	0.00959	0.306	1.92×10^{-6}	0.00033
photo TL HJ GI temp S	0.484	0.316	0.344	16.7
photo TL HJ iso	0.41	0.307	1.42×10^{-6}	0.000258
photo TL HJ iso S	0.382	0.452	0.43	16.0
photo TL HJ iso temp	0.323	0.26	5.37×10^{-7}	0.000151
photo TL HJ iso temp S	0.448	0.325	0.362	15.5
photo TL bio	0.0354	0.29	0.000149	0.017
photo TL bio S	0.252	0.495	0.423	14.9
photo TL bio temp	0.0406	0.291	0.000116	0.0252
photo TL bio temp S	0.33	0.376	0.413	20.9
photo TL bio GI	0.317	0.0329	0.00121	0.0717
photo TL bio GI S	0.104	0.413	0.385	17.9
photo TL bio GI temp	0.371	0.0357	0.00086	0.14
photo TL bio GI temp S	0.196	0.283	0.397	27.3
photo TL bio iso	0.087	0.43	0.000325	0.0495
photo TL bio iso S	0.249	0.5	0.393	20.4
photo TL bio iso temp	0.0982	0.432	0.00026	0.07
photo TL bio iso temp S	0.353	0.383	0.366	25.1
photo TL bio HJ	0.0347	0.289	0.000141	0.0167
photo TL bio HJ S	0.218	0.456	0.449	16.0
photo TL bio HJ temp	0.0399	0.29	0.000112	0.025
photo TL bio HJ temp S	0.311	0.354	0.424	21.6
photo TL bio HJ GI	0.316	0.033	0.00121	0.0715
photo TL bio HJ GI S	0.103	0.411	0.385	17.9
photo TL bio HJ GI temp	0.37	0.0359	0.000859	0.14
photo TL bio HJ GI temp S	0.196	0.282	0.398	27.2
photo TL bio HJ iso	0.0854	0.429	0.000318	0.0488

Table 5. *Cont.*

Criteria	$\mathbb{P}(\alpha_{obs})$	$\mathbb{P}(\beta_{obs})$	$\mathbb{P}(\gamma_{obs})$	$\mathbb{L}$
photo TL bio HJ iso S	0.243	0.49	0.397	20.7
photo TL bio HJ iso temp	0.0966	0.43	0.000255	0.0693
photo TL bio HJ iso temp S	0.348	0.375	0.367	25.3
yellow	0.486	0.162	8.78×10^{-7}	9.64×10^{-5}
yellow temp	0.492	0.161	2.31×10^{-7}	2.61×10^{-5}
yellow GI	0.00555	0.292	6.88×10^{-5}	0.0127
yellow GI temp	0.00329	0.272	2.81×10^{-5}	0.00588
yellow iso	0.391	0.219	2.03×10^{-6}	0.000336
yellow iso temp	0.387	0.218	5.35×10^{-7}	9.14×10^{-5}
yellow HJ	0.486	0.162	8.64×10^{-7}	9.55×10^{-5}
yellow HJ temp	0.492	0.161	2.27×10^{-7}	2.58×10^{-5}
yellow HJ GI	0.00554	0.292	6.83×10^{-5}	0.0126
yellow HJ GI temp	0.00329	0.272	2.79×10^{-5}	0.00584
yellow HJ iso	0.391	0.219	1.99×10^{-6}	0.000333
yellow HJ iso temp	0.387	0.218	5.26×10^{-7}	9.06×10^{-5}
yellow bio	0.0351	0.102	1.72×10^{-5}	0.00178
yellow bio S	0.123	0.47	0.38	2.88
yellow bio temp	0.0324	0.0912	5.21×10^{-6}	0.000555
yellow bio temp S	0.126	0.406	0.215	2.05
yellow bio GI	0.0503	0.0133	0.000736	0.13
yellow bio GI S	0.177	0.25	0.367	6.9
yellow bio GI temp	0.0316	0.00818	0.000315	0.0632
yellow bio GI temp S	0.299	0.177	0.474	7.21
yellow bio iso	0.081	0.187	5.23×10^{-5}	0.00821
yellow bio iso S	0.174	0.386	0.498	5.37
yellow bio iso temp	0.0741	0.168	1.62×10^{-5}	0.00262
yellow bio iso temp S	0.206	0.424	0.324	4.38
yellow bio HJ	0.0351	0.102	1.69×10^{-5}	0.00176
yellow bio HJ S	0.112	0.46	0.37	2.91
yellow bio HJ temp	0.0324	0.0912	5.14×10^{-6}	0.00055
yellow bio HJ temp S	0.12	0.4	0.209	2.05
yellow bio HJ GI	0.0503	0.0134	0.000732	0.13
yellow bio HJ GI S	0.177	0.251	0.368	6.9
yellow bio HJ GI temp	0.0316	0.00819	0.000313	0.0629
yellow bio HJ GI temp S	0.299	0.177	0.473	7.21
yellow bio HJ iso	0.0809	0.187	5.13×10^{-5}	0.00814
yellow bio HJ iso S	0.169	0.392	0.493	5.4
yellow bio HJ iso temp	0.074	0.168	1.59×10^{-5}	0.0026
yellow bio HJ iso temp S	0.202	0.428	0.319	4.39
yellow TL	0.0303	0.0308	1.63×10^{-6}	0.000763
yellow TL S	0.457	0.377	0.431	32.6
yellow TL temp	0.0229	0.0251	4.55×10^{-7}	0.000253
yellow TL temp S	0.328	0.281	0.242	26.5
yellow TL GI	0.00345	0.38	4.84×10^{-5}	0.0256
yellow TL GI S	0.35	0.299	0.496	43.7
yellow TL GI temp	0.00196	0.372	2.04×10^{-5}	0.0129
yellow TL GI temp S	0.481	0.311	0.326	44.9
yellow TL iso	0.0249	0.0412	3.78×10^{-6}	0.00193
yellow TL iso S	0.488	0.424	0.469	44.7
yellow TL iso temp	0.0186	0.0335	1.06×10^{-6}	0.000636
yellow TL iso temp S	0.353	0.353	0.294	39.2
yellow TL HJ	0.0303	0.0308	1.58×10^{-6}	0.000756
yellow TL HJ S	0.474	0.351	0.407	33.8
yellow TL HJ temp	0.0229	0.0251	4.43×10^{-7}	0.000251
yellow TL HJ temp S	0.334	0.266	0.227	26.9

Table 5. Cont.

Criteria	$\mathbb{P}(\alpha_{obs})$	$\mathbb{P}(\beta_{obs})$	$\mathbb{P}(\gamma_{obs})$	$\mathbb{L}$
yellow TL HJ GI	0.00345	0.38	4.82×10^{-5}	0.0254
yellow TL HJ GI S	0.35	0.298	0.495	43.6
yellow TL HJ GI temp	0.00196	0.372	2.03×10^{-5}	0.0128
yellow TL HJ GI temp S	0.481	0.31	0.325	44.7
yellow TL HJ iso	0.0249	0.0412	3.69×10^{-6}	0.00191
yellow TL HJ iso S	0.493	0.416	0.463	45.1
yellow TL HJ iso temp	0.0186	0.0335	1.04×10^{-6}	0.00063
yellow TL HJ iso temp S	0.355	0.346	0.289	39.4
yellow TL bio	0.324	0.335	0.0114	5.54
yellow TL bio S	0.373	0.496	0.323	51.3
yellow TL bio temp	0.332	0.335	0.00786	4.56
yellow TL bio temp S	0.399	0.474	0.425	63.1
yellow TL bio GI	0.472	0.446	0.0126	6.64
yellow TL bio GI S	0.126	0.278	0.323	59.4
yellow TL bio GI temp	0.452	0.444	0.0093	5.88
yellow TL bio GI temp S	0.183	0.283	0.433	77.6
yellow TL bio iso	0.435	0.446	0.0158	8.4
yellow TL bio iso S	0.349	0.492	0.326	64.5
yellow TL bio iso temp	0.444	0.445	0.0108	6.82
yellow TL bio iso temp S	0.404	0.499	0.423	77.2
yellow TL bio HJ	0.322	0.333	0.0111	5.5
yellow TL bio HJ S	0.332	0.461	0.343	54.5
yellow TL bio HJ temp	0.331	0.333	0.00769	4.53
yellow TL bio HJ temp S	0.369	0.447	0.444	66.0
yellow TL bio HJ GI	0.472	0.446	0.0126	6.61
yellow TL bio HJ GI S	0.126	0.277	0.323	59.3
yellow TL bio HJ GI temp	0.453	0.445	0.00928	5.85
yellow TL bio HJ GI temp S	0.183	0.281	0.434	77.4
yellow TL bio HJ iso	0.433	0.444	0.0155	8.36
yellow TL bio HJ iso S	0.339	0.499	0.33	65.5
yellow TL bio HJ iso temp	0.442	0.443	0.0106	6.8
yellow TL bio HJ iso temp S	0.395	0.492	0.428	78.4

Funding: This research received no external funding.

Acknowledgments: I would like to thank Cullen Blake, Diana Dragomir, Scott Kenyon, Jabran Zahid, and Li Zeng for useful discussions.

Conflicts of Interest: The author declares no conflict of interest.

Appendix A. Planetary Parameters

In this appendix, we collect results on how various quantities relevant to our estimates in this work depend on orbital parameters of the stellar system, as well as fundamental quantities. To begin, we display the typical molecular binding energy:

$$T_{mol} = \sqrt{\frac{\alpha}{m_{mol} r_{mol}^3}} = 0.037 \frac{\alpha^2 m_e^{3/2}}{m_p^{1/2}}. \tag{A1}$$

This also defines the temperature required for liquid water.

Planets: The mass of a terrestrial planet, based on the criteria that carbon dioxide but not helium is gravitationally bound to the surface at these temperatures, is

$$M_{terr} = 9.2 - 202.2 \frac{\alpha^{3/2} m_e^{3/4} M_{pl}^3}{m_p^{11/4}}. \tag{A2}$$

References

1. Sandora, M. Multiverse Predictions for Habitability I: The Number of Stars and Their Properties. *arXiv* **2019**, arXiv:1901.04614.
2. Vilenkin, A. Predictions from Quantum Cosmology. *Phys. Rev. Lett.* **1995**, *74*, 846–849. [CrossRef] [PubMed]
3. Frank, A.; Sullivan, W.T., III. A New Empirical Constraint on the Prevalence of Technological Species in the Universe. *Astrobiology* **2016**, *16*, 359–362. [CrossRef] [PubMed]
4. Maschberger, T. On the function describing the stellar initial mass function. *Mon. Not. R. Astron. Soc.* **2012**, *429*, 1725–1733. [CrossRef]
5. Johnson, J.L.; Li, H. The first planets: The critical metallicity for planet formation. *Astrophys. J.* **2012**, *751*, 81. [CrossRef]
6. Petigura, E.A.; Marcy, G.W.; Howard, A.W. A plateau in the planet population below twice the size of Earth. *Astrophys. J.* **2013**, *770*, 69. [CrossRef]
7. Owen, J.E.; Wu, Y. The evaporation valley in the Kepler planets. *Astrophys. J.* **2017**, *847*, 29. [CrossRef]
8. Ginzburg, S.; Schlichting, H.E.; Sari, R. Core-powered mass loss sculpts the radius distribution of small exoplanets. *arXiv* **2017**, arXiv:1708.01621.
9. Zeng, L.; Jacobsen, S.B.; Sasselov, D.D.; Vanderburg, A. Survival function analysis of planet size distribution with Gaia Data Release 2 updates. *Mon. Not. R. Astron. Soc.* **2018**, *479*, 5567–5576. [CrossRef]
10. Adams, F.C. Constraints on Alternate Universes: Stars and habitable planets with different fundamental constants. *J. Cosmol. Astropart. Phys.* **2016**, *2016*, 042. [CrossRef]
11. Adams, F.C.; Coppess, K.R.; Bloch, A.M. Planets in other universes: Habitability constraints on density fluctuations and galactic structure. *J. Cosmol. Astropart. Phys.* **2015**, *2015*, 030. [CrossRef]
12. Weinberg, S. Anthropic bound on the cosmological constant. *Phys. Rev. Lett.* **1987**, *59*, 2607–2610. [CrossRef] [PubMed]
13. Thielemann, F.K.; Nomoto, K.; Hashimoto, M.A. Core-collapse supernovae and their ejecta. *Astrophys. J.* **1996**, *460*, 408. [CrossRef]
14. Rees, M.J.; Ostriker, J. Cooling, dynamics and fragmentation of massive gas clouds: Clues to the masses and radii of galaxies and clusters. *Mon. Not. R. Astron. Soc.* **1977**, *179*, 541–559. [CrossRef]
15. Padmanabhan, T. *Theoretical Astrophysics: Volume 2, Stars and Stellar Systems*; Cambridge University Press: Cambridge, UK, 2001.
16. Tremonti, C.A.; Heckman, T.M.; Kauffmann, G.; Brinchmann, J.; Charlot, S.; White, S.D.; Seibert, M.; Peng, E.W.; Schlegel, D.J.; Uomoto, A.; et al. The origin of the mass-metallicity relation: Insights from 53,000 star-forming galaxies in the sloan digital sky survey. *Astrophys. J.* **2004**, *613*, 898. [CrossRef]
17. Press, W.H.; Schechter, P. Formation of galaxies and clusters of galaxies by self-similar gravitational condensation. *Astrophys. J.* **1974**, *187*, 425–438. [CrossRef]
18. Dayal, P.; Ward, M.; Cockell, C. The habitability of the Universe through 13 billion years of cosmic time. *arXiv* **2016**, arXiv:1606.09224.
19. Woosley, S.E.; Heger, A.; Weaver, T.A. The evolution and explosion of massive stars. *Rev. Mod. Phys.* **2002**, *74*, 1015–1071. [CrossRef]
20. Burrows, A.S.; Ostriker, J.P. Astronomical reach of fundamental physics. *Proc. Natl. Acad. Sci. USA* **2014**, *111*, 2409–2416. [CrossRef]
21. Shakura, N.I.; Sunyaev, R.A. Black holes in binary systems. Observational appearance. *Astron. Astrophys.* **1973**, *24*, 337–355.
22. Alexander, R.; Pascucci, I.; Andrews, S.; Armitage, P.; Cieza, L. The dispersal of protoplanetary disks. *arXiv* **2013**, arXiv:1311.1819.
23. Apai, D.; Lauretta, D.S. *Protoplanetary Dust: Astrophysical and Cosmochemical Perspectives*; Cambridge University Press: Cambridge, UK, 2010; Volume 12.
24. Ercolano, B.; Clarke, C. Metallicity, planet formation and disc lifetimes. *Mon. Not. R. Astron. Soc.* **2010**, *402*, 2735–2743. [CrossRef]
25. Zahid, H.J.; Dima, G.I.; Kudritzki, R.P.; Kewley, L.J.; Geller, M.J.; Hwang, H.S.; Silverman, J.D.; Kashino, D. The universal relation of galactic chemical evolution: The origin of the mass-metallicity relation. *Astrophys. J.* **2014**, *791*, 130. [CrossRef]

26. Schellekens, A.N. Life at the Interface of Particle Physics and String Theory. *Rev. Mod. Phys.* **2013**, *85*, 1491–1540. [CrossRef]
27. Tegmark, M.; Rees, M.J. Why Is the Cosmic Microwave Background Fluctuation Level 10^{-5}? *Astrophys. J.* **1998**, *499*, 526–532. [CrossRef]
28. Fischer, D.A.; Valenti, J. The Planet-Metallicity Correlation. *Astrophys. J.* **2005**, *622*, 1102–1117. [CrossRef]
29. Batygin, K.; Bodenheimer, P.H.; Laughlin, G.P. In situ formation and dynamical evolution of hot Jupiter systems. *Astrophys. J.* **2016**, *829*, 114. [CrossRef]
30. Dawson, R.I.; Johnson, J.A. Origins of Hot Jupiters. *arXiv* **2018**, arXiv:1801.06117.
31. Raymond, S.N.; Mandell, A.M.; Sigurdsson, S. Exotic Earths: Forming Habitable Worlds with Giant Planet Migration. *Science* **2006**, *313*, 1413–1416. [CrossRef]
32. Smallwood, J.L.; Martin, R.G.; Lepp, S.; Livio, M. Asteroid impacts on terrestrial planets: The effects of super-Earths and the role of the ν 6 resonance. *Mon. Not. R. Astron. Soc.* **2017**, *473*, 295–305. [CrossRef]
33. Buchhave, L.A.; Bitsch, B.; Johansen, A.; Latham, D.W.; Bizzarro, M.; Bieryla, A.; Kipping, D.M. Jupiter Analogues Orbit Stars with an Average Metallicity Close to that of the Sun. *arXiv* **2018**, arXiv:1802.06794.
34. Ndugu, N.; Bitsch, B.; Jurua, E. Planet population synthesis driven by pebble accretion in cluster environments. *Mon. Not. R. Astron. Soc.* **2017**, *474*, 886–897. [CrossRef]
35. Fabrycky, D.; Tremaine, S. Shrinking binary and planetary orbits by Kozai cycles with tidal friction. *Astrophys. J.* **2007**, *669*, 1298. [CrossRef]
36. Becker, J.C.; Vanderburg, A.; Adams, F.C.; Rappaport, S.A.; Schwengeler, H.M. WASP-47: A hot Jupiter system with two additional planets discovered by K2. *Astrophys. Lett.* **2015**, *812*, L18. [CrossRef]
37. Chatterjee, S.; Ford, E.B.; Matsumura, S.; Rasio, F.A. Dynamical outcomes of planet-planet scattering. *Astrophys. J.* **2008**, *686*, 580. [CrossRef]
38. Spalding, C.; Batygin, K. A Secular Resonant Origin for the Loneliness of Hot Jupiters. *Astron. J.* **2017**, *154*, 93. [CrossRef]
39. Johansen, A.; Lambrechts, M. Forming Planets via Pebble Accretion. *Annu. Rev. Earth Planet. Sci.* **2017**, *45*, 359–387. [CrossRef]
40. Johnson, J.A.; Aller, K.M.; Howard, A.W.; Crepp, J.R. Giant planet occurrence in the stellar mass-metallicity plane. *Publ. Astron. Soc. Pac.* **2010**, *122*, 905. [CrossRef]
41. Taubner, R.S.; Pappenreiter, P.; Zwicker, J.; Smrzka, D.; Pruckner, C.; Kolar, P.; Bernacchi, S.; Seifert, A.H.; Krajete, A.; Bach, W.; et al. Biological methane production under putative Enceladus-like conditions. *Nat. Commun.* **2018**, *9*, 748. [CrossRef]
42. Bains, W. Many chemistries could be used to build living systems. *Astrobiology* **2004**, *4*, 137–167. [CrossRef] [PubMed]
43. Schulze-Makuch, D.; Irwin, L.N. The prospect of alien life in exotic forms on other worlds. *Naturwissenschaften* **2006**, *93*, 155–172. [CrossRef] [PubMed]
44. Rogers, L.A. Most 1.6 Earth-radius planets are not rocky. *Astrophys. J.* **2015**, *801*, 41. [CrossRef]
45. Ward, P.D.; Brownlee, D. *Rare Earth: Why Complex Life Is Uncommon in the Universe*; Copernicus Books: New York, NY, USA, 2003.
46. Owen, J.E.; Lai, D. Photoevaporation and high-eccentricity migration created the sub-Jovian desert. *Mon. Not. R. Astron. Soc.* **2018**, *479*, 5012–5021. [CrossRef]
47. Fulton, B.J.; Petigura, E.A.; Howard, A.W.; Isaacson, H.; Marcy, G.W.; Cargile, P.A.; Hebb, L.; Weiss, L.M.; Johnson, J.A.; Morton, T.D.; et al. The California-Kepler survey. III. A gap in the radius distribution of small planets. *Astron. J.* **2017**, *154*, 109. [CrossRef]
48. Fressin, F.; Torres, G.; Charbonneau, D.; Bryson, S.T.; Christiansen, J.; Dressing, C.D.; Jenkins, J.M.; Walkowicz, L.M.; Batalha, N.M. The false positive rate of Kepler and the occurrence of planets. *Astrophys. J.* **2013**, *766*, 81. [CrossRef]
49. Raymond, S.N.; Boulet, T.; Izidoro, A.; Esteves, L.; Bitsch, B. Migration-driven diversity of super-Earth compositions. *Mon. Not. R. Astron. Soc. Lett.* **2018**, *479*, L81–L85. [CrossRef]
50. Kokubo, E.; Kominami, J.; Ida, S. Formation of terrestrial planets from protoplanets. I. Statistics of basic dynamical properties. *Astrophys. J.* **2006**, *642*, 1131. [CrossRef]
51. Williams, J.P.; Cieza, L.A. Protoplanetary disks and their evolution. *Annu. Rev. Astron. Astrophys.* **2011**, *49*, 67–117. [CrossRef]

52. Pascucci, I.; Testi, L.; Herczeg, G.; Long, F.; Manara, C.; Hendler, N.; Mulders, G.; Krijt, S.; Ciesla, F.; Henning, T.; et al. A steeper than linear disk mass–stellar mass scaling relation. *Astrophys. J.* **2016**, *831*, 125. [CrossRef]
53. Bate, M.R. On the diversity and statistical properties of protostellar discs. *Mon. Not. R. Astron. Soc.* **2018**, *475*, 5618–5658. [CrossRef]
54. Morbidelli, A.; Lambrechts, M.; Jacobson, S.; Bitsch, B. The great dichotomy of the Solar System: Small terrestrial embryos and massive giant planet cores. *Icarus* **2015**, *258*, 418–429. [CrossRef]
55. Kennedy, G.M.; Kenyon, S.J. Planet formation around stars of various masses: The snow line and the frequency of giant planets. *Astrophys. J.* **2008**, *673*, 502. [CrossRef]
56. Kennedy, G.M.; Kenyon, S.J. Planet formation around stars of various masses: Hot Super-Earths. *Astrophys. J.* **2008**, *682*, 1264. [CrossRef]
57. Ida, S.; Guillot, T.; Morbidelli, A. The radial dependence of pebble accretion rates: A source of diversity in planetary systems-I. Analytical formulation. *Astron. Astrophys.* **2016**, *591*, A72. [CrossRef]
58. Adams, F.C.; Hollenbach, D.; Laughlin, G.; Gorti, U. Photoevaporation of circumstellar disks due to external far-ultraviolet radiation in stellar aggregates. *Astrophys. J.* **2004**, *611*, 360. [CrossRef]
59. Morbidelli, A.; Lunine, J.I.; O'Brien, D.P.; Raymond, S.N.; Walsh, K.J. Building terrestrial planets. *Annu. Rev. Earth Planet. Sci.* **2012**, *40*, 251–275. [CrossRef]
60. Youdin, A.N.; Kenyon, S.J. From disks to planets. In *Planets, Stars and Stellar Systems*; Springer: New York, NY, USA, 2013; pp. 1–62.
61. Izidoro, A.; Raymond, S.N. Formation of Terrestrial Planets. In *Handbook of Exoplanets*; Springer: New York, NY, USA, 2018; pp. 1–59.
62. Schlichting, H.E. Formation of close in super-Earths and mini-Neptunes: Required disk masses and their implications. *Astrophys. J. Lett.* **2014**, *795*, L15. [CrossRef]
63. Sinukoff, E.; Fulton, B.; Scuderi, L.; Gaidos, E. Below One Earth: The Detection, Formation, and Properties of Subterrestrial Worlds. *Space Sci. Rev.* **2013**, *180*, 71–99. [CrossRef]
64. Wu, Z.N.; Li, J.; Bai, C.Y. Scaling relations of lognormal type growth process with an extremal principle of entropy. *Entropy* **2017**, *19*, 56. [CrossRef]
65. Cumming, A.; Butler, R.P.; Marcy, G.W.; Vogt, S.S.; Wright, J.T.; Fischer, D.A. The Keck planet search: Detectability and the minimum mass and orbital period distribution of extrasolar planets. *Publ. Astron. Soc. Pac.* **2008**, *120*, 531. [CrossRef]
66. Zeng, L.; Jacobsen, S.B.; Sasselov, D.D.; Vanderburg, A. Survival Function Analysis of Planet Orbit Distribution and Occurrence Rate Estimate. *arXiv* **2018**, arXiv:1801.03994.
67. Simon, J.B.; Armitage, P.J.; Li, R.; Youdin, A.N. The mass and size distribution of planetesimals formed by the streaming instability. I. The role of self-gravity. *Astrophys. J.* **2016**, *822*, 55. [CrossRef]
68. Mordasini, C. Planetary population synthesis. *Handbook of Exoplanets*; Springer: New York, NY, USA, 2018; pp. 1–50.
69. Raymond, S.N.; Scalo, J.; Meadows, V.S. A decreased probability of habitable planet formation around low-mass stars. *Astrophys. J.* **2007**, *669*, 606. [CrossRef]
70. Tasker, E.; Tan, J.; Heng, K.; Kane, S.; Spiegel, D.; Brasser, R.; Casey, A.; Desch, S.; Dorn, C.; Hernlund, J.; et al. The language of exoplanet ranking metrics needs to change. *Nat. Astron.* **2017**, *1*, 0042. [CrossRef]
71. Kopparapu, R.K.; Ramirez, R.; Kasting, J.F.; Eymet, V.; Robinson, T.D.; Mahadevan, S.; Terrien, R.C.; Domagal-Goldman, S.; Meadows, V.; Deshpande, R. Habitable zones around main-sequence stars: New estimates. *Astrophys. J.* **2013**, *765*, 131. [CrossRef]
72. Yang, J.; Boué, G.; Fabrycky, D.C.; Abbot, D.S. Strong dependence of the inner edge of the habitable zone on planetary rotation rate. *Astrophys. J. Lett.* **2014**, *787*, L2. [CrossRef]
73. Kasting, J.F.; Whitmire, D.P.; Reynolds, R.T. Habitable zones around main sequence stars. *Icarus* **1993**, *101*, 108–128. [CrossRef]
74. Leconte, J.; Forget, F.; Charnay, B.; Wordsworth, R.; Pottier, A. Increased insolation threshold for runaway greenhouse processes on Earth-like planets. *Nature* **2013**, *504*, 268. [CrossRef]
75. Walker, J.C.; Hays, P.; Kasting, J.F. A negative feedback mechanism for the long-term stabilization of Earth's surface temperature. *J. Geophys. Res. Ocean.* **1981**, *86*, 9776–9782. [CrossRef]
76. Rushby, A.J.; Johnson, M.; Mills, B.J.; Watson, A.J.; Claire, M.W. Long-Term Planetary Habitability and the Carbonate-Silicate Cycle. *Astrobiology* **2018**, *18*, 469–480. [CrossRef]
77. Barnes, R.; Quinn, T. The (in) stability of planetary systems. *Astrophys. J.* **2004**, *611*, 494. [CrossRef]

78. Dawson, R.I. Tightly Packed Planetary Systems. In *Handbook of Exoplanets*; Springer: New York, NY, USA, 2017; pp. 1–18.
79. Holman, M.J.; Wisdom, J. Dynamical stability in the outer solar system and the delivery of short period comets. *Astron. J.* **1993**, *105*, 1987–1999. [CrossRef]
80. Fang, J.; Margot, J.L. Are planetary systems filled to capacity? A study based on Kepler results. *Astrophys. J.* **2013**, *767*, 115. [CrossRef]
81. Raymond, S.N.; Barnes, R.; Veras, D.; Armitage, P.J.; Gorelick, N.; Greenberg, R. Planet-planet scattering leads to tightly packed planetary systems. *Astrophys. J. Lett.* **2009**, *696*, L98. [CrossRef]
82. Borucki, W.J.; Koch, D.; Basri, G.; Batalha, N.; Brown, T.; Caldwell, D.; Caldwell, J.; Christensen-Dalsgaard, J.; Cochran, W.D.; DeVore, E.; et al. Kepler planet-detection mission: Introduction and first results. *Science* **2010**, *327*, 977–980. [CrossRef] [PubMed]
83. Snellgrove, M.; Papaloizou, J.; Nelson, R. On disc driven inward migration of resonantly coupled planets with application to the system around GJ876. *Astron. Astrophys.* **2001**, *374*, 1092–1099. [CrossRef]
84. Unterborn, C.T.; Desch, S.J.; Hinkel, N.R.; Lorenzo, A. Inward migration of the TRAPPIST-1 planets as inferred from their water-rich compositions. *Nat. Astron.* **2018**, *2*, 297–302 . [CrossRef]
85. Trilling, D.E.; Lunine, J.I.; Benz, W. Orbital migration and the frequency of giant planet formation. *Astron. Astrophys.* **2002**, *394*, 241–251. [CrossRef]
86. Baruteau, C.; Masset, F. Recent developments in planet migration theory. In *Tides in Astronomy and Astrophysics*; Springer: New York, NY, USA, 2013; pp. 201–253.
87. Morbidelli, A.; Crida, A. The dynamics of Jupiter and Saturn in the gaseous protoplanetary disk. *Icarus* **2007**, *191*, 158–171. [CrossRef]
88. Masset, F.; Snellgrove, M. Reversing type II migration: Resonance trapping of a lighter giant protoplanet. *Mon. Not. R. Astron. Soc.* **2001**, *320*, L55–L59. [CrossRef]
89. Ida, S.; Lin, D.N. Toward a deterministic model of planetary formation. I. A desert in the mass and semimajor axis distributions of extrasolar planets. *Astrophys. J.* **2004**, *604*, 388. [CrossRef]
90. Tanaka, H.; Takeuchi, T.; Ward, W.R. Three-dimensional interaction between a planet and an isothermal gaseous disk. I. Corotation and Lindblad torques and planet migration. *Astrophys. J.* **2002**, *565*, 1257. [CrossRef]
91. Ormel, C.W.; Liu, B.; Schoonenberg, D. Formation of TRAPPIST-1 and other compact systems. *Astron. Astrophys.* **2017**, *604*, A1. [CrossRef]
92. Gillon, M.; Triaud, A.H.; Demory, B.O.; Jehin, E.; Agol, E.; Deck, K.M.; Lederer, S.M.; De Wit, J.; Burdanov, A.; Ingalls, J.G.; et al. Seven temperate terrestrial planets around the nearby ultracool dwarf star TRAPPIST-1. *Nature* **2017**, *542*, 456. [CrossRef] [PubMed]
93. Tamayo, D.; Rein, H.; Petrovich, C.; Murray, N. Convergent Migration Renders TRAPPIST-1 Long-lived. *Astrophys. J. Lett.* **2017**, *840*, L19. [CrossRef]
94. Sandora, M. Multiverse Predictions for Habitability III: Fraction of Planets That Develop Life. *arXiv* **2019**, arXiv:1903.06283.
95. Sandora, M. Multiverse Predictions for Habitability IV: Fraction of Life that Develops Intelligence. *arXiv* **2019**, arXiv:1904.11796.
96. Schulze-Makuch, D.; Méndez, A.; Fairén, A.G.; Von Paris, P.; Turse, C.; Boyer, G.; Davila, A.F.; António, M.R.d.S.; Catling, D.; Irwin, L.N. A two-tiered approach to assessing the habitability of exoplanets. *Astrobiology* **2011**, *11*, 1041–1052. [CrossRef]
97. Cockell, C.S.; Bush, T.; Bryce, C.; Direito, S.; Fox-Powell, M.; Harrison, J.; Lammer, H.; Landenmark, H.; Martin-Torres, J.; Nicholson, N.; et al. Habitability: A review. *Astrobiology* **2016**, *16*, 89–117. [CrossRef]
98. Press, W.H.; Lightman, A.P. Dependence of macrophysical phenomena on the values of the fundamental constants. *Philos. Trans. R. Soc. Lond. Ser. A* **1983**, *310*, 323–334. [CrossRef]
99. Tegmark, M.; Aguirre, A.; Rees, M.J.; Wilczek, F. Dimensionless constants, cosmology, and other dark matters. *Phys. Rev. D* **2006**, *73*, 023505. [CrossRef]
100. Shu, F.H. Self-similar collapse of isothermal spheres and star formation. *Astrophys. J.* **1977**, *214*, 488–497. [CrossRef]

© 2019 by the author. Licensee MDPI, Basel, Switzerland. This article is an open access article distributed under the terms and conditions of the Creative Commons Attribution (CC BY) license (http://creativecommons.org/licenses/by/4.0/).

Article

Multiverse Predictions for Habitability: Fraction of Planets that Develop Life

McCullen Sandora [1,2]

1. Institute of Cosmology, Department of Physics and Astronomy, Tufts University, Medford, MA 02155, USA; mccullen.sandora@gmail.com
2. Center for Particle Cosmology, Department of Physics and Astronomy, University of Pennsylvania, Philadelphia, PA 19104, USA

Received: 14 May 2019; Accepted: 12 July 2019; Published: 14 July 2019

Abstract: In a multiverse context, determining the probability of being in our particular universe depends on estimating its overall habitability compared to other universes with different values of the fundamental constants. One of the most important factors in determining this is the fraction of planets that actually develop life, and how this depends on planetary conditions. Many proposed possibilities for this are incompatible with the multiverse: if the emergence of life depends on the lifetime of its host star, the size of the habitable planet, or the amount of material processed, the chances of being in our universe would be very low. If the emergence of life depends on the entropy absorbed by the planet, however, our position in this universe is very natural. Several proposed models for the subsequent development of life, including the hard step model and several planetary oxygenation models, are also shown to be incompatible with the multiverse. If any of these are observed to play a large role in determining the distribution of life throughout our universe, the multiverse hypothesis will be ruled out to high significance.

Keywords: multiverse; habitability; life

1. The Fraction of Habitable Planets that Develop Life

In this paper, we continue our investigation from References [1,2] into the probabilities of measuring our observed values of the fundamental physical constants $\alpha = e^2/4\pi$, $\beta = m_e/m_p$ and $\gamma = m_p/M_{pl}$ within a multiverse framework. Focusing on these three quantities supplements the traditional treatment of the cosmological parameters, and, as we will show, has the potential to go much further in terms of predictive power. The overarching goal of this investigation is to elevate the status of the multiverse to a traditional scientific theory, capable of making testable predictions that are verifiable on reasonable timescales.

The framework has been to use the principle of mediocrity [3,4], wherein the probability of measuring a given set of constants is directly proportional to the number of observers in universes with those constants. Since this is often a strong function of these physical constants, we expect to be in a universe that nearly optimally reflects what life needs. It is usually easier to tell what types of environments our universe is good at producing rather than determining the exact requirements for life, since the former relies only on physics, while the latter involves extrapolating from biology. This approach relies on a notion of habitability: that is, the precise conditions required for the emergence of observers, here identified with life that is 'complex enough'. Herein lies the predictive nature of this enterprise: as there is no strong consensus on the conditions for life to arise and survive long enough to develop intelligence, we investigate a multitude of possibilities. We then tabulate which are compatible with our existence in this universe, and which would imply the existence of much more fecund universes that host the majority of observers. In the latter case, our universe would be a backwater, so much so that the probability of being one of those few observers here is low, sometimes to an extreme degree. This provides a test of these notions of habitability themselves: once we accrue large amounts of exoatmoshperic measurements from a diverse array of planetary environments, which is poised to happen in the coming decades, we will be able to correlate which, if any, are capable of hosting life. This will allow us to finally determine whether the true habitability condition matches the multiverse predictions. Because there are dozens of potentially important planetary and stellar characteristics, and the multiverse would be falsified if just one of them were deemed to be incompatible, this will serve as a very efficient method for putting this overarching framework to a rigorous test.

To estimate the number of observers in the universe, we use the Drake equation [5], which factors this question into subcomponents that are more or less capable of functioning in isolation. If this prescription is taken, then the probability of observing our values of the physical constants is

$$P(\alpha, \beta, \gamma) \propto p_{\text{prior}}(\alpha, \beta, \gamma) N_\star(\alpha, \beta, \gamma) \int d\lambda \, p_{\text{IMF}}(\lambda, \alpha, \beta, \gamma) h(\lambda, \alpha, \beta, \gamma)$$
$$h = f_p \times n_e \times f_{\text{bio}} \times f_{\text{int}} \times N_{\text{obs}} \quad (1)$$

Here, $N_\star$ is the number of stars in the universe, p_{IMF} is the initial mass function, and h defines the notion of habitability of a given environment, defined as the likelihood that it gives rise to the emergence of observers. It may be worth bearing in mind that this differs from the definition that astrobiologists typically use, who are more focused on the occurrence of unicellular life. This quantity naturally factorizes into a product of separate factors, which are: the fraction of stars that have planets f_p, the number of habitable planets per planet-hosting star n_e, the fraction of planets that develop life f_{bio}, the fraction of life-bearing worlds that develop intelligence f_{int}, and the total number of observers on intelligence-bearing worlds N_{obs}. In addition the factor $p_{\text{prior}} \sim 1/(\beta\gamma)$ is used to account for the relative frequency of occurrence of each type of universe, derived from high energy physics considerations. All of these quantities may depend on the fundamental constants, as well as the local environment, which in our analysis is restricted to the stellar mass λ (made dimensionless by comparing to the natural scale $(8\pi)^{3/2} M_{pl}^3/m_p^2$). Our strategy has been to estimate the overall probability by working our way through these factors from left to right, incorporating our previous findings for various habitability proposals into a cohesive analysis.

In Reference [1] we considered the number of habitable stars in a universe as a function of our physical constants, for various definitions of habitability. The main stumbling block is the strength of gravity, which is capable of being 2 orders of magnitude larger without affecting habitability in any obviously adverse way. If gravity were stronger, stars would be smaller, and so there would be more of them, and if each represents an independent opportunity for life to evolve, there would be more observers in those universes. We also incorporated various other potential habitability criteria, such as the requirement that planets not be tidally locked, that stars be not fully convective, that starlight be photosynthetic, and that stars last for biological timescales. While none of these were capable of rescuing the multiverse hypothesis, the tidally locked and photosynthesis conditions will be crucial components of our discussions here.

This was extended to the study of planets in Reference [2]. There are two separate terms related to this in the Drake equation, the first of which being the fraction of stars with planets. While recent results indicate that a minimum metallicity is required for protoplanetary disks to form planets, the constraints on parameters from this condition are quite mild, indicating that planets themselves are generic features throughout a range of alternate universes.

The second planet-related factor is the average number of habitable planets found around stars that do possess them. Here, we address various planetary habitability criteria. It is pointed out that our universe seems to preferentially produce roughly Earth mass planets, out of the eight orders of magnitude it could have chosen. If we take the assumption that Earthlike planets are necessary to host life, this fits in quite well with the multiverse hypothesis, almost regardless of the specific planet formation scenario one employs. Also of note is that the width of the temperate zone is roughly equivalent to the interplanetary spacing (around sunlike stars), ensuring that some planet will be capable of supporting liquid water on its surface in essentially every planetary system.

In this work, we extend our previous analysis to the fraction of planets that develop simple life. This is of course unknown at the moment, and many of the guesses we make about this will be utterly incompatible with the multiverse hypothesis. However, we demonstrate one that is fully capable of making our values typical, and discuss the associated distribution of life that we expect based off this. We first consider that the emergence of life depends on the amount of time, the planetary size, and the entropy production of the host star, and find that only the last is compatible with the multiverse, and further that it is fully able to bring the predictions into alignment with observation. In other words, we have discovered what our universe is truly good at: producing lots of entropy.

An important aspect of our analysis in this paper is not just the probability of observing our constants, but also our particular position within our universe. Many environmental parameters should be assessed, including planetary mass, orbit, metallicity, amount of water, carbon to oxygen ratio, and so forth, but we restrict our analysis to stellar size here. This becomes especially important when considering that habitability is proportional to entropy production: taken blindly, this leaves unanswered why we do not live around a much smaller star at a much later point in the future. This may be used to argue that the cutoff mass for stellar habitability is not too much below the solar value. This consideration favors several of our previously proposed habitability criteria, including the tidal locking and yellow light conditions. Synthesizing this with the analysis we perform on the physical constants is able to more powerfully constrain the viable habitability criteria.

After this, we consider a few more geologically inspired conditions: the notion that the size of the biosphere is limited by the amount of nutrient flux on a planet is investigated, and shown to fare worse than the entropy limited case. The question of whether radiogenic plate tectonics is necessary for life is addressed, and this condition is shown to also do worse than the case where it is ignored, though in some cases not terribly so.

We extend our analysis to more sophisticated accounts of the emergence of complex life to determine which of these are compatible with the multiverse. Chiefly, we examine the hard step model, the bated breath model, and the easy stroll model. We find reason to strongly disfavor the hard step model, both within the multiverse context and also on a biological basis. Multiple scenarios for atmospheric oxygenation are investigated, all sufficiently explaining our appearance toward the end of our planet's habitable phase within our universe, but all failing to explain this coincidence in the multiverse context. We determine that only the last is fully compatible with the multiverse.

2. What Factors Influence the Emergence of Life?

As explicitly incorporated into the Drake equation, we specialize our discussion of the emergence of life to planets orbiting stars. We now wish to estimate the fraction of habitable planets which develop life. This is likely to depend, at least to some extent, on the properties of the planet under consideration, and the distinction between factors is not always as clear cut as it may first appear. For instance, in Reference [2] we investigated different notions of the definition of habitability, such as the size and temperature of a planet, which could just as easily have been classified as affecting f_{bio}. Here we specify to temperate, terrestrial planets, and ask what may further influence the emergence of primitive, that is, microscopic, life. There are a number of conceivable factors that may influence this rate: the time in the temperate zone, planetary size, the amount of entropy and nutrients processed, the presence or absence of plate tectonics, and so forth. These will in turn be considered here, but this is far from an exhaustive list. We will succeed in showing that many of the reasonable expectations for the factors dictating where life can emerge will turn out to be incompatible with the multiverse hypothesis. In turn, this will lead us to some definite predictions for where life should be found in our universe.

2.1. Is Habitability Proportional to Stellar Lifetime?

As a first trial, we assign a habitability that is linearly proportional to the lifetime of the star, $h(\lambda) \propto t_\star(\lambda)$. The reasoning behind this is that if life typically takes very long to develop, then the chance of it arising around any given star will be small, but will grow with the star's total lifetime. This stands in contrast to our naive treatment in Reference [1], where we treated all stars as equihabitable, $h(\lambda) \propto 1$. (We also crudely accommodated stellar lifetime in this setup by optionally considering h to be a simple step function of the lifetime of the star: here, a star was deemed habitable if its lifespan exceeded a certain number of 'ticks of the molecular clock', which we took to be $N_{bio} \sim 10^{30}$ to equate to several billion years, and uninhabitable otherwise. This allowed us to place an absolute upper limit on the strength of gravity, $\gamma < 134 \gamma_{obs}$, as above this value no star would have a suitable lifetime.) Let us also note that a more general time dependence, along the lines of Reference [6], may be expected: our analysis in this section can represent a first order Taylor expansion, valid when probabilities are always small. Generalizations to this are considered below.

We use that the stellar lifetime is $t_\star(\lambda) = 110\alpha^2 M_{pl}^2/(m_e^2 m_p \lambda^{5/2}) \equiv \hat{t}_\star/\lambda^{5/2}$ [7]. To make a comparison to other universes, this needs to be divided by another timescale to define a dimensionless ratio: here we use the molecular timescale given by the expression $t_{mol} = 27 m_p^{1/2}/(\alpha^2 m_e^{3/2})$. This ratio counts the total number of interactions any given molecule experiences throughout the star's lifetime. Then, using the fact that $\int d\lambda \, p_{IMF}(\lambda) \lambda^q \sim \lambda_{min}^q$ because λ_{min} is the only scale in the initial mass function, we arrive at

$$P_{t_\star} \propto p_{prior} N_\star \frac{\hat{t}_\star}{N_{bio} t_{mol}} \frac{1}{\lambda_{min}^{5/2}} \propto \frac{\beta^{9/8}}{\alpha^{5/4}} \qquad (2)$$

One interesting aspect of this expression is that the dependence on γ entirely drops out. This indicates that the total habitable time in a universe is independent of this quantity, as although there are more stars in universes with stronger gravity, they last longer in universes with weaker gravity, and there are more of these universes to exactly compensate any preference. What does change is the number of stars this total time is divided among, but because of the simple linear relationship, life would be indifferent to this partitioning. Clearly, this indifference must break down at extreme values of this parameter, that would either create a small number of nearly indefinite stars, or else a cornucopia of exceedingly briefly shining objects. However, with this criterion we would not expect to be situated as we are, more than two orders of magnitude away from the upper boundary used in Reference [1]. The distribution of observers throughout the multiverse is plotted in Figure 1. The probabilities for this habitability criterion, defined as the smaller of P and $1 - P$, are[1]:

$$\mathbb{P}(\alpha_{obs}) = 0.251, \quad \mathbb{P}(\beta_{obs}) = 0.196, \quad \mathbb{P}(\gamma_{obs}) = 0.007 \qquad (3)$$

These can be compared to the values $\mathbb{P}(\alpha_{obs}) = 0.20$, $\mathbb{P}(\beta_{obs}) = 0.44$, and $\mathbb{P}(\gamma_{obs}) = 4.2 \times 10^{-7}$ that were found by simply taking $h(\lambda) = 1$. Though the probability for observing our γ in particular is orders of magnitude better than what we had found without weighting by stellar lifetime, it is still disquietingly small. We conclude that habitability can not be a simple linear function of stellar lifetime, otherwise we would be in a universe where gravity was stronger.

This conclusion has an important corollary: if habitability cannot depend on stellar lifetime, we can conclude that older stars should not be more likely to host biospheres. This expectation makes explicit use of the multiverse hypothesis, and so if a future catalog of biospheres displays a correlation with stellar age, it will constitute evidence against the multiverse, at the level of 2.7σ.

---

[1] The code to compute all probabilities discussed in the text is made available at https://github.com/mccsandora/Multiverse-Habitability-Handler.

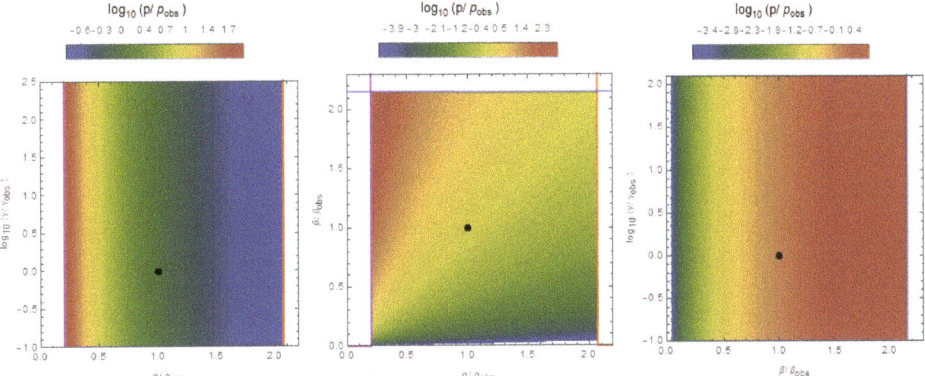

Figure 1. Distribution of observers from imposing the stellar lifetime condition. The black dot denotes the values in our universe, and the orange, blue, and purple lines are the hydrogen stability, stellar fusion, and galactic cooling thresholds, respectively, discussed in Reference [1].

2.2. Does Habitability Depend on Planetary Size?

The previous treatment of habitability was indifferent to the aspects of the planet in question. In general the habitability will depend on a great variety of factors, including size, mineral and volatile composition, amount of atmosphere and ocean, irradiance, source of internal heat, spin, orbit, obliquity, eccentricity, presence of any moons, possible secular resonances, and overall solar system architecture. Here we neglect all of these potentially important factors aside from size: the others constitute habitability hypotheses in their own right which can be incorporated into this analysis in the future. For the moment we restrict our attention to terrestrial worlds: that is, worlds capable of retaining a marginal atmosphere. In Reference [2] we discussed how this selects a relatively narrow range of planetary masses, characterized in terms of physical constants as $R_{terr} = 3.6 M_{pl}/(\alpha^{1/2} m_e^{3/4} m_p^{5/4})$. For the present purposes we assert that the fraction of stars that will have a planet of this size as independent of both the underlying parameters and stellar mass (evidence supporting this assumption can be found in, for example, Reference [8]). This will be paired with the more sophisticated treatment we undertook in Reference [2] in Section 4, but this will not affect our qualitative results.

Changing the physical constants will change the size of terrestrial planets. It is not difficult to imagine that the larger a planet, the greater the chances life will arise, essentially because of the greater number of experiments its chemical soup would be able to carry out [9]. This is bolstered by the observation that Earth, the one planet we know of that possesses life, is the largest terrestrial planet in our solar system [10]. For this, we define the habitability of a planet to be $h \propto N_{interactions} = N_{sites} t_\star / t_{mol}$, the number of chemical interactions that occur over the planet's lifetime. This weights the previous estimates based solely off lifetime by the number of active sites a planet contains.

Of course, this is a highly simplistic method of taking size into account. Much work has been done on what are termed superhabitable worlds recently [10], which asks the question of how the habitability properties may scale with, among other things, planetary size. There it was pointed out that larger worlds may very well be less habitable, because plate tectonics may not be operational, or because continents may be larger, yielding proportionally more desert regions, and so forth. Likewise, smaller planets may be expected to be less habitable because they cannot retain their atmospheres, cool more quickly, and may not possess a protective magnetic field. How planetary properties scale with size in our universe is a different question than how they scale with values of fundamental parameters, however, though the one may potentially inform the other.

The number of sites will not scale as simply as $(R_{\text{terr}}/L_{\text{mol}})^2$, however; a more nuanced analysis must be carried out. To estimate the total number of reaction sites we follow Reference [11], where the number of sites is estimated as

$$N_{\text{sites}} \sim \frac{V_{\text{clay}} \rho_A}{L_{\text{mol}}^2} \sim \alpha^{3/2} \beta^{3/4} \gamma^{-3} \quad (4)$$

Here, several quantities were used: the total amount of clay upon which chemical reactions can take place is given roughly by the average depth of clay times the surface area of the Earth. This depth is set by the same physics that yields the size of mountains, as we detail in the Appendix: it is set by equating the gravitational energy to the molecular energy, though the average depth of clay is several orders of magnitude smaller than a typical mountain, on account of the chemical bonds being much weaker. Then we have $H_{\text{clay}} \sim 0.01 H_{\text{mountain}} \sim 0.01 E_{\text{mol}}/(g m_p)$, and $V_{\text{clay}} \sim 4\pi R_{\text{terr}}^2 H_{\text{clay}}$. Note that the height of mountains scales inversely with the planetary radius, so that the number of sites is actually linear in radius. We also need the 'surface area per volume' ρ_A of typical clay, which takes into account the high fractal dimension of the mineral surface: in Reference [11] this was estimated to be 10^{-6}cm^{-1}, which in terms of physical constants we take to be set by the size of molecules, given by the Bohr radius.

Taking this hypothesis yields

$$P_{\text{size}} \propto \frac{\alpha^{1/4} \beta^{15/8}}{\gamma^3} \quad (5)$$

The distribution of observers for this is plotted in Figure 2. This gives the probabilities

$$\mathbb{P}(\alpha_{obs}) = 0.37, \quad \mathbb{P}(\beta_{obs}) = 0.11, \quad \mathbb{P}(\gamma_{obs}) = 0.01 \quad (6)$$

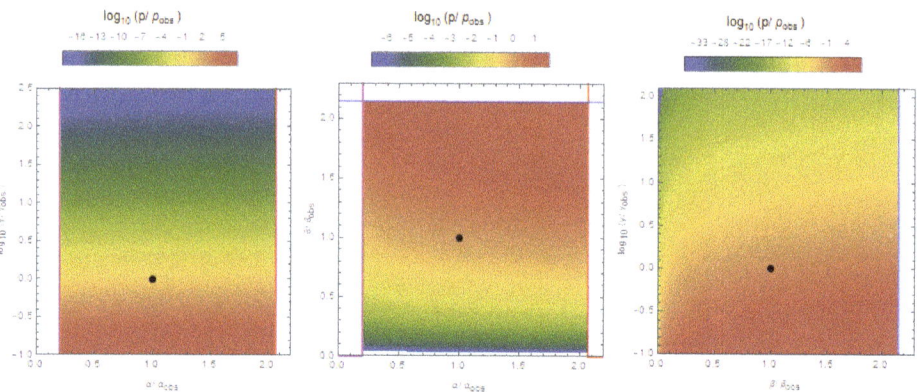

Figure 2. Distribution of observers from imposing the size condition.

This notion that the habitability of a planet should scale with its size is in conflict with what we observe, because our universe does not favor particularly large habitable planets. Here the main conflict is again due to the strength of gravity, though the dependence that plagued the criteria of our previous analyses has now been inverted, so that now extremely small values are preferred. The estimate we presented was taken using $\gamma_{\min} = 0.1 \gamma_{\text{obs}}$, though we do not find any lower bounds on this quantity from any of our considerations. From this, we find that if planet size does dictate habitability, the multiverse hypothesis will have made a wrong prediction.

However, including some of the criteria we have discussed in previous papers can ameliorate this situation. Insisting that tidally locked planets are uninhabitable raises the probability for observing our value of γ, giving

$$\mathbb{P}(\alpha_{obs}) = 0.10, \quad \mathbb{P}(\beta_{obs}) = 0.09, \quad \mathbb{P}(\gamma_{obs}) = 0.06 \tag{7}$$

This is because for small γ, the dependence is tempered by a factor of $(\lambda_{\min}/\lambda_{\text{TL}})^{4.85}$. This makes the probability proportional to $\gamma^{-1.24}$, alleviating the strong preference for smaller values. Similar results hold if the photosynthesis condition is applied as well. However, the increase in the probability of γ here is compensated by a decrease in the probability of α.

Then, the size condition may be kept with certain caveats: the size of a planet may in fact be important, but only if tidally locked planets are uninhabitable and/or photosynthesis is required. It is certainly not as clean as being able to discard this notion of habitability in its entirety, but it illustrates the state of affairs we hope to achieve. Of the laundry list of potential conditions necessary for life, only certain combinations will be compatible with the multiverse hypothesis, and if complicated conditionals must be employed in order to check consistency, then so be it.

2.3. Is Habitability Dependent on Entropy Production?

One further quantity that the development of complex life may depend on is the entropy produced on the planet per unit time, which serves as an upper limit for the rate of information processing a biosphere can hypothetically manage. On Earth, entropy production is dominated by the downconversion of sunlight to lower frequencies, which yields approximately

$$\dot{S} \sim \frac{L_\star}{T_\star} \frac{R_{\text{terr}}^2}{4 a_{\text{temp}}^2} \sim 10^{-3} \frac{\alpha^{13/2} \beta^4}{\lambda^{19/40} \gamma^{9/4}} \sim 10^{36} \frac{\text{bits}}{\text{sec}} \tag{8}$$

We have made use of the estimates for all these quantities from the appendix of Reference [1], which are stellar luminosity $L_\star$, stellar surface temperature $T_\star$, and temperate orbit a_{temp}. Note that here, we have specified to planets that orbit within the temperate zone, at which liquid water can exist on the surface. We also assume that stellar temperature is much greater than that of the planet, which holds for all main sequence stars; to extend this analysis to systems such as brown dwarfs, refinements such as found in Reference [12] should be used.

This can be compared to estimates for the total information processed by the biosphere, which was estimated as $\dot{S}_{\text{biosphere}} = 10^{39}$ bits/sec in Reference [13]. The fact that this is higher than the incident entropy production is not an indication of the violation of the second law of thermodynamics, as the authors admit that their figure is likely to be an overestimate, based off of rates measured in metabolically active bacteria cultured in the lab. If we try ourselves by using the entropy of a single bacterium $S_{\text{bact}} = 2 \times 10^{11}$ from Reference [14] and the cell turnover rate of 1.7×10^{30} cells/year from Reference [15], we find $\dot{S}_{\text{biosphere}} = 10^{34}$ bits/s, which is 1% of the total information processing available. This is in line with the result that biological information processing systems universally converge to several orders of magnitude below the theoretical limit [16], the rest being converted into waste heat. What is of note, however, is that these two numbers are indeed comparable, signaling that the ultimate size of the biosphere [17] (and ultimately technosphere [18]) is foremost limited by the amount of possible information that can be processed in its environment. If the emergence of life were dependent on the amount of information processed, rather than the number of ticks of the molecular clock, we would expect this quantity to be selected for.

The entropy production rate can also be used to determine the size of the biosphere by considering the amount of entropy produced per molecular time, $\Delta S \sim \dot{S} t_{\text{mol}}$. This was considered in References [19] and [20], where it was shown that the requirement that planets be large enough to host biospheres of sufficient complexity to contain conscious societies did not serve as a very strong constraint on physical parameters. This constraint will not be considered further here.

More appropriately, we may take the presence of complexity to be dependent on the total amount of entropy delivered to the system (as before, this obviously breaks down in its extreme limits, such as if all the entropy were delivered within a single minute). This is actually a more natural choice than just considering the amount of time a planet spends in the habitable zone, as the rate of evolution should be weighted by the overall size of the system doing the exploration [21]. In this case, we have

$$\Delta S_{\text{tot}} \sim \dot{S} \, t_\star \sim \frac{\alpha^{17/2} \beta^2}{\lambda^{119/40} \gamma^{17/4}} \sim 10^{54} \tag{9}$$

Let us also note that we have been purposefully vague as to what we are trying to encapsulate with this criterion: how can the probability of the emergence of life depend on the size of biosphere? This is a major presupposition. Rather, what we are actually computing represents the probability that a given biosphere can attain some given state, be that intelligent observers, multicellularity or whatever else. As such, this may more naturally be classified under one of the other Drake factors, such as f_{int}. Our unwillingness to commit to a definite interpretation of this quantity justifies including it in the current discussion instead.

Before using this to estimate probabilities, an important caveat must be made: the total entropy itself should not be important unless it can be utilized by living organisms. This is achieved on our planet through the process of photosynthesis, whereby sunlight is converted into chemical energy. The size of the biosphere must be conditioned on the fact that the star's light be within the chemically absorptive range, a feature that was discussed originally in Reference [22] and at length in Reference [1]. Due to this fact, the estimate for the probability of observing certain values of the constants does not attain as simple a form as our estimates above, but nevertheless can be computed,

$$\mathbb{P}_S \propto \frac{\alpha^{2.54} \beta^{3.98}}{\gamma^{2.25}} \left(\min\left\{1, 0.45 \frac{L_{\text{fizzle}}}{1100 \text{ nm}} Y^{1/4}\right\}^{9.11} - \min\left\{1, 0.16 \frac{L_{\text{fry}}}{400 \text{ nm}} Y^{1/4}\right\}^{9.11} \right) \tag{10}$$

Here, $Y = 3.19 \alpha^{-63/20} \beta^{137/40} \gamma$ and the length scales that appear delimit the wavelengths of photosynthetic light. Here, we take the optimistic upper bound taken from Reference [23] on the basis that the light be above the thermal background and the lower bound from Reference [24] to avoid photodissociation. The distribution of observers for this criterion is displayed in Figure 3.

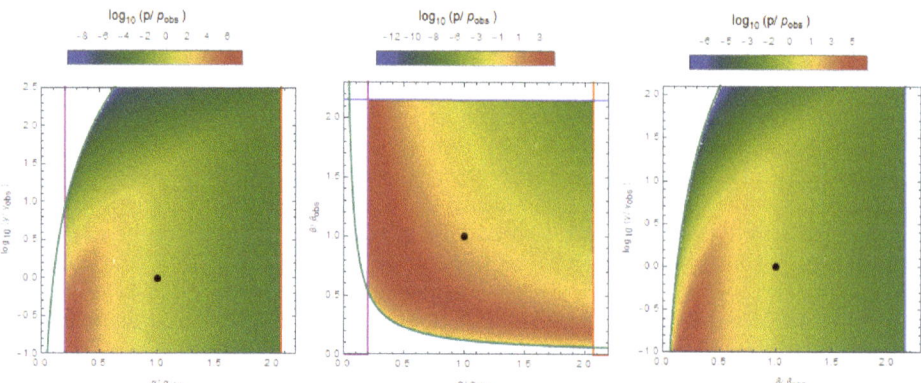

Figure 3. Distribution of observers from imposing the entropy condition.

For optimistic values of the potential photosynthetic range, $400 \text{ nm} < L < 1100 \text{ nm}$, the corresponding probabilities are

$$\mathbb{P}(\alpha_{obs}) = 0.24, \quad \mathbb{P}(\beta_{obs}) = 0.38, \quad \mathbb{P}(\gamma_{obs}) = 0.38 \tag{11}$$

This will be referred to as the 'photosynthesis condition'. For more pessimistic values of the photosynthetic range, 600 nm < L < 750 nm, which we refer to as the 'yellow condition', we have

$$\mathbb{P}(\alpha_{obs}) = 0.19, \quad \mathbb{P}(\beta_{obs}) = 0.44, \quad \mathbb{P}(\gamma_{obs}) = 0.32 \tag{12}$$

Suffice it to say, this habitability criteria is fully consistent with the multiverse hypothesis, and not very sensitive to the photosynthetic range used. It has implications for the distribution of observers that may be eventually tested: we should expect to find complex life in those locales with the most amount of entropy production. While fully determining the places this distinguishes will rely on an in-depth analysis, this would include planets that orbit more active stars, for longer, and able to collect more incident radiation. This last criterion would include planets which orbit closer to their host star and are perhaps as large as can be, within the ranges compatible with life.

These predictions bring a certain amount of subtlety, however: at first glance they seem to be in direct conflict with the results of the previous two sections, that we should expect no correlation of life with stellar lifetime or planetary size. The distinction here is that life's presence should only depend on these quantities inasmuch as they determine the entropy collected. This is not degenerate with the criteria of before, though the number of samples needed to distinguish these two scenarios is left for future work.

However well this criterion may do in explaining the observed values of our constants, it fails to account for our location within our universe on its own. This is an equally powerful test of which habitability criteria are compatible with observations, and so we now turn our attention to this as well.

2.4. Why Are We Around a Yellow Star?

With the inclusion of the entropy production criteria, we have a notion of habitability that makes our observed values of the three microphysical parameters we focused on consistent with the multiverse model. Now, it is necessary to include local observables to further test the consistency of this criterion: namely, if the probability of life arising around a star is proportional to the total amount of entropy it produces over its entire lifetime, we must ensure that this is compatible with our presence around a star such as our sun. This consideration is capable of yielding extra information about where we should expect life to be in our universe: since smaller stars produce significantly more entropy over their lifetimes, we should expect some cutoff not too far below 1 solar mass where complex life cannot develop.

Restricting our attention to within our universe, then, we may ask what the probability is that we find ourselves around a star of one solar mass. This has been the subject of recent investigation, for instance in References [25–27]. Of course, this places us in the undesirable situation of trying to conduct a statistical analysis based off of a sample size of one, and with heavy selection effects at that. A more robust question would be to predict the distribution of biospheres as a function of stellar mass, which can test the model more concretely. Until we are technologically able to perform such measurements, however, we focus on the immediately accessible question. For a generic definition of the habitability of a system, the probability of being around a solar mass star or larger is

$$P(M_\odot) = \frac{\int_{\lambda_\odot}^{\infty} d\lambda \, p_{\text{IMF}}(\lambda) h(\lambda)}{\int_{\lambda_{\min}}^{\infty} d\lambda \, p_{\text{IMF}}(\lambda) h(\lambda)} \tag{13}$$

For the simplest habitability hypothesis that all stars are equally habitable, we find that $P(M_\odot) = 0.14$, since approximately 14% of stars are larger than the sun (in agreement with contemporary surveys and Reference [25]). This is a perfectly reasonable account for our current position within our universe. However, we remind the reader that it failed miserably at accounting for the values of the constants themselves. If we instead use the entropy condition, the probability of being around a smaller star is weighted much higher: $h(\lambda) \sim \lambda^{-3}$. This is a direct consequence of the lower temperature and especially the longer lifetime of small mass stars. If this habitability hypothesis is used, we instead find that $P(M_\odot) = 0.02$, so that only 1 in 50 civilizations should expect to be around a star this large (not to mention this early [26]). Thus, neither habitability hypothesis can simultaneously explain our position in our universe and within the multiverse itself.

However, these are not the only two notions of habitability we have encountered- far from it. If we include the 4 new possibilities along with the 480 from References [1,2], this brings the total to 1920 separate habitability criteria to test. Since the aim is now to explain why we do not live around a smaller star as well as why we live in this universe, we focus on those criteria that penalize small mass stars. From before, we had three of these: considering tidally locked planets to be uninhabitable rules out stars below $0.85 M_\odot$, if convective stars are uninhabitable the minimum is $0.35 M_\odot$, and if only yellow light can be photosynthetic the minimum is also $0.85 M_\odot$. The probabilities for each of these are displayed in Table 1. Of these potential explanations, the convective criterion does nothing to alleviate the problem, since the cutoff is below even the most optimistic photosynthetic mass. The other two hypotheses are understandably similar, since they introduce the same low mass cutoff. They are not identical because the yellow criterion also introduces a high mass cutoff at $1.3 M_\odot$, but both of these work even better than the equihabitable criterion.

Table 1. Probability of orbiting a star larger than or equal to our sun with the various habitability hypotheses. Whenever the entropy condition is used, the photosynthesis condition is also employed, except for the 'none' and 'yellow' rows. Since the size and nutrient flux conditions have the same dependence on λ as the stellar lifetime condition, they all have the same probability of being around the sun.

Criteria	$P(M_\odot), h \propto 1$	$P(M_\odot), h \propto S$	$P(M_\odot), h \propto t_\star$
none	0.142	1.3×10^{-4}	4.3×10^{-4}
TL	0.835	0.528	0.570
convective	0.345	0.024	0.030
photo	0.308	0.024	0.038
yellow	0.585	0.424	0.449

When we considered the effects each of these criteria on the multiverse probabilities in Reference [1], we found no strong preference for whether to expect stars of these sorts to be habitable. Now that we incorporate additional criteria, however, they become crucial. This is due to the fact that because we place a strong preference on high entropy production, this favors stars that produce more than our sun. We need some sort of reason, then, why low mass stars are inhospitable. While the presence of convective flares, tidal locking, or absence of photosynthetic radiation are all reasonable hypotheses, only the latter two are coherent explanations. While we cannot uniquely specify the reason for the inhospitability of low mass stars, we end up with the prediction that either life cannot thrive on tidally locked planets or that photosynthesis is only possible with yellow light (or both). Flare stars may be uninhabitable too, but this does not constitute as good an explanation of our star's mass as it first appeared to.

We also explore the possibility that while a higher entropy production will be more conducive to the development of life, at some point the dependence must turn over, as the probability of development saturates to a near certainty. This may be encapsulated in the trial function $h(\lambda) = 1 - e^{-\Delta S(\lambda)/S_0}$. If S_0 is large compared to all produced stellar entropies considered, this recovers the analysis from before, whereas if S_0 is small the probability is essentially 1. Intermediate values interpolate between these extremes. One may think that if the value of S_0 is close to the solar value for whatever reason, this may naturally explain our presence in this universe without the need to invoke a large value for the smallest habitable star. The probabilities of our constants, as well as of being around a sunlike star, are plotted in Figure 4 as a function of S_0, where we find the interpolating behavior as advertised: for small S_0, it tends toward the photosynthesis criterion, which has the probabilities $\mathbb{P}(\alpha_{obs}) = 0.44$, $\mathbb{P}(\beta_{obs}) = 0.18$, and $\mathbb{P}(\gamma_{obs}) = 8.4 \times 10^{-7}$, whereas for large S_0 it tends toward the entropy condition. Intermediate values fail to simultaneously account for $P(M_\odot)$ and $P(\gamma_{obs})$, with one of these quantities being below 6% for any choice of S_0. So, while there may very well be some amount of entropy production that almost guarantees that life will arise, there is no reason to expect that it is anywhere close to the amount so far produced by the sun, and it plays no role in explaining the mass of our star.

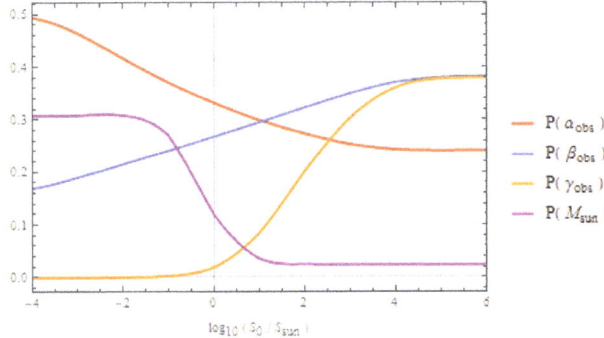

Figure 4. Probabilities with differing interpolation entropy, assuming the entropy production and photosynthesis criteria. This interpolates between the photosynthesis criterion for small S_0 and linear scaling for large S_0.

2.5. Is the Biosphere Entropy or Material Limited?

Above, we have shown that treating the biosphere as set by the total amount of entropy produced yielded the best account for our position within this universe, and have quoted several studies suggesting that this may indeed place the ultimate limit on biosphere size. However, there is plenty of reason to be skeptical of this claim: often ecosystems are instead resource limited on Earth [28], and are expected to be elsewhere as well [29]. Much of the discussion on the total primary productivity is centered on exactly which nutrient is the limiting factor for growth [30].

This being said, there are a few indications that it is indeed entropy that sets the ultimate limit of the size of the biosphere. While if one nutrient is found to be scarce it can be recycled many times, there is no real method for recycling light energy. Indeed, phosphorus, which is often a limiting factor, can be recycled as many as 500 times before leaving the biosphere [31]. Plankton have the ability to substitute many of the trace metals for each other in their various enzymes to take advantage of any local imbalance, and it has been found that the availabilities of each nutrient are roughly equal [32]. This colimitation is a natural outcome of life adjusting its activity to bolster its utilization of one resource, until the point where such optimization would no longer be beneficial. The fact that one of the colimiting factors can be light has been demonstrated to occur in subarctic ecosystems, where a concomitant increase in irradiance and iron flux lead to the largest amount of phytoplankton growth [33]. This indicates that though the tropics have more than enough light energy to sustain the same level of productivity year round, the balance between nutrients and light are roughly comparable. Thus, the biosphere seems to push recycling capacity until it hits the hard limit, dictated by the total amount of entropy that can be harnessed.

That being said, here we adopt the traditional stance that the size of the biosphere is limited by nutrient flux. In this scenario, the total mass of living organisms is set by the rate at which material that is weathered. The details of how to estimate this are relegated to the Appendix, but the result is

$$\Delta C_{tot} \sim 10^3 \, \epsilon_C \, \frac{\alpha^{9/2} \beta^{1/2}}{\lambda^{5/2} \gamma^3} \tag{14}$$

The biosphere size that can be supported depends on the actual residence times of each nutrient, which depend on geochemical and hydrological factors that we do not attempt to model here. Instead, we must content ourselves with parameterizing this in the efficiency factor ϵ_C for the time being, trusting that the overall scaling will not be altered by too much. The scalings in this quantity are not too different from the entropy limited case in Equation (9). However, this criterion does not perform as well: if we impose that photosynthesis is necessary in this case as well, to facilitate comparison (as well as ensuring that the probability of orbiting a sunlike star is not too low), the probabilities are

$$\mathbb{P}(\alpha_{obs}) = 0.09, \quad \mathbb{P}(\beta_{obs}) = 0.15, \quad \mathbb{P}(\gamma_{obs}) = 0.07 \tag{15}$$

These are uniformly worse than the entropy limited scenario. Using a 'law of the minimum' criterion, with $f_{bio} \propto \min(\Delta S_{tot}, \Delta C_{tot})$ interpolates between these two scenarios, based on the value of ϵ_C. The best fit for this class of models is for ϵ_C to be large enough that the biosphere is entropy limited, which is precisely what we have argued that life would strive for anyway.

2.6. Does Life Need Plate Tectonics?

We now turn to another planetary property that may be crucial for life: plate tectonics. Though it may come as a surprise to those who have not encountered it before, plate tectonics is considered by many geologists to be essential for life on Earth for at least three reasons: first, subduction is responsible for creating the granite which comprises the continents today, and so is ultimately responsible for producing practically all land surface on Earth [34]. Secondly, even life that does not live on land ultimately is built out of materials that are eroded from the Earth's mountain ranges [35]. Thirdly, and perhaps most importantly, the silicate weathering that takes place as a result provides an additional negative feedback loop for the amount of carbon in the atmosphere, which regulates the temperature over geological timescales to a much higher precision than would have occurred otherwise [36]. In short, plate tectonics provides a "living rock" for which the stage of life is set. However, some authors, such as those of Reference [37], consider that plate tectonics may not be crucial for maintaining planetary habitability over long timescales, making this habitability criterion subject to debate.

There is every indication that plate tectonics is 'hard' to achieve, and does not seem to be the typical state for rocky planets. Firstly, none of the other rocky bodies in our solar system have plate tectonics [38]. Additionally, its presence seems to have been facilitated by a number of compounding factors: the presence of liquid water greatly increases the ductility of the crust and mantle, enabling subduction [39]. Life itself may play a critical role in speeding up the process by enhancing erosion and deposition of carbonate [40]. The fact that it relies on two different sources of internal heat, both primordial and radioactivity, of roughly equal contribution [41], could be construed as the hallmark of a selection pressure. Taken together, these strongly argue that plate tectonics is the exception rather than the rule, and is accomplished on Earth only by a plethora of independent helping factors.

All this is further compounded by the interesting coincidence: there is a narrow range of planetary masses for which plate tectonics exists, between $0.7 - 2R_\oplus$ [42,43]. This is determined by the tuning that the convective stress of the mantle appropriately balances the lithospheric yield stress of the crust. Since mantle convection is dictated by the amount of heat contained, it is a function of the planetary mass: too small, and the crust is locked in a stagnant lid regime, too large, and it is molten. This becomes all the more intriguing when it is noted that this narrow range happens to precisely coincide with the equally narrow range of planetary masses permitting an Earthlike atmosphere, between $0.7 - 1.6R_\oplus$ [44]. This mass range is usually taken to be important as well, under the auspices that both Marslike and Neptunelike planets are inhospitable to complex life requiring the presence of liquid water, so far as we can tell. It seems a remarkable coincidence that these two narrow windows just so happen to coincide, given the many orders of magnitude of planetary masses.

This coincidence was investigated in detail in Reference [45], where it was found that, since the radioactive heat is generated by alpha decays, which are tunneling processes, their lifetime depends exponentially on the fine structure constant (and to a lesser degree on the other parameters). If α were increased to a value of 1/136, all possibly relevant alpha decays would occur with half lives of less than Gyr timescales, and so would have decayed by this point, leaving the Earth cool and stagnant. If decreased beyond a value of 1/153, the typical timescale would be much larger, making all radioactive compounds effectively stable. Thus, if radiogenic plate tectonics is deemed important for life, the range of allowable α is considerably narrowed. What is more, the observed value of 1/137 is extremely close the the maximal value, indicating a strong preference for large α. If plate tectonics is crucial, we expect a habitability criterion that reflects this, by exhibiting strong preference for large α.

We have systematically combined the plate tectonics condition with all our previous habitability criteria to determine which combinations are consistent with the multiverse hypothesis. We report a few: if the yellow and entropy conditions are included along with the plate tectonics condition, we have the probabilities

$$\mathbb{P}(\alpha_{obs}) = 0.064, \quad \mathbb{P}(\beta_{obs}) = 0.38, \quad \mathbb{P}(\gamma_{obs}) = 0.20 \tag{16}$$

If we include the photosynthesis, entropy and tidal locking condition, we have

$$\mathbb{P}(\alpha_{obs}) = 0.063, \quad \mathbb{P}(\beta_{obs}) = 0.50, \quad \mathbb{P}(\gamma_{obs}) = 0.29 \tag{17}$$

This is not an exhaustive list: we are reaching a point where it becomes untenable to report every criterion in table format, even when restricting to those above some threshold, and so we will release the full list online as supplemental material at publication. Rather, these three representatives form germs: combinations of habitability criteria that all additional successful hypotheses will contain. If one additionally includes the convective, biological timescale, terrestrial mass, or temperate conditions to any of the above, the probabilities will be shifted slightly but the overall conclusions will still hold. This class of criteria can be said to be indifferent to these additional hypotheses.

The first thing to note is that the probability of observing our value of the fine structure constant is always diminished, since it is still rather close to the anthropic boundary. This makes the Bayesian evidence for the necessity of plate tectonics around 3–6 times weaker than for the hypothesis that it is unnecessary, which is not quite low enough to exclude this scenario.

There are a number of subtleties in the interpretation of this, however. Firstly, it is unclear whether this indicates that plate tectonics itself should be unimportant for complex life, or whether radioactivity is ultimately unimportant for plate tectonics. If the former, then we should expect to find just as much life of planets that do not support plate tectonics, be they too dry, small, large, or stiff. If the latter, then we will no doubt discover planets with perfectly active plate tectonics that are not as enriched in radioactive isotopes as ours, be that from the circumstances of their birth environment, or possibly their age.

We stress again that it will be impossible to ultimately derive a version of habitability that is uniquely compatible with the multiverse hypothesis, robust against the future inclusion of additional considerations. What we can hope to achieve is the enumeration of all possible notions that are compatible with the multiverse, and the eventual determination of which is true. Should the single true condition match any of these, it can be taken as compatible with a multiverse, and should any of the independent components of this ultimate criterion fail, it will be strong evidence against.

3. Why Did Life Procrastinate So Long?

In the previous section, we found habitability criteria that make our observed values of the constants typical. In order to be fully consistent, however, it was necessary to include an additional ingredient, the probability that we find ourselves in our particular location within our universe. Likewise, we may ask a similar question, the probability of finding ourselves at our particular time. Maintaining that our notion of habitability ought to account for this as well is shown to be equally constraining, and can allow us to make inferences about the distribution and frequency of life throughout our universe that we would not have been able to deduce without the added input of the multiverse hypothesis.

Namely, the conundrum we address now is the question of why we find ourselves so close to the end of our star's habitable phase, when the timescales of biological and stellar evolution are not obviously related. Several different models have been put forward to account for this coincidence, all of which recast it as an artefact of a selection effect. The hard step model posits that the evolutionary path to intelligence required a half dozen or so incredibly difficult innovations which individually each have a very small probability of occurring within a stellar lifetime [46]. The bated breath model, on the other hand, allows that the ratio of these two timescales is a steep function of stellar mass, and so naturally most observers would arise around the smallest stars capable of giving rise to intelligent observers [47]. The easy stroll model holds that intelligence is rather reliably developed after a certain period of time, but that local planetary conditions cause the distribution of habitable lifetimes be be very steep [48].

These three models will be considered in turn. While there is no way to distinguish which is right on an observational basis at the moment, we take a different approach and ask what the compatibility of each is with the multiverse hypothesis. The first two will be shown to be incompatible with the multiverse, causing us to greatly favor the third. These will ostensibly be tested in the conventional sense eventually, allowing us to compare the prediction we make with observation.

Note that strictly speaking, the contents of this section deal more directly with the f_{int} term in the Drake equation, which is the probability that a planet that has already developed life gives rise to intelligent observers. Though this factor will be the main subject of our follow-up work [49], we include this discussion here anyway.

3.1. Would We Live in This Universe if the Hard Step Model Is True?

The first hypothesis we consider is the hard step model, originally proposed in Reference [46]. Its tenet is that the emergence of intelligence requires a small number of very hard evolutionary innovations, each of which typically take much longer than the 5–10 Gyr timescale of stellar evolution. This scenario was studied in Reference [50], where it was found that because we are roughly 4/5 of the way through the Earth's habitable phase, the best fit value of the number of steps is 4, though within 95% confidence the possible range is between 1–16 [27]. A biological perspective was applied to try to identify what these steps could be in Reference [51] on the basis of reorganizations of information processing, and is consistent with this number, and including the distribution of these other purported hard steps in time bolsters the agreement with this model. The hard step model was combined with stellar activity models to deduce that life should be most probable around K dwarfs in Reference [52]. One important consequence of this model is that intelligent life should be quite rare in the universe, since it relies on a sequence of improbable events.

According to this model, the probability of intelligence arising on a planet after a time t is

$$f_{int}(t) = \left(\frac{t}{T}\right)^{n_{hard}} \tag{18}$$

For definiteness we take $n_{hard} = 4$ throughout, and the timescale T is taken to be much larger than any other that appears in the evolution of the system. In the following we may use in this expression any of the parameterizations for time which we considered: either strictly proportional to time elapsed, or else weighted by the size of the planet in question, or the size of the biosphere. This defines the pure hard step model, but a more general version will be considered after.

It is simple to see that this is incompatible with the multiverse: roughly speaking, since it greatly favors stars with the longest possible lifetimes, we would be ten thousand times more likely to inhabit a universe where stars last just ten times as long. This only exacerbates the problems we found when we considered the probability of the emergence of life to be linearly dependent on the total time. Without weighting by entropy, the probabilities we find are

$$\mathbb{P}(\alpha_{obs}) = 0.49, \quad \mathbb{P}(\beta_{obs}) = 0.009, \quad \mathbb{P}(\gamma_{obs}) = 1.1 \times 10^{-5} \tag{19}$$

Additionally, it was pointed out in Reference [27] that if the hard step model is employed, our chances of orbiting a yellow star are greatly reduced except in the case where tidally locked planets are considered uninhabitable. Accordingly, we find that $P(M_\odot) = 1.4 \times 10^{-12}$ without the tidal locking criterion, and $P(M_\odot) = 0.18$ including it. Our numbers differ from their analysis because there a more sophisticated measure of how the stellar lifetime scales with mass was used, but our simplified parameterization is sufficient to prove the point.

Since previously we had more success considering that the emergence of life should be not just dependent on the time available, but also weighted by the size of the biosphere, we may try this here, to see if it fares any better. We find that this modified version of the hypothesis $f_{int} \propto \Delta S_{tot}^{n_{hard}}$ is even more problematic, yielding

$$\mathbb{P}(\alpha_{obs}) = 0.12, \quad \mathbb{P}(\beta_{obs}) = 0.044, \quad \mathbb{P}(\gamma_{obs}) = 2.2 \times 10^{-9} \tag{20}$$

The size condition does even worse than these, giving probabilities which are indistinguishable from 0 to the 16-point numerical precision to which we work. This is so far the worst suite of hypotheses considered.

3.1.1. Can the Hard Step Model Work if We Are Close to the Turnover Scale?

We have seen that the hard step model as specified is drastically incompatible with the framework we are employing. The other extreme, the equihabitable condition, works much better, but this fails to explain the coincidence that it has taken approximately the full duration of the habitable time for intelligent life to arise on Earth. Before discarding this model completely, we may ask whether these two failures can be reconciled by acknowledging that they both are limiting cases of a more general probability distribution for life to arise.

Let us illustrate this in the 1 step case first, for simplicity: then the probability for life to arise on a suitable planet after a time t is

$$c_1(t) = 1 - e^{-t/t_1} \tag{21}$$

As can be seen, for times much shorter than the intrinsic timescale of this distribution t_1, this recovers the linear dependence we saw previously. As t becomes larger, the probability that life would have emerged at some point becomes more certain, eventually saturating at 1. This more general distribution then interpolates between the two cases we considered before, with the expense of adding an additional parameter.

This can be generalized to n steps, by noting that the probability density function (giving the chances of a step to occur at a given moment in time) is given recursively by the formula $p_n(t) = \int_0^t dt' \, p_1(t-t') p_{n-1}(t')$. This yields an expression for the cumulative probability:

$$c_n(t) = 1 - \sum_{i=1}^{n} t_i^{n-1} Z_i e^{-t/t_i}, \quad Z_i = \frac{1}{\prod_{j \neq i}(t_i - t_j)} \tag{22}$$

In the limit that the time is much shorter than all timescales in this expression, this asymptotes to

$$c_n(t) \to \frac{t^n}{n! \prod_{i=1}^{n} t_i}, \tag{23}$$

which reproduces the hard step model in Equation (18) (where the factorial had been absorbed into the definition of T), and asymptotes to 1 in the opposite limit. It has the additional feature that it approximates an *m* step model if *m* of the times are much greater than the timescale in question, the others much shorter. This is a consequence of the mathematical formulae

$$\sum_i Z_i t_i^k = \begin{cases} (\prod_i t_i)^{-1} & k = -1 \\ 0 & 0 \leqslant k < n-1 \\ 1 & k = n-1 \\ \sum_i t_i & k = n \end{cases} \tag{24}$$

and allows us to treat the number of critical steps as a sliding scale that depends on the time frame in question. Thus, the probability for life to emerge, for instance, on a Mars or Venus like planet, which went through a brief habitable phase in the beginning of the solar system, would not be a simple extrapolation of the 4 step model that appears to govern life on Earth, but instead would be given by a much larger number of steps. Evolutionary innovations that are trivial on the scale of millions of years become insurmountable when you only have an afternoon.

Before discussing the complication of how the timescales in this distribution are chosen, we make the simplification that they are all equal to a common timescale T. Then, the probability attains a highly simplified form

$$c_n(t) = \frac{\gamma(n, t/T)}{\Gamma(n)} \tag{25}$$

where $\gamma(n, x) = \int_0^x ds\, s^{n-1} e^{-s}$ is the lower incomplete gamma function. With this distribution, the expected time for the emergence of intelligent life on a planet with habitable duration t_{hab}, conditioned on the fact that the event does occur, is

$$t_{\text{int}} = \frac{\gamma(n+1, t_{\text{hab}}/T)}{\gamma(n, t_{\text{hab}}/T)} T \to \begin{cases} \frac{n}{n+1} t_{\text{hab}} & t_{\text{hab}} \ll T \\ n\, T & t_{\text{hab}} \gg T \end{cases} \tag{26}$$

Let us discuss the behavior of this model: its features can be roughly summarized by saying that for stars with lifetimes greater than T, the probability of life is a constant, whereas for those with lifetimes less than T it is suppressed. We then must integrate over the distribution of stars to arrive at the probabilities for this habitability hypothesis. However, this always yields inconsistent results, which interpolate between the pure hard step model of Equation (20) and the photosynthesis criterion, both of which are in conflict with the multiverse. So perhaps unsurprisingly, given the results of the previous section (which would correspond to the $n_{\text{hard}} = 1$ model), taking the entropy produced by the sun as close to the threshold to guarantee that life arises does nothing to rescue the hard step model's incompatibility with the multiverse hypothesis.

3.1.2. Disparate Timescales

Previously, we made the simplification that all critical step timescales were the same, in order to simplify the expressions needed. This is certainly an unwarranted approximation; here we rectify this, and show that there is even more reason to disfavor this model.

What is needed is the underlying distribution of timescales for biological innovations to take place. This can be determined by extrapolation: since life on Earth has developed a whole suite of innovations throughout its history, statistics can be performed on the relatively more mundane ones that took place, and used to determine the underlying probability for an innovation to take a given amount of time. We use the list compiled in Reference [53], where the origination of 60ish innovations of higher organisms are tabulated. Taking rank-order statistics of the time difference between successive innovations, as displayed in Figure 5, yields a cumulative distribution function consistent with a power law of slope 1/4,

$$c(t) = \left(\frac{t_{\text{cut}}}{t}\right)^{1/4} \tag{27}$$

Restricting to our lineage instead leads to a slope of 1/2, but the difference between these two is inconsequential, so we specify to the former. This also requires a cutoff timescale t_{cut}, which specifies what is to be considered a 'hard' innovation. In the following, we take $t_{\text{cut}} = 50$ Myr, but the results are rather insensitive to the exact number used. With these choices, the number of hard steps will be given by a binomial distribution

$$p(N_{\text{hard}}) = \text{Binomial}\left(\frac{4}{c(t_\oplus)}, c(t_\oplus)\right) \quad (28)$$

So that the expected number of hard steps will be $\langle N_{\text{hard}} \rangle = 4 \pm 2\sqrt{1 - c(t_\oplus)}$, the variance being $\sigma_{N_{\text{hard}}} = 1.65$ for our choices. We have normalized the mean to be the most likely value for our Earth system, for definiteness.

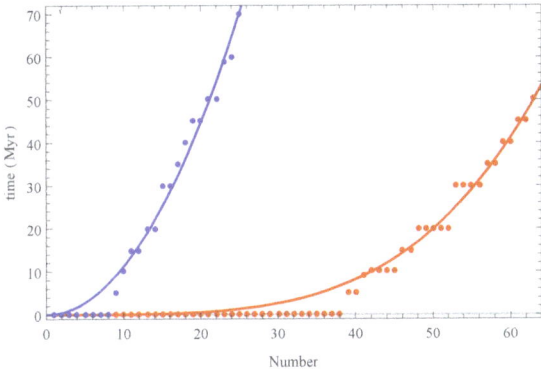

Figure 5. The observed distribution of innovation timescales, along with the best fit curves. Here, blue specifies to innovations occurring in our lineage, and orange to innovations across all lineages.

This raises an important criticism of the hard step model: if the timescales are all independent identically distributed variables drawn from this random distribution, then it is fairly likely for a random instantiation to have relatively fewer hard steps. This must then be coupled to the steep suppression of the emergence of life on systems with a larger number of critical steps. In this view, we would be overwhelmingly likely to have developed on a planet with an unusually small number of critical steps, rather than arising as one of the extraordinarily lucky representatives with an average number of steps. This argument is independent of multiverse reasoning: as long as there is any distribution for the hardness of evolutionary innovations, it is much easier for a planet to be accidentally easier at facilitating the development of intelligent species than it is for intelligence to arise on an average planet. Based off this consideration alone, we should not expect the hard step model to be a good description of the distribution of intelligent life within our universe. Of course, some innovations, such as the evolution of oxygenic photosynthesis and eukaryogenesis, may belong to a qualitatively more difficult class, in which case the hard step model preserves its explanatory character.

3.2. Was Complex Life Waiting for Earth to Oxygenate?

A second popular explanation for why complex life only started around 600 Mya is that it was necessary for the Earth's atmosphere to fully oxygenate first [54]. This increased the energy available for metabolic processes, which subsequently became indispensable for animal life, and also created an ozone shield that made the colonization of land possible. In this framework, the reason complex life took billions of years to get started is simply due to the fact that that is how long it took the Earth to oxygenate. Upon inspection, though, this explanation raises the secondary conundrum as to why it takes almost precisely the stellar lifetime for this to occur; this has not been answered conclusively.

Here we detail several leading mechanisms for what triggered this, and how they fit in with anthropic selection pressures.

Though photosynthetic life is usually implicated in the oxygenation of Earth, this alone would not contribute a significant amount of oxygen if counterbalancing oxygen sinks adjusted to absorb it [55]. Additionally, the innovation of oxygenic photosynthesis cannot be the complete mechanism, as there is evidence [56] that this evolved at least 500 Myr before the first increase in oxygen levels 2.2 Gya. Even after the original oxygenation event, levels stalled at perhaps 1% of current levels until 600 Mya, when oxygen levels reached roughly their present values [57], so the long delay must have been due to significant oxygen depletion that kept pace with the biological production rates [58]. This depletion is to be expected, as at the time both the Earth's crust and atmosphere would have been strongly reducing, soaking up any oxygen that was produced. It was only after enough had been injected into the system that these oxygen sinks would have depleted, finally allowing for a buildup of the gas to form. These sinks can be classified into two broad types: either the reductants would have been depleted by being drawn down, into the Earth's mantle, or up, into space. The fact that the relative rates of each of these processes is uncertain to within an order of magnitude [59] leads to uncertainty over which was dominant. We consider each in turn.

3.2.1. Drawdown

The most standard explanation for the oxygenation time is that the Archean Earth was reducing, with plenty of raw iron and sulfur in the crust that would have immediately neutralized any excess oxygen that appeared in the atmosphere [60]. It was only once this initial reservoir was depleted that oxygen could build up to its current state. In order to estimate the time it would take for this transition to take place, we can compare the rate of drawdown to the total mass of the atmosphere to find

$$t_{O_2\downarrow} \sim \frac{M_{atm}}{\Gamma_{down}} \qquad (29)$$

If we know how both of these quantities scale with parameters, we can use this to determine whether this timescale is naturally of the order of the stellar lifetime. Of the two, the drawdown rate is on relatively firmer footing; this is because there is no clear explanation for the mass of the Earth's atmosphere.

There are various schools of thought as to why the mass of our atmosphere should be $M_{atm} \sim 10^{-6} M_\oplus$. Even within our own solar system, there is a very large spread in the ratio of atmospheric mass to planet mass, so atmosphere is likely to be highly variable and sensitive to local conditions [61]. Given that the source of Earth's atmosphere seems to have important contributions both from planetary outgassing and delivery from asteroids or comets [62], the distribution is highly uncertain. This uncertainty is compounded by the fact that we still do not know what the range of habitable atmospheric masses is. Here, we will follow the expectation that there is a narrow window of habitable atmospheric masses, based off the pressure at the surface.

The total mass of the atmosphere can be related to the pressure at the surface by the expression $M_{atm} = 4\pi R_{terr}^2 P_{atm}/g$. Noting that the surface pressure of Earth is two orders of magnitude larger than the minimum required for the presence of stable liquid water, $P_{trip} = 0.006$ atm, we use this as a guiding value to set the atmospheric mass necessary in alternate universes. In fact, this could easily be a natural state of affairs: the closer the atmosphere is to the triple point, the more susceptible it would be to climate fluctuations that could lead to a runaway icehouse or greenhouse scenario. We do not pursue this line of reasoning quantitatively here, but use it to argue that it is plausible that the smallest possible atmosphere would be an order of magnitude or two above the triple point of water.

It still needs to be explained why we would be situated near the minimal value, if larger atmospheres are favored on climate stability grounds. One possibility would be that the atmospheric mass distribution could be very steep, greatly favoring smaller values. Alternatively, it may be that smaller atmospheres are more conducive to life. Earth's usual biochemistry ceases to operate in regions of extreme pressure. This is especially true of the lipid chemistry that is essential for the functioning of cell membranes- at high pressures, the membrane stops behaving as a two dimensional fluid surface [63]. On the other hand, the piezosphere, the regions of the Earth with unusually high pressure such as deep in the ocean and within the crust, is not completely devoid of life. There, piezophiles have adapted to their environment by using unsaturated fatty acids, which are more 'knobby' that the simple rod shaped saturated ones normally employed, as to resist jamming [64]. These types of adaptations even allow for complex animal life deep within the Mariana trench. There is certainly less biological activity in these realms, but it is challenging to attribute this to the extreme pressures, as these regions also have reduced nutrient flux, lower light, and lower temperature [65]. In light of this, it is not clear that a significantly larger atmosphere would pose much of an evolutionary challenge to life. However, perhaps it would lead to a significant lengthening of evolutionary timescales, and this is the explanation for the atmospheric mass we observe. In this case, the range of habitable atmospheric masses would be quite sharp, and so we would be justified in fixing $P_{atm} \sim 100 P_{trip}$. We follow this route in this paper, though it certainly would be interesting to explore these ideas about atmospheric mass more fully, and the implications they have for the distribution of both atmospheres and life throughout the universe.

This then raises the question of how the triple point of water depends on fundamental constants. The solid-liquid transition is practically independent of pressure, and occurs when the temperature is high enough to excite vibrational modes of the water molecules. This energy is given by $E_{vib} \sim T_{mol} = 0.037 \alpha^2 m_e^{3/2}/m_p^{1/2}$. The water-gas phase curve is given by the Clausius-Clapeyron equation as $P(T) = P_0 e^{-L/T}$. The latent heat of evaporation is equal to the intermolecular binding energy, $L \sim \alpha/r_{H_2O}$. The coefficient P_0 is an integration constant that in principle can be derived from a statistical mechanical treatment capable of yielding an exact expression for the chemical potential of liquid water, but the author is not aware of progress in this direction, so the phenomenological value $P_0 = 1.3 \times 10^5 T_{mol}/r_{H_2O}^3$ will be used instead. This aesthetic choice will not strongly affect the calculation, since the exponential dependence plays the dominant role in the scaling with parameters. Using $r_{H_2O} = 5.9 a_0$, where a_0 is the Bohr radius, we find

$$P_{atm} = 22.8 \frac{\alpha^5 m_e^{9/2}}{m_p^{1/2}} e^{-0.44 \sqrt{\frac{m_p}{m_e}}} \tag{30}$$

From here, the rate of drawdown of O_2 from the oxidation of eroded material must be calculated to determine the timescale. We use Equation (A10) from the Appendix, which for our current atmospheric mass gives several Gyr as an oxidation time.

In relation to the lifetime of the host star, this yields

$$\frac{t_{O_2 \downarrow}}{t_\star} = 341 \lambda^{5/2} \alpha^{-3} \beta^{1/4} e^{-0.44 \beta^{-1/2}} \tag{31}$$

This is normalized to be 1 Gyr for sunlike stars for definiteness, giving a ratio of about 0.2, in agreement with Reference [66]. Demanding that this quantity be less than 1 for a planet to be habitable gives an upper bound on the mass of the star, which is in contrast to the requirement that arises if the oxidation were due to escape to space.

The aim of this model was to explain the ratio of the two timescales, despite their drastically different physical origins. Even though this scenario favors large mass stars, the scaling is very close to the Salpeter slope in the initial mass function, so that in the equihabitable scenario the observed value is not so unlikely. Even in the entropy-weighted scenario, this ratio can usually be close to 1, provided that there is a cutoff in habitable stellar masses close to the solar value. We quantify this in two ways: the first is $P(r_t > 0.2)$ restricting to our observed values of the physical constants, and the second is $\mu_{r_t} \pm \sigma_{r_t}$, again within our constants. In Table 2, we find that with either statistic these ratios are quite compatible with observation. Of note is that the first is essentially equivalent to the probability of orbiting a star of our sun's mass, as long as no high mass cutoff is introduced, as in the biological timescale criterion.

Table 2. Statistics for the ratio of oxygenation time to the stellar lifetime, both within our universe (subscript U) and in the multiverse (subscript M). All figures are computed with the entropy+yellow criteria. Even though they all satisfactorily explain our observed ratio of about 0.2 when restricted to our universe, none of them explain why we do not live in another universe where the average ratio is much smaller. Note that these values are optimistic: if the observed ratio is taken as being any higher, the probabilities will be even lower.

Mechanism	$P(r_t > 0.2)_U$	$(\mu_{r_t} \pm \sigma_{r_t})_U$	$P(r_t > 0.2)_M$	$(\mu_{r_t} \pm \sigma_{r_t})_M$
drawdown	0.38	0.22 ± 0.07	0.03	0.03 ± 0.08
drawup	0.49	0.24 ± 0.14	0.05	0.04 ± 0.11
combined	0.56	0.20 ± 0.04	0.05	0.004 ± 0.10

This appears to be a valid explanation of why life took so long to develop. However, if we employ the multiverse hypothesis, we can add additional statistics by letting the physical constants vary. There, we find that this ratio is actually much smaller than one in a sizable fraction of universes. The majority of universes, in fact, we find will oxygenate their planets much more quickly than ours, producing a paradox of why we would have ended up in this one.

In contrast, the oxygenation delay does not appreciably alter the probabilities of being in our universe, as displayed in Table 3. So, in contrast to the other hypotheses considered in this paper, this one is incompatible with the multiverse on the basis that its purported explanatory powers are undermined when the two ideas are combined. This will hold for the alternative oxygenation mechanism as well, as we now discuss.

Table 3. The probabilities with various oxygenation mechanisms. All utilize the entropy + yellow criteria.

Mechanism	$\mathbb{P}(\alpha_{\text{obs}})$	$\mathbb{P}(\beta_{\text{obs}})$	$\mathbb{P}(\gamma_{\text{obs}})$	$P(M_\odot)$
drawdown	0.187	0.487	0.317	0.384
drawup	0.279	0.172	0.430	0.507
combined	0.190	0.494	0.322	0.443

3.2.2. Drawup

An explanation for the coincidence of these two timescales was proposed in Reference [47], on the basis that their ratio $t_{\text{life}}/t_\star$ may decrease with stellar mass. Then, since the stellar mass distribution is so steep, we would naturally expect to be situated around a star just large enough to barely satisfy the requirement $t_{\text{life}} \sim t_\star$. If the oxygenation of the Earth were dependent on stellar activity this would naturally fit with this explanation, as this process would then take longer around smaller stars.

This mechanism was proposed in Reference [59], where atmospheric reductants are eventually lost to space. Here, geochemical processes would have dissociated methane and water molecules, followed by the escape of the hydrogen. This leads to an imbalance of carbon, which combines into carbon dioxide that is then drawn down into the mantle. The escape process is limited by the UV flux of the star, which depends on stellar mass, and only matches the stellar timescale around sunlike stars.

We are not in a position to judge this hypothesis based off its geological merit, but we may consider the implications it has on the distribution of observers throughout the multiverse. If this process is the limiting factor for where life can arise, then we would expect to be unable to find universes that can oxygenate planets much faster than our own. In the following we specify to main sequence stars, disregarding any enhanced atmospheric erosion that would occur during flares of young stars. This has recently been a topic of intense interest [67], and may play an important role in determining the habitability of planets around red dwarfs [68].

The UV process is dominated by photons that can just barely cause hydrogen to escape the atmosphere, since those with higher energies are exponentially suppressed. Then, to estimate the time required to completely oxygenate a planetary atmosphere, the amount of flux in this range must be inferred. For a blackbody at temperature T, this would be

$$f_{XUV} \approx \frac{1}{2}\left(\frac{E_{XUV}}{T}\right)^2 e^{-E_{XUV}/T}, \tag{32}$$

where $E_{XUV} \sim 10$ eV. Stars are not blackbodies in this energy range, but the flux can be estimated as being a factor of 20 higher than the black body flux [69]. For the early Earth, the sun's flux implied a photolysis rate of 10^{12-13} mol/year [59]. (It is worth noting that the oxidation rate is much smaller currently, due to the presence of cold traps and other concentration effects that prevent dissociable compounds from reaching the Earth's exobase. Additionally, planetary characteristics such as a magnetic field can prevent atmospheric loss, leading to uncertainties in the overall rate [70,71].) To first approximation, about one hydrogen ion escapes for every UV photon incident, and so the oxygenation timescale is then given by

$$t_{O_2\uparrow} = \frac{M_{atm}/m_p}{\Phi_{XUV} R_{terr}^2} \tag{33}$$

Here M_{atm} is the mass of the atmosphere and the flux is given by $\Phi_{XUV} = 20\, T^3 f_{XUV}$, so that the ratio of timescales is

$$\frac{t_{O_2\uparrow}}{t_\star} = 1.1 \times 10^9 \frac{\beta^{3/4} \gamma^{3/4}}{\alpha^4} \lambda^2 \hat{e}\left(-\frac{0.44}{\sqrt{\beta}} + 841 \frac{\alpha^{3/2} \beta}{\gamma^{1/4} \lambda^{1/2}}\right) \tag{34}$$

Though the exponential dependence on mass differs from the usual power law form found in the literature, around solar mass values this function behaves with effective power law index $p \equiv d\log(t_{O_2}/t_\star)/d\log(\lambda)$, which ranges from -8.1 to -3.2 within a factor of two of solar mass. This can be compared to the estimates of $p = -3.4$ from Reference [72] and $p = -6.6$ of Reference [47], and a seeming -4.25 from Reference [73].

For life to develop, this ratio must necessarily be less than 1. This will be the case for intermediate values of λ, constants permitting. For small masses, the stellar temperature is so low that photons capable of ejecting hydrogen from the atmosphere are infrequent, leaving the gas trapped and unable to oxidize. For large masses, the stellar lifetime is very short, leading the star to burn out well before enough hydrogen has escaped. To first approximation, these delineating masses are given by

$$\lambda_{min} \approx 3.7 \times 10^6 \frac{\alpha^3 \beta^3}{\gamma^{1/2}}, \quad \lambda_{max} \approx 3.0 \times 10^{-5} \frac{\alpha^2}{\beta^{3/8} \gamma^{3/8}} e^{.22/\sqrt{\beta}} \tag{35}$$

The former corresponds to $0.66 M_\odot$ in our universe, and the latter is irrelevantly large. (The exact expressions can be given in terms of Lambert productlogs, but the error from these approximations are only a few percent.) The minimum of the ratio of timescales will be larger than 1 when

$$\frac{\alpha^2 \beta^{19/4}}{\gamma^{1/4}} e^{-0.44/\sqrt{\beta}} < 8.5 \times 10^{-21}. \tag{36}$$

Finally, let us comment on the possibility that both drawup and drawdown are relevant for setting the oxygenation timescale. In this case, we would have

$$\frac{1}{t_{O_2}} = \frac{\epsilon}{t_{O_2\downarrow}} + \frac{1-\epsilon}{t_{O_2\uparrow}} \tag{37}$$

where the two timescales are given by Equations (31) and (34) above, and ϵ is a free parameter between 0 and 1 that dictates the relative importance of each process. Taking both contributions to be equal is problematic, for as we have seen the rate of drawdown is higher for low mass stars, and the rate of drawup is higher for large mass stars. In this case, then, every star's oxygenation time is smaller than its lifetime, with a switchover in the dominant mode occurring for intermediate masses, as shown in Figure 6. Generically, the presence of drawdown serves to spoil the explanation for the observed coincidence of timescales, as it removes the low mass cutoff. The only way for this not to occur is for sufficiently small values of ϵ: for $t_{O_2\downarrow} \gtrsim 18 t_\star$, a range of stellar masses has a ratio which is larger than 1. However, when considered from a multiverse perspective, most observers would still not expect these two timescales to coincide on their planet. Therefore, the multiverse gives reason to disfavor planetary oxygenation as the mechanism for the delay of complex life. This is an otherwise perfectly viable hypothesis, and if it does turn out to be true, we will have strong evidence against the multiverse.

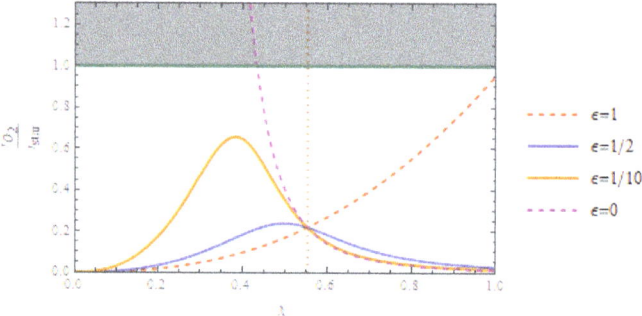

Figure 6. Ratio of oxygenation time to stellar lifetime as a function of stellar mass. For this criterion, stars with ratio above 1 are uninhabitable. Here, ϵ parameterizes the relative importance of drawdown and drawup. The dotted line corresponds to 1 $M_\odot$.

3.3. Easy Stroll

The remaining account for the coincidence of solar and biological timescales is known as the easy stroll model [48]. In this model the distribution of planetary habitable durations is relatively decoupled from the lifetime of the host star, with a steep preference for very short duration. If we take the planetary histories of Venus and Mars into account, for example, it becomes easy to imagine that the majority of planets, born even within the temperate zone, will through runaway processes lose their clement nature on the order of a geological timescale [74]. It has been proposed that the presence of life itself can temper some of the most dangerous negative feedbacks, leading to a bottleneck amongst worlds where this large scale alteration initially takes hold [75].

In this model, developing complex life is relatively easy, being perhaps proportional to the total lifetime (weighted appropriately) as in Section 2, but the planetary maintenance of a habitable phase becomes the significant bottleneck. Since the majority of planets will have habitable lifetimes much too short for complex life to develop, we would then naturally expect to arise on a planet that is quite close to its expiration date.

This state of affairs effectively adds a hidden variable that we do not consider in our simplistic account, which dwells only on stellar mass. This may be planetary mass, composition, volatile abundance, orbital parameters, or any number of other things. The prediction of this model is that the distribution for at least one of these will be important for habitability, and also sufficiently steep so that the ratio of timescales is typically on the order of the threshold value. Exploring this prediction in detail will need to remain for future work.

4. Discussion: Comparing 10,560 Hypotheses

In the multiverse setting, we expect to live in a universe that is good at producing observers. There are undoubtedly many conditions that must be met for a universe to be able to achieve this feat, but at the moment we do not know what they are. As such, there are a vast number of hypotheses in the literature, some complementary, others mutually contradictory. By tabulating all of these, it will be possible to delineate which are compatible with the multiverse expectation that our universe is an exceptional observer factory, and which are not. As of now, each habitability criterion is strictly hypothetical, but one day we will definitively know what life needs. Once this is established, it can be used as either negative evidence against the multiverse, or positive evidence for the multiverse. This rather baroque procedure is necessary to circumvent the major charges against this scientific paradigm, that it does not lead to any directly observable consequences. This reliance on indirect means in no way diminishes this framework, as in fact the act of doing so allows many predictive and explanatory statements to be made.

Of course, if I tell you that the multiverse dictates that complex life should need photosynthesis, and we later verify that this is in fact true, it would be a tremendous overstatement to say that the multiverse is anywhere close to being proven right. The opposite situation is much less forgiving, mind you: if the prediction is wrong, we can safely forget the idea of other universes, and focus on explaining why ours is unique. However, the key to overcoming this rather weak positive evidence, which on the face of things boils down to something like a 50/50 chance of being right, is the fact that statements can be made about a very large variety of potentially relevant conditions for life. Taken all together, this can amount to perhaps a dozen independent predictions for what life needs. By marshalling these various predictions, the even split from a lone condition turns into a 1 in a 1000 chance of getting all of them right. This could be compared to the scenario where the multiverse is a fiction, in which case we would expect roughly half the predictions to be false.

The challenge here is that the different predictions are not exactly all independent: as they add various anthropic boundaries and preferences for specific parameter values, there can be a nontrivial interplay between the different habitability conditions. We have seen some of this already, as for example when the plate tectonics condition lead to acceptable probabilities for only a subset of the hypotheses. Because of this, it becomes necessary to check each combination individually, to ensure that no acceptable combinations are being overlooked. The number of possible combinations quickly proliferates, however: even with the dozen or so conditions we have considered in the three papers on the subject, there are over 10,000 combinations. What is more, there are many more relevant habitability conditions we have not even attempted to incorporate yet. Though incorporating each new criterion has already been made as streamlined as deemed possible through the python code available at https://github.com/mccsandora/Multiverse-Habitability-Handler, it still takes a laptop about a minute to check each condition to a reasonable accuracy. Even with a drastic increase in computing power, it will soon become infeasible to check every individual hypothesis.

It is important to extract general, qualitative features about how each notion of habitability affects the distribution of observers throughout the multiverse. Armed with these, it becomes possible to build an intuition as to how they will combine to deliver the ultimate figures of merit in our framework, those probabilities of observing our values of the constants.

Though we have focused on the fraction of planets which develop life in this work, we are very far from having done this topic justice. Our estimates of which planets can give rise to life has been very rudimentary and broad-brush, neglecting very many features that are probably extremely important. However, we hesitate to be too apologetic for this omission: what has been provided is a framework that can be readily extended to include arbitrary habitability criteria. The reasons for including more are twofold: first, the greater the input, the greater the output. Each habitability condition represents a potential check of the multiverse framework, and though not all will give rise to concrete predictions, incorporating as many as possible will lead to the strongest possible attempt at testing this idea. Secondly, it will be necessary to ensure that nothing is overlooked. If some habitability condition is passed over because it greatly favors some region of the multiverse which later turns out to be sterile for completely unrelated reasons, we will be in danger of drawing false conclusions. This is the challenge that presents itself, and if we want to be able to utilize this reasoning effectively it will need to be dealt with to the best of our ability.

We regard the results we have found so far as encouraging. Out of the various potentially reasonable habitability hypotheses we have discussed, the linear dependence on entropy was a clear winner. It is worth reiterating that in our assessment we used not only the probabilities of observing the values of the constants we see, but also several other local conditions, namely the mass of our star and our current moment most of the way through the Earth's habitable phase. Utilizing these additional pieces of information highlights a complementarity in our approach: some criteria that have no trouble explaining our position within our universe are dramatically incompatible with the multiverse, and others that are compatible with the multiverse cannot be reconciled with our position within our universe. It is necessary, within a multiverse framework, to be able to explain both simultaneously, and this more rigorous standard is capable of more effectively pruning the potential habitability criteria than either alone. Though we have focused our attention on a relatively few number of parameters, this procedure can eventually be done for the entire suite of both physical constants and environmental variables, which will ultimately be necessary to fully test this framework. This is going to be a tremendous challenge that will require synthesizing knowledge from a great variety of fields, but the promise of being able to fully determine whether there are other universes out there beyond our horizon will make this herculean task well worth the effort.

Funding: This research received no external funding.

Acknowledgments: I would like to thank Ileana Pérez-Rodríguez for useful discussions.

Conflicts of Interest: The author declares no conflict of interest.

Appendix A. Some Geology

Many calculations throughout the text rely on the characteristic size and timescales of terrestrial planets. Here we estimate these, and determine how they depend on the physical constants.

Firstly, the mass and radius of a terrestrial planet are given by the condition that the escape velocity must be slightly greater than the thermal velocity [22]:

$$M_{\text{terr}} = 91.9 \, \frac{\alpha^{3/2} m_e^{3/4} M_{pl}^3}{m_p^{11/4}}, \quad R_{\text{terr}} = 3.6 \, \frac{M_{pl}}{\alpha^{1/2} m_e^{3/4} m_p^{5/4}} \tag{A1}$$

These have been normalized to Earth's values and we have used that the density of matter is $\rho \sim \alpha^3 m_e^3 m_p$.

The typical mountain height can be estimated by equating the molecular and gravitational energies, $H_{\text{mountain}} \sim E_{\text{mol}}/(g m_p)$ [76], to give

$$H_{\text{mountain}} = 0.0056 \, \frac{M_{pl}}{\alpha^{1/2} m_e^{3/4} m_p^{5/4}} \tag{A2}$$

This scales in the same way as the terrestrial planet radius, which is also set by balancing gravitational and molecular energy, but is 600 times smaller (10 km).

Many properties on Earth, such as the speed of continental drift and erosion rates, are determined by the planet's internal heat. A rough estimate of this can be given by dimensional analysis as

$$Q \sim G M_{\text{terr}} \rho \, \kappa_{\text{heat}} = 92.5 \, \frac{\alpha^{9/2} m_e^{7/2} M_{pl}}{m_p^{5/2}} \tag{A3}$$

Which is normalized to the observed value of 47 TW. Here, we made use of the thermal diffusivity of rock $\kappa_{\text{heat}} \sim c_s L$, where L is the size of a cell in the solid, which in rock just scales with the Bohr radius. and $c_s \sim \sqrt{E_{\text{vib}}/m_p} \sim 6$ km/s is the speed of sound, yielding $\kappa_{\text{heat}} = 2/(m_e^{1/4} m_p^{3/4})$.

A more sophisticated estimate of the Earth's heat, including its time dependence, will be necessary for some applications. This problem is somewhat muddied because there are two relevant sources: the primordial heat of formation, and that released by radioactive decay. The relative importance of each was estimated from other bodies in the solar system in Reference [77], and a measured from geoneutrino flux in Reference [41]. Remarkably, these indicate that the Earth's heat budget is split almost equally between radioactivity and primordial heat. This itself is a startling coincidence, but its only relevant consequence for our present purposes is that we must calculate both contributions to the heat.

First, the primordial heat: the internal temperature is set by the gravitational energy of the planet's formation, which is given by

$$T_i \sim \frac{G M_{\text{terr}} m_p}{R_{\text{terr}}} \sim 7600 \text{ K} \tag{A4}$$

Crucially, this is above the melting temperature of rock, which led to an initially molten Earth, and its subsequent differentiation into mantle and core. This is generic for terrestrial planets: since the gravitational energy is a bit lower than the molecular bond energy to retain gases and liquids, we always have $T_i \propto T_{\text{mol}}$ for any values of the fundamental constants. Since T_i is an order of magnitude larger, terrestrial planets will always be predisposed to differentiation. This carries many potential benefits, such as a magnetic field and the sequestration of highly reducing iron minerals.

The mantle's heat conductivity is much higher than garden variety rocks, on account of the dominant method of heat transfer being by convection rather than conduction. It is the conductivity of the lithosphere, the Earth's rigid outer skin, which serves as the last line of defense against the emanation of heat to space, and so it will be crucial to determine what sets the lithosphere's thickness and conductivity. From Reference [78], If the Earth is modeled as an inner mantle with infinite conductivity, and the upper lithosphere as having having diffusivity κ_{heat}, then the total heat radiating through the surface at time t will be

$$Q_{\text{form}} = 4\pi R_{\text{terr}}^2 \frac{\kappa_{\text{heat}} T_i}{L_{\text{lith}}(t)} \hat{e}\left(\frac{-3 \kappa_{\text{heat}} t}{R_{\text{terr}} L_{\text{lith}}(t)}\right) \tag{A5}$$

The depth of the lithosphere increases with time as $L_{\text{lith}}(t) = 2\sqrt{\kappa_{\text{heat}} t}$. We then arrive at

$$L_{\text{lith}} = 1.56 \frac{t^{1/2}}{m_e^{1/8} m_p^{3/8}} \tag{A6}$$

In terms of constants, we find

$$Q_{\text{form}} = 237 \frac{\alpha^{9/2} m_e^{7/2} M_{pl}}{m_p^{5/2}} \frac{1}{s} e^{-s}, \quad s = 10 \frac{\alpha^{1/2} m_e^{5/8} m_p^{7/8} t^{1/2}}{M_{pl}} = \left(\frac{t}{2.5 \times 10^9 \text{ yr}}\right)^{1/2} \tag{A7}$$

If we neglect the s dependence of this expression, we find the scaling from above based off simple dimensional analysis.

Now, the radiogenic component: for a planet with multiple radioactive species, the heat generated by decay is given by a sum of their individual contributions. The total is then

$$Q_{\text{rad}} = \sum_i \frac{f_i \epsilon_i M_{\text{terr}}}{\tau_i} e^{-t/\tau_i} \tag{A8}$$

where f_i is the fraction and ϵ_i the binding energy (in units of m_p) of species i.

The rate of continental drift can be found in Reference [79]. If we use the rough estimate for the planet's heat, we have

$$v_{\text{drift}} = \frac{Q}{4\pi R_{\text{terr}}^2 n (L + c_p \Delta T)} \sim 1376 \, \alpha^{1/2} \beta^{1/2} \gamma \sim \frac{\text{cm}}{\text{yr}} \qquad (A9)$$

Here n is the number density of mantle, L is the latent heat, and ΔT is the difference between melting and surface temperatures, which can both be estimated as $L \sim \Delta T \sim T_{\text{mol}}$.

The weathering rate can be estimated as

$$\Gamma_{\text{drawdown}} \sim v_{\text{drift}} H_{\text{mountain}} R_{\text{terr}} n = 8.9 \left(\frac{\alpha \, m_e}{m_p}\right)^{5/2} M_{pl} = 5.4 \times 10^{13} \frac{\text{mol}}{\text{yr}} \qquad (A10)$$

which agrees with observed rates.

References

1. Sandora, M. Multiverse Predictions for Habitability I: The Number of Stars and Their Properties. *arXiv* **2019**, arXiv:1901.04614.
2. Sandora, M. Multiverse Predictions for Habitability II: Number of Habitable Planets. *arXiv* **2019**, arXiv:1902.06784.
3. Vilenkin, A. Predictions from Quantum Cosmology. *Phys. Rev. Lett.* **1995**, *74*, 846–849. [CrossRef] [PubMed]
4. Bostrom, N. *Anthropic Bias: Observation Selection Effects in Science and Philosophy*; Routledge: London, UK, 2013.
5. Gleiser, M. Drake equation for the multiverse: From the string landscape to complex life. *Int. J. Mod. Phys. D* **2010**, *19*, 1299–1308. [CrossRef]
6. Lingam, M.; Loeb, A. Reduced diversity of life around Proxima Centauri and TRAPPIST-1. *Astrophys. J. Lett.* **2017**, *846*, L21. [CrossRef]
7. Hansen, C.J.; Kawaler, S.D.; Trimble, V. *Stellar Interiors: Physical Principles, Structure, and Evolution*; Springer Science & Business Media: New York, NY, USA, 2012.
8. Kaltenegger, L. How to characterize habitable worlds and signs of life. *Annu. Rev. Astron. Astrophys.* **2017**, *55*, 433–485. [CrossRef]
9. Scharf, C.; Cronin, L. Quantifying the origins of life on a planetary scale. *Proc. Natl. Acad. Sci. USA* **2016**, *113*, 8127–8132. [CrossRef]
10. Heller, R.; Armstrong, J. Superhabitable worlds. *Astrobiology* **2014**, *14*, 50–66. [CrossRef]
11. Hazen, R.M. Chance, necessity and the origins of life: A physical sciences perspective. *Philos. Trans. R. Soc. A* **2017**, *375*, 20160353. [CrossRef]
12. Scharf, C. Exoplanet Exergy: Why useful work matters for planetary habitabilty. *Astrophys. J.* **2019**, *876*, 16. [CrossRef]
13. Landenmark, H.K.; Forgan, D.H.; Cockell, C.S. An estimate of the total DNA in the biosphere. *PLoS Biol.* **2015**, *13*, e1002168. [CrossRef]
14. Popovic, M. Negentropy concept revisited: Standard thermodynamic properties of 16 bacteria, fungi and algae species. *arXiv* **2019**, arXiv:1901.00494.
15. Whitman, W.B.; Coleman, D.C.; Wiebe, W.J. Prokaryotes: The unseen majority. *Proc. Natl. Acad. Sci. USA* **1998**, *95*, 6578–6583. [CrossRef]
16. Kempes, C.; Wolpert, D.; Cohen, Z.; Pérez-Mercader, J. The thermodynamic efficiency of computations made in cells across the range of life. *Philos. Trans. Ser. A Math. Phys. Eng. Sci.* **2017**, *375*. [CrossRef]
17. Wolpert, D.H. The free energy requirements of biological organisms; implications for evolution. *Entropy* **2016**, *18*, 138. [CrossRef]
18. Gillings, M.R.; Hilbert, M.; Kemp, D.J. Information in the biosphere: Biological and digital worlds. *Trends Ecol. Evolut.* **2016**, *31*, 180–189. [CrossRef]
19. Dyson, F.J. Time without end: Physics and biology in an open universe. *Rev. Mod. Phys.* **1979**, *51*, 447. [CrossRef]

20. Adams, F.C. Constraints on Alternate Universes: Stars and habitable planets with different fundamental constants. *J. Cosmol. Astropart. Phys.* **2016**, *2016*, 042. [CrossRef]
21. Fisher, D.S. Course 11 evolutionary dynamics. *Les Houches* **2007**, *85*, 395–446.
22. Press, W.H.; Lightman, A.P. Dependence of macrophysical phenomena on the values of the fundamental constants. *Philos. Trans. R. Soc. Lond. Ser. A* **1983**, *310*, 323–334. [CrossRef]
23. Krishtalik, L.I. Energetics of multielectron reactions. Photosynthetic oxygen evolution. *Biochim. Biophys. Acta (BBA) Bioenerg.* **1986**, *849*, 162–171. [CrossRef]
24. Kiang, N.Y.; Segura, A.; Tinetti, G.; Govindjee; Blankenship, R.E.; Cohen, M.; Siefert, J.; Crisp, D.; Meadows, V.S. Spectral Signatures of Photosynthesis. II. Coevolution with Other Stars And The Atmosphere on Extrasolar Worlds. *Astrobiology* **2007**, *7*, 252–274. [CrossRef]
25. Haqq-Misra, J.; Kopparapu, R.K.; Wolf, E.T. Why do we find ourselves around a yellow star instead of a red star? *Int. J. Astrobiol.* **2018**, *17*, 77–86. [CrossRef]
26. Loeb, A.; Batista, R.A.; Sloan, D. Relative likelihood for life as a function of cosmic time. *J. Cosmol. Astropart. Phys.* **2016**, *2016*, 040. [CrossRef]
27. Waltham, D. Star masses and star-planet distances for Earth-like habitability. *Astrobiology* **2017**, *17*, 61–77. [CrossRef]
28. Tyrrell, T. The relative influences of nitrogen and phosphorus on oceanic primary production. *Nature* **1999**, *400*, 525–531. [CrossRef]
29. Chyba, C.F.; Phillips, C.B. Possible ecosystems and the search for life on Europa. *Proc. Natl. Acad. Sci. USA* **2001**, *98*, 801–804. [CrossRef]
30. Sigman, D.M.; Hain, M.P. The biological productivity of the ocean. *Nat. Educ. Knowl.* **2012**, *3*, 21.
31. Ward, L.M.; Rasmussen, B.; Fischer, W.W. Primary Productivity was Limited by Electron Donors Prior to the Advent of Oxygenic Photosynthesis. *J. Geophys. Res. Biogeosci.* **2019**, *124*, 211–226. [CrossRef]
32. Morel, F.M. The co-evolution of phytoplankton and trace element cycles in the oceans. *Geobiology* **2008**, *6*, 318–324. [CrossRef]
33. Maldonado, M.T.; Boyd, P.W.; Harrison, P.J.; Price, N.M. Co-limitation of phytoplankton growth by light and Fe during winter in the NE subarctic Pacific Ocean. *Deep Sea Res. Part II Top. Stud. Oceanogr.* **1999**, *46*, 2475–2485. [CrossRef]
34. Labrosse, S.; Jaupart, C. Thermal evolution of the Earth: Secular changes and fluctuations of plate characteristics. *Earth Planet. Sci. Lett.* **2007**, *260*, 465–481. [CrossRef]
35. Valentine, J.W.; Moores, E.M. Global Tectonics and the Fossil Record. *J. Geol.* **1972**, *80*, 167–184. [CrossRef]
36. Walker, J.C.; Hays, P.; Kasting, J.F. A negative feedback mechanism for the long-term stabilization of Earth's surface temperature. *J. Geophys. Res. Oceans* **1981**, *86*, 9776–9782. [CrossRef]
37. Foley, B.J. Habitability of Earth-like stagnant lid planets: Climate evolution and recovery from snowball states. *Astrophys. J.* **2019**, *875*, 72. [CrossRef]
38. Korenaga, J. Initiation and evolution of plate tectonics on Earth: Theories and observations. *Annu. Rev. Earth Planet. Sci.* **2013**, *41*, 117–151. [CrossRef]
39. Regenauer-Lieb, K.; Yuen, D.A.; Branlund, J. The initiation of subduction: Criticality by addition of water? *Science* **2001**, *294*, 578–580. [CrossRef]
40. Rosing, M.T.; Bird, D.K.; Sleep, N.H.; Glassley, W.; Albarede, F. The rise of continents—An essay on the geologic consequences of photosynthesis. *Palaeogeogr. Palaeoclimatol. Palaeoecol.* **2006**, *232*, 99–113. [CrossRef]
41. Gando, A.; Gando, Y.; Ichimura, K.; Ikeda, H.; Inoue, K.; Kibe, Y.; Kishimoto, Y.; Koga, M.; Minekawa, Y.; Mitsui, T.; et al. Partial radiogenic heat model for Earth revealed by geoneutrino measurements. *Nat. Geosci.* **2011**, *4*, 647–651.
42. Valencia, D.; O'connell, R.J.; Sasselov, D.D. Inevitability of plate tectonics on super-Earths. *Astrophys. J. Lett.* **2007**, *670*, L45. [CrossRef]
43. Alibert, Y. On the radius of habitable planets. *Astron. Astrophys.* **2014**, *561*, A41. [CrossRef]
44. Rogers, L.A. Most 1.6 Earth-radius planets are not rocky. *Astrophys. J.* **2015**, *801*, 41. [CrossRef]
45. Sandora, M. The fine structure constant and habitable planets. *J. Cosmol. Astropart. Phys.* **2016**, *8*, 048. [CrossRef]
46. Carter, B. The anthropic principle and its implications for biological evolution. *Philos. Trans. R. Soc. Lond. A* **1983**, *310*, 347–363. [CrossRef]

47. Livio, M. How rare are extraterrestrial civilizations, and when did they emerge? *Astrophys. J.* **1999**, *511*, 429–431. [CrossRef]
48. Simpson, F. The longevity of habitable planets and the development of intelligent life. *Int. J. Astrobiol.* **2017**, *16*, 266–270. [CrossRef]
49. Sandora, M. Multiverse Predictions for Habitability IV: Fraction of Life that Develops Intelligence. *arXiv* **2019**, arXiv:1904.11796.
50. Watson, A.J. Implications of an anthropic model of evolution for emergence of complex life and intelligence. *Astrobiology* **2008**, *8*, 175–185. [CrossRef]
51. Szathmáry, E.; Smith, J.M. The major evolutionary transitions. *Nature* **1995**, *374*, 227–232. [CrossRef]
52. Lingam, M.; Loeb, A. Role of stellar physics in regulating the critical steps for life. *arXiv* **2018**, arXiv:1804.02271.
53. Vermeij, G.J. Historical contingency and the purported uniqueness of evolutionary innovations. *Proc. Natl. Acad. Sci. USA* **2006**, *103*, 1804–1809. [CrossRef]
54. Catling, D.C.; Glein, C.R.; Zahnle, K.J.; McKay, C.P. Why O_2 Is Required by Complex Life on Habitable Planets and the Concept of Planetary "Oxygenation Time". *Astrobiology* **2005**, *5*, 415–438. [CrossRef]
55. Langmuir, C.H.; Broecker, W. *How to Build a Habitable Planet: The Story of Earth from the Big Bang to Humankind-Revised and Expanded Edition*; Princeton University Press: Princeton, NJ, USA, 2012.
56. Brocks, J.J.; Logan, G.A.; Buick, R.; Summons, R.E. Archean Molecular Fossils and the Early Rise of Eukaryotes. *Science* **1999**, *285*, 1033–1036. [CrossRef]
57. Holland, H.D. The oxygenation of the atmosphere and oceans. *Philos. Trans. R. Soc. Lond. B Biol. Sci.* **2006**, *361*, 903–915. [CrossRef]
58. Konhauser, K.O. *Introduction to Geomicrobiology*; John Wiley & Sons: Hoboken, NJ, USA, 2009.
59. Catling, D.C.; Zahnle, K.J.; McKay, C.P. Biogenic methane, hydrogen escape, and the irreversible oxidation of early Earth. *Science* **2001**, *293*, 839–843. [CrossRef]
60. Lenton, T.; Watson, A. *Revolutions that Made the Earth*; Oxford University Press: Oxford, UK, 2011.
61. Del Genio, A.D.; Brain, D.; Noack, L.; Schaefer, L. The Inner Solar System's Habitability Through Time. *arXiv* **2018**, arXiv:1807.04776.
62. Pepin, R.O. Atmospheres on the terrestrial planets: Clues to origin and evolution. *Earth Planet. Sci. Lett.* **2006**, *252*, 1–14. [CrossRef]
63. Winter, R.; Jeworrek, C. Effect of pressure on membranes. *Soft Matter* **2009**, *5*, 3157–3173. [CrossRef]
64. Simonato, F.; Campanaro, S.; Lauro, F.M.; Vezzi, A.; D'Angelo, M.; Vitulo, N.; Valle, G.; Bartlett, D.H. Piezophilic adaptation: A genomic point of view. *J. Biotechnol.* **2006**, *126*, 11–25. [CrossRef]
65. Picard, A.; Daniel, I. Pressure as an environmental parameter for microbial life—A review. *Biophys. Chem.* **2013**, *183*, 30–41. [CrossRef]
66. Lingam, M.; Loeb, A. Is life most likely around Sun-like stars? *J. Cosmol. Astropart. Phys.* **2018**, *2018*, 020. [CrossRef]
67. Luger, R.; Barnes, R. Extreme water loss and abiotic O2 buildup on planets throughout the habitable zones of M dwarfs. *Astrobiology* **2015**, *15*, 119–143. [CrossRef]
68. Airapetian, V.S.; Glocer, A.; Khazanov, G.V.; Loyd, R.O.P.; France, K.; Sojka, J.; Danchi, W.C.; Liemohn, M.W. How hospitable are space weather affected habitable zones? The role of ion escape. *Astrophys. J. Lett.* **2017**, *836*, L3. [CrossRef]
69. Pierrehumbert, R.T. *Principles of Planetary Climate*; Cambridge University Press: Cambridge, UK, 2010.
70. Dong, C.; Lingam, M.; Ma, Y.; Cohen, O. Is Proxima Centauri b Habitable? A Study of Atmospheric Loss. *Astrophys. J. Lett.* **2017**, *837*, L26. [CrossRef]
71. Dong, C.; Jin, M.; Lingam, M.; Airapetian, V.S.; Ma, Y.; van der Holst, B. Atmospheric escape from the TRAPPIST-1 planets and implications for habitability. *Proc. Natl. Acad. Sci. USA* **2018**, *115*, 260–265. [CrossRef]
72. Lingam, M.; Loeb, A. Physical constraints on the likelihood of life on exoplanets. *Int. J. Astrobiol.* **2018**, *17*, 116–126. [CrossRef]
73. Rugheimer, S.; Segura, A.; Kaltenegger, L.; Sasselov, D. UV surface environment of Earth-like planets orbiting FGKM stars through geological evolution. *Astrophys. J.* **2015**, *806*, 137. [CrossRef]
74. Ramirez, R. A more comprehensive habitable zone for finding life on other planets. *Geosciences* **2018**, *8*, 280. [CrossRef]

75. Chopra, A.; Lineweaver, C.H. The case for a Gaian bottleneck: The biology of habitability. *Astrobiology* **2016**, *16*, 7–22. [CrossRef]
76. Weisskopf, V.F. Of atoms, mountains, and stars: A study in qualitative physics. *Science* **1975**, *187*, 605–612. [CrossRef]
77. McDonough, W.F.; Sun, S.S. The composition of the Earth. *Chem. Geol.* **1995**, *120*, 223–253. [CrossRef]
78. Fowler, C. *The Solid Earth: An Introduction to Global Geophysics*; Royal Holloway, University of London: Egham, Surrey, UK, 2004.
79. O'Reilly, T.C.; Davies, G.F. Magma transport of heat on Io: A mechanism allowing a thick lithosphere. *Geophys. Res. Lett.* **1981**, *8*, 313–316. [CrossRef]

© 2019 by the author. Licensee MDPI, Basel, Switzerland. This article is an open access article distributed under the terms and conditions of the Creative Commons Attribution (CC BY) license (http://creativecommons.org/licenses/by/4.0/).

Article

Multiverse Predictions for Habitability: Fraction of Life That Develops Intelligence

McCullen Sandora [1,2]

1. Institute of Cosmology, Department of Physics and Astronomy, Tufts University, Medford, MA 02155, USA; mccullen.sandora@gmail.com
2. Center for Particle Cosmology, Department of Physics and Astronomy, University of Pennsylvania, Philadelphia, PA 19104, USA

Received: 14 May 2019; Accepted: 14 July 2019; Published: 17 July 2019

Abstract: Do mass extinctions affect the development of intelligence? If so, we may expect to be in a universe that is exceptionally placid. We consider the effects of impacts, supervolcanoes, global glaciations, and nearby gamma ray bursts, and how their rates depend on fundamental constants. It is interesting that despite the very disparate nature of these processes, each occurs on timescales of 100 Myr-Gyr. We argue that this is due to a selection effect that favors both tranquil locales within our universe, as well as tranquil universes. Taking gamma ray bursts to be the sole driver of mass extinctions is disfavored in multiverse scenarios, as the rate is much lower for different values of the fundamental constants. In contrast, geological causes of extinction are very compatible with the multiverse. Various frameworks for the effects of extinctions are investigated, and the intermediate disturbance hypothesis is found to be most compatible with the multiverse.

Keywords: multiverse; habitability; mass extinctions

---

1. Introduction

This is a continuation of the work initiated in [1–3] aimed at advancing progress on the multiverse hypothesis by connecting it to biological and geological notions of habitability to generate predictions which will be testable within the timespan of several decades. Our technique has been to use detailed criteria for what life needs to count the number of environments suitable for life in the universe, and to check how this depends on the fundamental constants of physics. This counting depends on the assumptions we make about what constitutes a habitable environment, and several of the choices we made imply the existence of much more fertile universes, which would host the majority of observers. When the use of a habitability criterion leads to a very small probability of our existence in this universe, we conclude that this criterion is incompatible with the multiverse hypothesis. Thus, the existence of the multiverse can be used to predict which notions of habitability are right or wrong. While there is currently no way to distinguish among competing notions of habitability, our knowledge on this front is advancing rapidly, and future telescopes and space missions are slated to greatly elucidate what conditions are required for life. We will ultimately determine the exact conditions for habitability, and when this occurs, we may check how this compares to the predictions the multiverse has made. Since there are a great number of factors to consider, there are several independent testable predictions that will serve to test this otherwise almost untestable hypothesis.

The main tool for estimating the number of observers within a universe is the Drake equation, which is a product of factors associated with stellar, planetary and biological habitability. As such, our analysis was (relatively) neatly split by this compartmentalization, with each of our previous papers devoted to a separate domain. While the number and properties of stars, and as of recently the planets orbiting them, are quite well known, as we progress through the factors the analysis becomes more speculative. In this paper, we focus our attention on the fraction of biospheres that develop intelligent societies. Without wading too much into the debate of what exactly constitutes an intelligent observer, we adopt the self-aggrandizing view that humanlike intelligence is somewhat representative. The author adopts the rather common view that language may represent an excellent proxy for general purpose intelligence, but the results of this paper do not depend too much on the details of this assumption.

Determining the fraction of life bearing planets that develop intelligence is a monumental task, and we make no claim to doing it justice in this letter. We can begin to estimate the physical effects that can preclude this event, adopting the viewpoint that $f_{int} = 1 - f_\dagger$, where $f_\dagger$ is the fraction of biospheres that are so affected by cataclysm that intelligence cannot develop. This treats the emergence of intelligence (noogenesis) as otherwise inevitable, though there are many additional factors that could contribute to this that should eventually be folded into the full analysis. Even in our restricted setting, we cannot tabulate all possible catastrophes and runaway processes that can occur on a planet in order to provide a true estimate of the fraction that survive sufficiently long, but we instead highlight a few for which we are able to readily encapsulate the dependence on physical parameters.

Mass Extinctions

The history of life on Earth is punctuated by several episodes of mass extinction, where a large fraction of the species present at the time did not survive. In the 540 Myr since the advent of complex life, there have been five of these absolutely catastrophic episodes, as well as 20–30 distinguishable lesser episodes. Although for the majority of the study of natural history these great extinctions were treated as 'different in size but not kind' that were not in need of explanation, it is by now generally understood that they are a product of catastrophic changes in the Earth's environment [4]. The single piece of evidence that shifted public perception so radically was the discovery of the iridium anomaly at the onset of the Cretaceous extinction, which indicated that it was triggered by a massive impactor [5]. While this has subsequently been thoroughly confirmed by the discovery of the associated impact crater at Chicxulub (see [6] for a review), the causes of the earlier mass extinctions remain under intense debate to this day.

Several salient features of the observed mass extinctions are worth pointing out for our purposes. Firstly, they all manifest differently. The fossil records of each indicate that the characteristics of the species that went extinct, in which order, the abruptness, severity, and recovery pattern, were all markedly different from event to event [4]. This indicates that the change in conditions that ultimately triggered the extinction did not arise from the same underlying cause, but instead, each event represents a unique flavor of catastrophe. The list of possible causes includes glaciation, impact, supervolcano, biological innovation, anoxia, sea level change, gamma ray burst, and climate change. Many of these causes can lead to additional other causes, leading to a domino effect of extinctions. They may not be mutually exclusive, and some of the extinctions could well have been the product of several concurrent factors. Secondly, there is no strong consensus for the causes of the earlier events, so that there is still room for speculation on the ultimate cause(s). In fact, every possible environmental trigger for mass extinctions has, at some point or another, been applied to explain every mass extinction. Nevertheless, the differences now known lead to a better understanding of the probable causes for each individual extinction, which we go through in detail now.

The Ordovician extinction (444 Mya), the first to occur, is the one which most clearly occurred in two separate waves, separated by a period of 0.5 Myr [4]. There is strong evidence that this was accompanied by a global cooling, which may either have occurred through the glaciation of a continent drifting over a pole [7], or a nearby gamma ray burst [8]. The Devonian extinction (360 Mya) shows clear signs of ocean anoxia on the ocean shelf, and the primary culprit is the innovation of land plants, which lead to a drastically increased weathering rate on the continents [9]. The Permian (251 Mya), which was the largest extinction, is noted to coincide with the supervolcano that essentially created modern Siberia [10], as well as the development of the chemical pathways necessary to decompose organic matter that had accumulated on the sea floor for the eons prior, both of which could have drastically altered the climate [11]. The Triassic extinction (200 Mya) is possibly associated with sea level change from the breakup of Pangaea [12], but the cause of this extinction is particularly uncertain. As mentioned earlier, the Cretaceous extinction (66 Mya) was caused by the Chicxulub impact. The cause of the current mass extinction underway today is the most unambiguously established to be the result of a recent biological innovation within homo sapiens that has led to an unprecedented ability to alter the environment [13]. On average, the interval between mass extinctions is around 90 Myr.

In Section 2 we discuss several different models for the biological effect of mass extinctions and the time needed for the biosphere to recover. In Section 3, we discuss the rate of deadly comet impacts, and how this depends on fundamental parameters. In Section 4 we discuss the geological contributions to extinctions, including glaciations, sea level rise, and volcanoes. In Section 5 we discuss the rate of gamma ray bursts. In Section 6 we combine these rates into estimates for the fraction of biospheres that develop intelligence.

2. Rates

2.1. Catastrophes

Before we begin estimating the rates of various potential catastrophic processes, we first derive the fraction of biospheres that develop intelligence for a given rate of mass extinctions $\Gamma_\dagger$. The full rate can be found as the sum of all the various contributions as

$$\Gamma_\dagger = \Gamma_{\text{comets}} + \Gamma_{\text{glac}} + \Gamma_{\text{vol}} + \Gamma_{\text{grb}} + \ldots \tag{1}$$

where the displayed rates are ones we will discuss in the text, though there may potentially be more.

Let us make a brief comment on the implications of the relative rates of each of these, because they all seem to occur with a frequency more or less on the order of 100 Myr-Gyr. This itself is enough to suggest the presence of some selection effect that greatly favors the total rate to be as small as possible. This is reminiscent of a 'law of the minimum', wherein systems wishing to maximize some function of independent variables should only be willing to tune those variables inasmuch as they affect the outcome [14]: in this setting, it would do no good to expend effort making any of these rates arbitrarily small, when the sum will always be dominated by the largest term. Of course, this does not imply the presence of any agent doing the selecting, or indeed even a multiverse: there is plenty of variability within our universe to find a system that happens to be unusually quiet on several different fronts. The purpose of the present paper is to determine to what extent the multiverse is required, or even capable of, explaining this state of affairs.

Now, let us estimate the probability that a biosphere beset by random extinction events develops to the point of intelligence. This will depend on the assumptions for how evolutionary processes operate, and so we will end up with three separate functional forms to test for compatibility with the multiverse. These can be called the setback model, where extinctions cause a relatively short period of reduced biodiversity, the reset model, where extinctions result in a complete loss of progress toward intelligence, and the intermediate disturbance hypothesis model, where there is an optimum rate of extinction. Throughout most of this work we will favor the first model, saving discussion of the other two for Section 6.

Our first model is the setback model of extinctions: here, mass extinctions cause a dramatic reduction of biodiversity that leaves the planet ecologically impoverished for a period of time. After a few speciation timescales, however, niches become repopulated, ultimately reaching the complexity the system exhibited previously. The net effect in this scenario is that the biosphere spends this amount of time in a recovery phase, after which it proceeds as normal. This describes the fossil record well, with the recovery time set by $t_{\text{rec}} \sim 10$ Myr [4].

In this view, extinctions reduce the total amount of time the biosphere can spend exploring evolutionary strategies that may lead to intelligence. Because in [3] we favored a model where the probability of developing intelligent life is linearly dependent on total time (importantly, weighted by the biosphere size, set by total entropy production rate), the fraction of biospheres that develop intelligence would then be proportional to the average amount of undisturbed time. Here, we can model the total amount of habitable time for a system as its stellar lifetime, $t_{hab} \sim t_\star$ (it will actually only be some fraction of this, but this sets the rough timescale). Then the number of extinction events n will be given by a Poisson distribution $p(n)$ with average $\Gamma_\dagger t_{hab}$. A simple estimate for fraction of biospheres that develop intelligence is $f_{int}^{setback} = \mathbb{E}((1 - n/n_{max})\theta(n_{max} - n))$, where $n_{max} = t_{hab}/t_{rec}$ is the number of extinctions which would correspond to the system being bombarded, on average, more frequently than it can recover. This is a somewhat simplified prescription, since it neglects the case where all hits are concentrated in a small interval, after which the system settles down to let life proceed unhindered; however, this situation will be rare, and it suffices to provide a simple formula encapsulating these effects through which the dependence on physical parameters can be tracked. Then we find

$$f_{int}^{setback} = \frac{(\Gamma_\dagger t_{hab})^{n_{max}+1}}{n_{max}! \, n_{max}} \left(e^{-\Gamma_\dagger t_{hab}} + (n_{max} - \Gamma_\dagger t_{hab}) \, \mathrm{E}_{-n_{max}}(\Gamma_\dagger t_{hab}) \right) \qquad (2)$$

where $E_a(b)$ is the exponential integral. This somewhat cumbersome expression can be approximated to essentially indistinguishable precision by

$$f_{int}^{setback} \approx (1 - \Gamma_\dagger t_{rec}) \, \theta(1 - \Gamma_\dagger t_{rec}) \qquad (3)$$

where $\theta(x)$ is the Heaviside step function. The only difference from the full expression occurs past $\Gamma_\dagger \sim 1/t_{rec}$, so that this approximation discounts incredibly rare survivors. For instance, if $\Gamma_\dagger t_{rec} = 1.3$, the full expression gives 10^{-11} rather than 0. The essential feature here is that if the extinction rate exceeds the recovery rate, intelligence will never develop. Given the simplicity of the last expression and its conveyance of this key property, this will be the form we will use.

Next, we discuss a second model of how extinctions affect biospheres, which can be called the reset model. This takes the view that extinction events set the clock back to zero, so that life must start over from scratch each time. In this model, there is a noogenesis timescale $t_{noo} \sim$ 100 Myr that is required to develop intelligence, and any progress toward this outcome is erased in the course of a mass extinction. The probability of a biosphere developing intelligence is then simply equal to the probability that a time interval equal to t_{noo} exists at some point during the planet's history for which no mass extinctions occur. In favor of clarity, we opt for a simplified, but easier estimate of the probability: if we break up the total lifetime into $N = t_{hab}/t_{noo}$ intervals and ask what the probability is that at least one of these is undisturbed, we find

$$f_{int}^{reset} = 1 - \left(1 - e^{-\Gamma_\dagger t_{noo}}\right)^{\frac{t_{hab}}{t_{noo}}} \qquad (4)$$

This is approximately a step function enforcing $\Gamma_\dagger < 1/t_{noo}$, but with a somewhat large tail.

A third viewpoint is known as the intermediate disturbance hypothesis: this is the notion that instead of life faring best under the most placid circumstances, there is some value of the extinction rate that maximizes biodiversity [15]. This is a relatively standard idea in ecology, though it only holds for some ecosystems [16]. A method of determining which systems it is applicable to has recently been developed [17], though it is certainly still too premature to settle whether this idea holds for the biosphere as a whole. Adopting this view for the entire biosphere is no doubt fueled by the somewhat narcissistic observation that had the dinosaurs never died out, then we mammals would never have had a chance to radiate, and intelligent life may never have evolved. Nevertheless, it may have merit, and can readily be included in our analysis. For this model, we have

$$f_{\text{int}}^{\text{IDH}} = 4\, \Gamma_\dagger\, t_{\text{dist}}\, (1 - \Gamma_\dagger\, t_{\text{dist}}) \qquad (5)$$

This fraction is maximized at $\Gamma_\dagger = 1/(2 t_{\text{dist}})$.

The above three functions are plotted in Figure 1 as a function of $\Gamma_\dagger$. All of them have been normalized to 1 at their maxima, seemingly implying that if biospheres are left alone they are guaranteed to develop intelligence. However, there are surely other factors that may affect this: several that were discussed in [3] are the amount of time a planet spends in the habitable zone, the total entropy it processes, and planet size, though there are undoubtedly many more. We leave any additional considerations to future work, noting that we expect these effects to be largely treatable in a fashion that factorizes from the effects we deal with here.

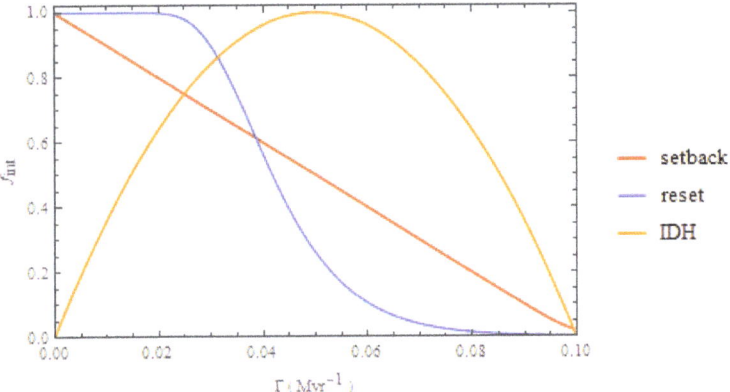

Figure 1. Different parameterizations for f_{int}. The red curve is Equation (2), the blue Equation (4), and the orange Equation (5).

2.2. What Sets the Recovery Time?

All our analysis relies not only on the extinction rate $\Gamma_\dagger$, but also the recovery time t_{rec} (or noogenesis time or disturbance time, resp.). Somewhat vaguely, this should be related to the rate of evolution or speciation, but it is far from clear how to connect this to the fundamental properties of physics. Here, we entertain several different educated guesses for what this dependence may be, and then track how the final outcome depends on our choice. The two scenarios are that the recovery time is set by the rate of chemical processes, which dictate the rate of mutations, or else that the recovery time is set by the length of the year, which may dictate the generational timescale for complex organisms.

Our first guess is that the recovery time is set by molecular evolution timescale. This should have some bearing on the rate of genetic mutations, which in turn will dictate how fast speciation can occur. In [18], the molecular timescale was found to be equal to the inverse of the typical molecular binding energy, so if the recovery time is proportional to this, we have

$$t_{\text{rec}}^{\text{mol}} = 3.0 \times 10^{29} \, \frac{m_p^{1/2}}{\alpha^2\, m_e^{3/2}} \qquad (6)$$

Here α is the fine structure constant, m_e is the electron mass, m_p is the proton mass, and below M_{pl} is the Planck mass and λ is a dimensionless measure of the stellar mass, as in [1]. The large coefficient, while somewhat off-putting when regarded as a supposed constant independent of physics, is necessary to bridge the gap between these two extremely disparate timescales.

Alternatively, we may take the view that the recovery time is set by the lifespan of macroscopic organisms. It was argued in [19] that this is set by the year length, as organisms would take advantage of nature's cyclic variations for feeding and reproductive purposes. If this is so, then the recovery rate is instead given by

$$t_{rec}^{year} = 5.9 \times 10^8 \frac{\lambda^{17/8} m_p^{7/4} M_{pl}^{1/4}}{\alpha^{15/2} m_e^3} \quad (7)$$

This again requires a rather large numerical coefficient. These may ultimately be related to the rate of mutations and the size of complex organisms, but we do not go further into detail on these points, because these are more likely dictated by the laws of complex systems rather than the underlying physical substrate.

Though the basis for evolutionary timescales is still not completely known, both of these can be used as hypotheses for our analysis. They lead to different effects: if organism lifespan is truly set by the length of the year, then planets which orbit further from their stars would possess life that lives on a much longer timescale, which would not be observed if it is instead determined by the molecular time. In the other case, however, hotter planets, which have shorter molecular timescales, would have faster evolution. While distinguishing these two scenarios experimentally is probably a ways off (we may like to find a second sample of life first), they do yield differences which are in principle observable.

In the following sections, we will use the setback model and the molecular recovery timescale for definiteness. In Section 6 we will fully explore the alternative choices as well.

3. Comets

We now wish to investigate the effect of fundamental parameters on Earth's impact rate. These disruptive events, though rare, can cause a significant effect on complex ecosystems worldwide, as evidenced by their contribution to at least one of the five mass extinctions in Earth's history [4,5]. To get a measure of the rate of impacts, it is essential to know the source of these impactors, as well as their properties.

In general, there are three sources of impacts that may affect our planet. These are asteroids, short period comets, and long period comets. Asteroids are the planetesimal scraps situated between Mars and Jupiter, situated at 2–3 AU. This region of the solar system is in one of Jupiter's orbital resonances, which lead to the conditions that prevent these bodies from condensing into a full planet. Their composition is highly differentiated as a result of primordial heating from a variety of sources, making them rocky. We do not focus on these in this paper, as their properties, such as total number, orbital distribution, and rate of perturbations is highly dependent on details of solar system architecture [20], and so will likely be highly environmentally variable.

There are two populations of comets: those with orbital periods the same order of magnitude as the planets in the outer solar system, and those with orbital periods much larger. Orbits that enter the inner solar system are unstable over geologic time, and so the very existence of such populations indicates a vast reservoir for both. For the short period comets, this is the Kuiper belt, that band of small bodies with orbits on the order of 100 AU that encompasses the dwarf planet Pluto. The long period comets have been inferred to come from an even greater reservoir that extends from 10,000–100,000 AU known as the Oort cloud. This inference is based on both the observed number of long period comets, as well as the orbits of all new comets being practically parabolic rather than hyperbolic, which would be indicative of an extrasolar source (see [21,22] for a reviews). Since these bodies extend so far out into the outer reaches of our solar system, they are routinely influenced by the galactic environment in a variety of ways, which in turn injects them into the inner solar system, with potential to collide with Earth.

The Oort cloud was initially formed out of material from the gas giant region of the solar system. Perturbations from these giant planets caused the originally nearly circular orbits of the comets to elongate secularly over time, increasing the semimajor axis, while preserving the perihelion distance [22]. If this process were allowed to continue indefinitely, most of these bodies would have ultimately been ejected from the solar system into interstellar space. However, once the orbits crossed a threshold of $a_{inner} \sim 1000$ AU, external perturbations from passing stars also perturb the orbits. These perturbations increase the perihelion out of the inner solar system, thereby preventing any further perturbations from the outer planets occurring. Once this happens, the object is stuck in a nearly stable orbit surrounding our sun until further perturbations either kick it back into the inner solar system, or out of the system completely. Further out, on the order $2-3 \times 10^4$ AU, the inclination or the orbit may also be changed [23]. This leads to a distinction between the inner and outer Oort cloud: inside this, comets orbit within the plane of the solar system, while beyond, the orbits splay into a sphere encompassing our sun.

From observed injection rates, it is estimated that there are approximately 10^{12} comets of diameter larger than 1 km within the Oort cloud [22]. At these densities, interactions between Oort cloud objects are negligible for our purposes.

3.1. Comet Dynamics

Once a comet is placed in the Oort cloud, its orbit is relatively stable. It is, however, subjected to perturbations that can eventually cause it to reenter the inner solar system, potentially causing an impact on Earth. The galactic tidal force is the dominant cause of reentry, but we first list the other forces for completeness.

Nongravitational forces are mostly caused by the sublimation of ices on the comet, causing the iconic comae that have been observed for thousands of years. This effect only happens once the comet crosses the ice line of the solar system, around 3 AU, and so plays no role while it is in the Oort cloud [21]. Once a comet does reenter the inner solar system, however, these forces play a significant effect on perturbing the orbit. Similarly, planetary perturbations also only play a role once a comet has entered the inner solar system. These were of prime importance for the creation of the Oort cloud [24,25], but not for its subsequent dynamics. The passage of our solar system through molecular clouds has the potential to significantly perturb cometary orbits. However, it was found in [26] that this effect is nonnegligible only for giant clouds, and so can only be expected to be operational once every Gyr or so. Encounters with other star systems can alter the orbits of comets; the density of stars in our galactic neighborhood is 0.185 $M_\odot/pc^3$, leading to a close encounter of a star within 1 pc of our system about every 100,000 years. The orientation of these encounters is practically random, so that in effect these cause the orbital parameters of the Oort cloud constituents to execute random walks.

The galactic tidal force is caused by the fact that the solar system has finite size, and so the Milky Way exerts a torque throughout the system. In contrast with the effect of close encounters, this effect is directed, causing the orbital parameters to decrease with time (for certain orbits) [22]. The magnitudes of these external influences are proportional to high powers of the semimajor axis a, and so these effects are utterly negligible on asteroids and short period comets, but are the most important perturbations on the Oort cloud. These also set the outer boundary of the Oort cloud as the Hill radius of the sun. A rough estimate of this can be found by making the approximation that all the galactic mass is concentrated at a point at the center of the galaxy [27], and is found to be

$$a_{outer} \sim \left(\frac{M_\odot}{\rho_{gal}}\right)^{1/3} = 0.0037 \frac{\lambda^{1/3} M_{pl}}{\kappa m_p^2} \qquad (8)$$

corresponding to 0.5 pc, or 10^5 AU[1]. It is possible to do a more sophisticated analysis by modeling the galactic disk: this leads to the conclusion that the Oort cloud is actually an ellipse, but does not change the size of the cloud [23].

The rate of injection into the inner solar system can be computed as [21]:

$$\Gamma_{comets} = N_{comets} f_{>d_\dagger} f_{inj} f_{hit} \frac{1}{P} \qquad (9)$$

The quantities in this equation are as follows: N_{comets} is the total number of comets, $f_{>d_\dagger}$ is the fraction of comets large enough to cause a mass extinction, f_{tide} is the rate of injection (per orbit, as to be dimensionless), f_{hit} is the fraction of injected comets that hit the Earth, and P is the period of the comets, given by $P = 2\pi a^{3/2}/\sqrt{GM_\odot}$. In truth these all depend on the radial distribution of comets within the Oort cloud. This is generally expected to decrease with semimajor axis a power law $N(a) \propto a^{-\gamma}$; depending on the processes involved, simulations favor $\gamma = 2.5 - 3.5$ [22] and $\gamma = 3.2 \pm 0.3$ is measured from observations in [28]. We make the simplification, however, that comets orbit at a characteristic radius a_{Oort}. A full treatment would integrate over the orbital radii of Oort cloud objects, weighted by the distribution of semimajor axes- however, the magnitude of the perturbing force depends quite strongly on orbital size, so the integral is dominated by the outermost orbits.

The fraction of comets injected into the inner solar system per period can be computed by considering the secular change in angular momentum, relating this to perihelion distance, and then conditioning on the perihelion to be within the inner system, as in [21]. The radius of the inner solar system is given by the orbits of the outer planets, which is about 15 AU. The giant planets' locations are set as a small multiple of the snow line, $a_{planets} = 5.6 \, a_{snow}$, where the snow line was found in [2] to be $a_{snow} = 30.4 \, \lambda M_{pl}/(\alpha^{5/3} m_e^{5/4} m_p^{3/4})$. However, the injection rate, as computed in [21], is actually independent of this quantity, being given by the ratio of kinetic to gravitational energy $f_{inj} = \Delta v^2/(2 \, GM_\odot/a_{Oort})$. This is because the perturbation in orbit is much smaller than the inner system size and is secular, leading to a steady supply of precarious comets right at threshold of entry: the typical change in speed per orbit due to the tidal perturbing force is given by $\Delta v = G\rho_{disk} aP$. The injection rate is dependent on the square of the perturbing force because for parabolic orbits, angular momentum is proportional to the square root of the perihelion distance $h = (2 \, GM_\odot q)^{1/2}$, and since the distribution of perihelia is uniform, the distribution of angular momentum in linear. When integrating over the entire 'loss cone', the fraction is then proportional to the square of the threshold angular momentum. Combining these elements gives

$$f_{inj} \sim \left(\frac{\rho_{disk} a_{Oort}^3}{M_\odot}\right)^2 \qquad (10)$$

This expression has the interesting feature that since the Oort cloud radius is set by the typical interplanet spacing in Equation (8), all dependences drop out, and f_{inj} becomes a pure number.

Once a comet is injected into the inner solar system, more likely than not, perturbations will alter its trajectory to a hyperbolic orbit, ejecting it from the system. The probability that it will impact on any of the rocky planets was found there to be $f_{hit} = 1.3 \times 10^{-7}$ [21]. This number is set as the ratio of the Earth's gravitational cross section to that of the sun's. Because comets' velocities are orbital, they are larger than the Earth's escape velocity[2] and so the Earth's cross section is given by its geometric value $\sigma_\oplus = \pi R_\oplus^2$. In contrast, the sun's escape velocity is larger than typical comet speeds, so its cross section

---

[1] Here, and in the following, the expressions from our previous appendices [1,2] are used. Here κ is a dimensionless measure of galactic density.

[2] This hierarchy does not actually hold for all parameters, but the full implications of this will be explored in future work.

is enhanced by gravitational focusing to be $\sigma_\odot = 2\pi G M_\odot R_\odot/v_{comet}^2$. The fraction that impinge on Earth is then given by

$$f_{hit} \sim \frac{R_\oplus^2}{2R_\odot a_{temp}} = 0.005 \frac{\alpha^6 \beta^{1/2}}{\lambda^{51/20} \gamma^{1/2}} \tag{11}$$

We use here $\beta = m_e/m_p$ and $\gamma = m_p/M_{pl}$.

Now, to estimate N_{comets}: the estimated size of the Oort cloud from formation scenarios is $3-4\,M_\oplus$, or about 3×10^{11} objects of km size or greater [22]. The total mass ejected during planet formation is dictated by the amount of material in the outer regions of the protoplanetary disk, and so is related to the disk density Σ by the expression $M_{ejected} \sim 0.01\,\pi a_{planets}^2 \Sigma(a_{planets})$. Most ejected material, however, was removed from the solar system completely. To estimate the amount placed on Oort cloud orbits, [29] found this to be reduced by the factor $a_{planets}/a_{Oort}$. This quantity is dominated by the outermost planets and the innermost Oort cloud orbits, so that Neptune is the most important perturbing agent.

From [30], the inner edge of the Oort cloud can be obtained by setting the timescale of tidal secular evolution equal to the ejection time, and solving for semimajor axis[3]:

$$a_{inner} \sim \frac{M_{Neptune}^{4/3}}{M_\star^{2/3} \rho_{gal}^{2/3} a_{Neptune}} = 1.1 \times 10^5 \frac{\lambda\, M_{pl}}{\alpha^{5/3} m_e^{5/4} m_p^{3/4}} \tag{13}$$

Here, we have used that the size of ice giant planets is set by the isolation mass evaluated a few times further out than the snow line, such that

$$M_{Neptune} = 3.5 \times 10^{10} \frac{\kappa^{3/2} \lambda^2 M_{pl}^3}{\alpha^{5/2} m_e^{15/8} m_p^{1/8}} \tag{14}$$

Equal to 17 $M_\oplus$.

Then the total Oort cloud mass is

$$M_{Oort} = 1.1 \times 10^{14} \frac{\kappa^2 \lambda^{7/3} m_p^{1/2} M_{pl}^3}{\alpha^{10/3} m_e^{5/2}} \tag{15}$$

To compute the number of comets, we now calculate the typical comet size d_{comet}, which is set by the accretion that can occur before ejection. The presence of the ice giants will impart a change in energy to all smaller bodies in their neighborhood; these will then diffuse outward with a timescale of 100 Myr, set by [30]

$$t_{eject} \sim 0.01 \frac{M_\star^2}{M_{Neptune}^2} P = 3.5 \times 10^{-16} \frac{\alpha^{5/2} m_e^{15/8} M_{pl}}{\kappa^3 \lambda\, m_p^{31/8}} \tag{16}$$

---

[3] The inner edge will exceed the outer for stars above the mass

$$\lambda_{none} = 5.7 \times 10^{-12} \frac{\alpha^{5/2} \beta^{15/8}}{\kappa^{3/2}} \tag{12}$$

so that stars larger than this value, corresponding to 17 $M_\odot$ in our universe, will not possess Oort clouds (assuming a similar planetary system architecture to the solar system). However, this bound is of little importance, as it only affects very massive stars unless α or β are several times smaller than their observed values.

From [29], the growth rate of a body is given by $\dot{M} = 2\pi G^2 M^2 \rho/c_s^3$. The ambient density can be expressed in terms of the disk surface density as $\rho \sim \Sigma/(c_s a)$, giving the characteristic size of Oort cloud objects to be $M_{\text{comet}} \sim T^2 a/(m_p^2 G^2 \Sigma t_{\text{eject}})$. This leads to a characteristic radius

$$d_{\text{comet}} = 2501 \frac{\kappa^{2/3} \lambda M_{pl}^{5/6}}{\alpha^{25/9} m_e^{3/2} m_p^{1/3}} \qquad (17)$$

This is set to be 1 km.

With all these, the rate of cometary impacts on Earth can be estimated. Altogether, this leads to

$$\Gamma_{\text{comets}} = 3.1 \frac{\kappa^{3/2} \alpha^8 m_p^{3/2}}{\lambda^{16/5} m_e^{1/2}} \min\left\{ \left(\frac{d_{\text{comet}}}{d_\dagger}\right)^p, 1 \right\} \qquad (18)$$

The normalization here is chosen to reproduce one mass extinction every 90 Myr (which would only be the desirable prescription if comets were the sole cause of extinctions). We have used that the fraction of comets large enough to cause a mass extinction is given by a power law, which holds over 16 orders of magnitude [31]. For the slope we use $p = 1.5$, in agreement with that found in [28], though there is considerable uncertainty in the measurement of this quantity. The only remaining quantity to estimate is the size of comets which will cause mass extinctions.

3.2. What Sets the Size of Deadly Comets?

From the geologic record, the size of the smallest impactor that can lead to global disruptions must be slightly more than 10 km, since the Chicxulub impact, at 14 km, caused a mass extinction, whereas the next largest impacts that occurred since the advent of complex life, the Popigai and Manicouagan, both 10 km, did not.

In fact, size is not strictly the sole determiner of the magnitude of the environmental perturbation. Other important factors include the mineral composition of the location of impact: the K-Pg crater location, being a shallow sea, was particularly laden with the mineral gypsum [32], making it an abnormally sulfate-rich site. Additionally, the kinetic energy of the impactor is more relevant, and there is a large spread in the speeds of comets relative to Earth[4]. Nevertheless, in terms of average conditions, the diameter of the impacting comet will dictate the strength of environmental response.

The limiting size depends on the exact mechanism of extinction during an impact event. This is not as straightforward to deduce as one might naively expect, essentially because many Earth systems are catastrophically affected during such a calamity, in many different ways [32]: first, a rather large portion of rock around the initial impact site is vaporized and thrown into the stratosphere. Dust can linger for months, darkening skies to the point where photosynthesis (and even vision) are impossible. Larger rocks can be strewn ballistically over the entire globe, the reentry of which may ignite forests worldwide. Nitrous oxide compounds are created in bulk during initial atmospheric deposition, potentially destroying the ozone layer. Sulfate compounds are liberated during the impact, leading to an intense global cooling that may last decades, and associated acidification that will result in widespread die-offs. If the impact occurs in the ocean, massive tsunamis will wreak havoc on shallow water and coastal ecosystems, and drastically alter the amount of water in the atmosphere. Given this litany of utterly brutal catastrophes, it is small wonder that there are differing ideas to the ultimate cause(s) of extinction.

---

[4] This is the reason comets can be deadlier than asteroids, even though they are less dense: since they can come from any direction rather than being roughly coorbital with the Earth, the average speed will be $\sqrt{3}v_\oplus = 52$ km/s, rather than $(\sqrt{3}-2\sqrt{2})v_\oplus = 12.4$ km/s, a factor of 17.5 more energy. Therefore, comets can be smaller and still impart more energy than asteroids, and so their rate of deadly impacts will be more numerous (which depends on their relative population sizes as well, of course).

The impact that triggered the K-Pg (end Cretaceous) extinction can be of great use here. Though all of these mechanisms are fiercely debated, by now the most plausible seems to be the climatic effect of sulfate injection. It was argued in [33] that not enough submicrometer dust was produced in the impact to cause significant attenuation of sunlight. In [34], it was found that the charcoal record, an indicator of forest fires, was not above the background level at the K-Pg layer, even for sites located in the Americas. It was argued in [35] that marine extinctions are consistent with acidification, rather than darkness-induced productivity collapse. In [36] it was argued that not enough NO_x was created to trigger a significant decrease in ozone layer. Climate modeling in [37] indicates that the amount of SO_x (x = 2,3) created is enough to cause 10–20° global cooling, depending on the uncertain residence time of these molecules. In the following we track the minimum size of an impactor for both the production of SO_x and of dust, especially since the argued relevance of both mechanisms indicates that the threshold diameters may be close in magnitude. Since these scale differently with physical parameters, this coincidence cannot be explained by selection effects operating purely within our universe, and may be a hallmark of anthropic selection.

3.2.1. Sulfate

We begin with the production of sulfate, a gas with the capacity to block sunlight and a residence time of years to decades. To find the minimum size of an impactor necessary for this effect to operate, we relate the optical depth of the sulfate material produced to the mass of the impactor.

Sulfate is generated by the vaporization of the surrounding crater at the impact site. Here, the initial kinetic energy is first converted into breaking molecular binding energy. The number of bonds that can be broken is given by

$$N_{SO_x} \sim \frac{m_i v_i^2}{E_{mol}} \quad (19)$$

Here, E_{mol} is set by the molecular binding energy, though the precise value is determined by assessing exactly how energy is distributed in the surrounding minimum during the initial shock wave [32,38].

Dust and debris created during the impact will rise as a plume high into the atmosphere, nearly vertically. Any that stays within the troposphere will only have a local effect, but the material that makes it to the stratosphere, where horizontal mixing is fast, will cover the Earth within a matter of hours. The energy needed for material to reach the stratosphere is orders of magnitude less than that needed to have a substantial effect on ecosystems, achieved even with paltry atomic bomb blasts [39], so it is a generic feature, of terrestrial planets in any universe, that distribution of material will be global. Here we assume the material will be dispersed relatively uniformly, and so the optical depth will be $\tau = N_{SO_x} \sigma_T / A_\oplus$, where σ_T is the molecular cross section. The critical optical depth is uncertain but close to 1, and a combination of this and the energy required per molecule of sulfate can simply be matched with the observed critical comet size in order to find the dependence on physical parameters. We find

$$d_{SO_x} \sim \left(\frac{E_{mol} A_\oplus}{\rho v_i^2 \sigma_T}\right)^{1/3} = 10.1 \frac{\lambda^{1/4} M_{pl}^{1/2}}{\alpha^3 m_e m_p^{1/2}} \quad (20)$$

This quantity depends most strongly on the electron mass and the fine structure constant: if either of these were significantly larger, the size of globally catastrophic comets would be much smaller, leading to an enhanced rate. The dependence of this quantity on the physical parameters is plotted in Figure 2, along with the other scales for comparison.

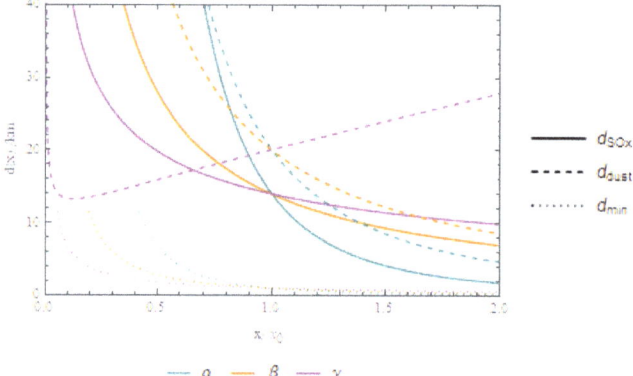

Figure 2. The dependence of the various cometary scales on physical parameters. Though these curves scale the same way with α and β, when varying with respect to γ a maximum size is reached close to our observed value. The typical comet size is the smallest of the three except for very small values of the parameters.

3.2.2. Dust

Though dust does not appear to have been the primary cause of extinction for the K-Pg event [33], it is still interesting to determine the mass of an impactor that would be required for this mechanism to go into effect, and not just for pure agnosticism toward differing environmental impact models. The reason is that the sensitivity of this hazard to fundamental parameters is much greater than the sulfate case, and so, for different values, dust is the main contribution to extinctions.

The main reason dust is not a major concern for our values is that when the crater is vaporized in the impact event, most of the dust produced is on the order of the grain size of crystals comprising the crust, which is $r_0 = 100$ µm [33]. Particles of this size have atmospheric residence times on the scale of days, and so any effect will quickly dissipate [32]. Submicrometer dust is required for any lasting impact on the environment, but very little is produced. We first determine how small a dust grain has to be in order to linger in the atmosphere, and then calculate the amount of dust smaller than this that is produced for an impact of a given size.

The residence time of a dust grain can be estimated by dividing the height at which it is deposited in the atmosphere by the rate at which it falls. Most of the plume from the impact is deposited at the base of the stratosphere, and so we simply use a multiple of the atmospheric scale height $H_{atm} = T/(m_a g)$, where m_a is average molecular weight of atmospheric gas. The terminal velocity is given by $v \sim \sqrt{gr_{dust}}$. The residence time is then $t_{res} \sim T/(m_a g^{3/2} r_{dust}^{1/2})$.

This can be used to determine the largest dust size that is capable of having a lasting impact on the environment, if the threshold residence time is known. However, this immediately becomes a very tricky quantity to define. Even on Earth, a planet we are relatively familiar with, the amount of time each photosynthesizing organism can go without sunlight is highly variable, and how many need to die for a complete ecosystem collapse is a difficult question to address. Nevertheless, several weeks seems to be a reasonable estimate. On other planets, we may assume roughly the same level of hardiness, though this may not be valid if the planet varies drastically, such as on planets which are tidally locked, or have extreme obliquity. On planets in another universe with different fundamental constants, any prediction for how long plants can survive without sunlight should be taken with extreme skepticism. Nevertheless, to make progress, we choose to take the threshold residence time to be several dozen days, where the length of a day is several times larger than the centrifugal limit, as on Earth. This sets a critical value for the size of dust particles that will affect the atmosphere for a prolonged period as

$$\frac{r_{\text{float}}}{r_0} \sim \frac{T^2}{m_a^2 g^3 (20 t_{\text{day}})^2 r_0} = 3.0 \times 10^{-19} \frac{\alpha^{1/2} \beta^{1/4}}{\gamma} \tag{21}$$

Notice that this is most sensitive to γ, and if it were much smaller, a significant fraction of the impact would stay in the atmosphere long enough to affect life. The main reason for this is that the pull of gravity will then be weaker, though this effect is partially compensated by the fact that then days will be longer as well. Because of this, smaller comets would be capable of having a drastic effect, and the overall rate would increase.

To relate this to the size of a comet needed, we find the optical depth of the dust injected into the atmosphere, as before. The key difference is that now, of the total amount of dust produced, we are only interested in the amount below this threshold value. The distribution of dust grains is observed to be lognormal distribution over microscopic ranges. Here, in accordance with [32,33,38], we take $r_0 = 100$ μm. To set the fraction of submicron dust to be 0.1% from [32], we take $\sigma_{\text{float}} = 1.24$ (whereas if we set the fraction to be 0.01%, as in [33], $\sigma_{\text{float}} = 1.08$). Then the number of particles that remain in the atmosphere is

$$N_{\text{dust}} = \frac{M_{\text{vap}}}{\frac{4\pi}{3} \rho r_0^3} f_{\text{float}} \tag{22}$$

where M_{vap} is the total vaporized mass, and the fraction of dust that stays in the atmosphere is

$$f_{\text{float}} = \frac{1}{2} + \frac{1}{2} \text{erfc} \left(\frac{\frac{1}{2} \sigma_{\text{float}}^2 + \log(r_{\text{float}}/r_0)}{\sqrt{2} \sigma_{\text{float}}} \right) \tag{23}$$

As with the sulfates above, the optical depth is given by $\tau = N_{\text{dust}} \sigma_M / A_\oplus$, except that here the cross section is given by the geometric size of the particles. The threshold values of the optical depth required for photosynthesis and human vision are given in [32] as $\tau = 29$ and $\tau = 143$, respectively. In practice, it matters very little which of these values we take, as the size of comet needed to produce the second effect is only a few times larger than the first. This is

$$d_{\text{dust}} \sim \left(\frac{E_{\text{mol}} A_\oplus r_0}{m_p v_i^2 f_{\text{float}}} \right)^{1/3} = \frac{1022}{f_{\text{float}} \left(3 \times 10^{-19} \alpha^{1/2} \beta^{1/4} \gamma^{-1}\right)^{1/3}} \frac{\lambda^{1/4} M_{pl}^{1/2}}{\alpha^{5/3} m_e m_p^{1/2}} \tag{24}$$

For definiteness, this has been normalized to 20 km. This length scale is compared to the size needed for sulfates and the typical comet size in Figure 2.

The size of dangerous comets is the smaller of these two scales, $d_\dagger = \min\{d_{\text{SO}_x}, d_{\text{dust}}\}$. Even though this has a maximum when varying the strength of gravity, when comparing to the dimensionless ratio given by dividing by the comet radius there is no turnover, just a change in slope. More importantly, the quantity $\Gamma_{\text{comet}} t_{\text{rec}}$ does not have a turnover either, as can be seen from Figure 3. There, the probability of observing any value of the constants α, β and γ is displayed, based solely on the number of stars and the fraction of those affected by comets. To give probabilities that are compatible with our observations, this analysis must be included with some of the other factors that affect habitability as discussed in the previous papers of this series. Throughout, we will take the entropy and yellow conditions as our baseline, as detailed in [1,3], and check how extinctions hinder this already viable criterion. A more thorough analysis would run through all the previous combinations to check which are affected by extinctions, but the results of this would be much too cumbersome to report on in this paper.

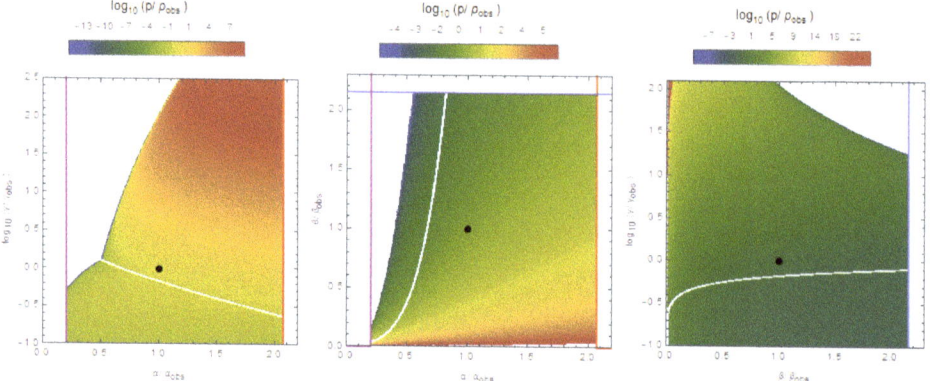

Figure 3. Distribution of observers taking comet impact extinctions into account. The black dot denotes values in our universe, and the orange, blue, and purple lines are the hydrogen stability, stellar fusion, and galactic cooling thresholds discussed in [1]. The white line corresponds to the discontinuity where $d_{\text{dust}} = d_{SO_x}$.

If just the sulfate radius is used, the probabilities of observing our values of the constants are

$$\mathbb{P}(\alpha_{obs}) = 0.07, \quad \mathbb{P}(\beta_{obs}) = 0.17, \quad \mathbb{P}(\gamma_{obs}) = 0.30 \tag{25}$$

where we use the setback model of mass extinctions and the molecular timescale parameterization of recovery time from Section 2. If just the dust radius is used, then

$$\mathbb{P}(\alpha_{obs}) = 0.14, \quad \mathbb{P}(\beta_{obs}) = 0.09, \quad \mathbb{P}(\gamma_{obs}) = 0.28 \tag{26}$$

while for both[5],

$$\mathbb{P}(\alpha_{obs}) = 0.13, \quad \mathbb{P}(\beta_{obs}) = 0.08, \quad \mathbb{P}(\gamma_{obs}) = 0.29 \tag{27}$$

The numbers shift by a factor of two, but there is little difference between these three scenarios. Interestingly, however, even though sulfate production seems to have been the dominant cause of extinction in Earth's previous episodes, the threat of dust more strongly dictates our position in this universe. These should be compared to the values when extinctions are not taken into account, $\mathbb{P}(\alpha_{obs}) = 0.19$, $\mathbb{P}(\beta_{obs}) = 0.44$, $\mathbb{P}(\gamma_{obs}) = 0.32$.

4. Volcanism, Glaciations, and Sea Level Change

Three of the other main impetuses of mass extinction, volcanism, glaciations, and sea level change, all depend directly on the amount of heat generated from the mantle. The rate of volcanoes perhaps explicitly, as the yearly output can be directly linked to internal heat, and the observed power law can then be used to extrapolate to the rate of biosphere-altering supervolcanoes. Glaciations and sea level change are set by this tempo as well, as their occurrence is a direct consequence of continental drift leading to a rearrangement of the Earth's land surface capable of triggering a climate instability into a secondary equilibrium. Glaciations as such can occur when an isolated continent is situated over one of the Earth's poles, as occurred with Antarctica 34 Mya, which triggered a glaciation and coincident cooling of the Earth by several degrees [40]. Though the subsequent global extinction was relatively minor, a similar event has been implicated in the Ordovician mass extinction [4], this time

---

[5] The code to compute all probabilities discussed in the text is made available at https://github.com/mccsandora/Multiverse-Habitability-Handler.

with the glaciation of Gondwana. Similarly, the original formation of Pangaea lead to a reduction of coastal area, triggering a marine dieoff in the Devonian (and Rodinia as well for stromatolites in the Proterozoic [41]). Since the majority of marine bioproductivity is situated close to continents, any rearrangements have the potential to trigger catastrophic instabilities. Coral reef ecosystems, for example, are extremely sensitive to the changes in available sunlight that accompany sea level change, even by a few meters. Whether by sea level change due to glaciations, or the eventual closing up of intercontinental seaways, both processes depend on the rate of continental rearrangement.

4.1. Glaciations

Let us estimate the rate of glaciations/sea level change first. A simple estimate of this is just given by

$$\Gamma_{\text{glac}} \sim \frac{v_{\text{drift}}}{R_\oplus} \tag{28}$$

In our previous paper ([3] and references therein), we used an expression for the continental drift rate $v_{\text{drift}} \sim Q/(A_\oplus n \Delta T)$, which derives this quantity in terms of the internal heat flow Q, number density n, area of Earth $A_\oplus$, and temperature difference ΔT. This expression then simplifies,

$$\Gamma_{\text{glac}} \sim \frac{Q \, m_p}{M_\oplus \, E_{\text{mol}}} = 29.6 \, \frac{Q \, m_p^{17/4}}{\alpha^{7/2} \, m_e^{9/4} \, M_{pl}^3} \tag{29}$$

Using $Q = 47$ TW, this indeed yields $\Gamma_{\text{glac}}^{-1} \sim 100$ Myr. Once an expression for Q is used, this can then be incorporated into f_{int} to determine which values of the constants are compatible with the emergence of intelligent observers.

The Earth's internal heat is somewhat subtle, though, since it depends on time and has multiple distinct sources. If we use the naive dimensional analysis estimate we first found in [3], $Q_{\text{naive}} \sim 92.5 \alpha^{9/2} m_e^{7/2} M_{pl}/m_p^{5/2}$, we find that $\Gamma_{\text{glac}} t_{\text{rec}} = 8.3 \times 10^{32} \gamma^2/(\alpha \beta^{1/4})$. If this naive estimate is used, then increasing the strength of gravity by a factor of 3 would increase the rate of glaciations to below the recovery timescale! This is certainly a very big departure from what we've been discussing to this point, where the strength of gravity could vary by two orders of magnitude. Since this was the basis of exclusion for many habitability criteria which favor larger values of this quantity, a drastic reduction in anthropically allowed space such as this would alter our previous conclusions.

However, this sharp boundary is spurious, and disappears if the time dependence of Earth's heat flux is taken into account. For the heat of formation, we found before that

$$Q_{\text{form}} = 2.6 \, Q_{\text{naive}} \frac{1}{s} e^{-s}, \quad s = 10 \, \alpha^{1/2} \beta^{5/8} \gamma \sqrt{m_p \, t} \tag{30}$$

When this is used, the exponential dependence on γ balances the prefactor, so that stronger gravity no longer obviates the development of complex life.

Above, we only took the heat of formation into account when estimating the rate of glaciations. However, an additional source of heat comes from radioactive elements in the mantle decaying. Indeed, these two sources of heat are comparable [42]. Given this intriguing fact, along with the high sensitivity of this latter form of heat to the fine structure constant uncovered in [43], it is worth thoroughly investigating the interplay between these two quantities for generic values of the parameters.

Generically, the heat generated by radioactivity in the mantle is given by

$$Q_{\text{rad}} = \sum_i f_i \frac{E_i}{t_i} e^{-t/t_i} \tag{31}$$

where f_i is the fraction of species i in the mantle, and the sum is performed over all radioactive species. This is a tall order, especially since the relevant species depend on the constants themselves.

For tractability, we note that in [43], when this sum was performed in earnest, the resultant heat resembled a Gaussian which peaked at $\alpha = 1/144$ at a value 2.7 times our observed heat. This allows us to use a simplified expression

$$Q_{rad} \sim 5.6 \times 10^{-51} \frac{\alpha^{3/2} m_e^{3/4} M_{pl}^3}{m_p^{7/4}} f_{rad}(\alpha), \quad f_{rad}(\alpha) = 2.7 \, e^{-383.88\left(\alpha - \frac{1}{144}\right)^2} \qquad (32)$$

This approximation yields at least 15% accuracy, though much greater for the majority of values[6].

When both sources of heat are used, the probabilities of observing our quantities become

$$\mathbb{P}(\alpha_{obs}) = 0.19, \quad \mathbb{P}(\beta_{obs}) = 0.42, \quad \mathbb{P}(\gamma_{obs}) = 0.30 \qquad (33)$$

These are almost indistinguishable from the baseline case, so that including glaciations and other related geological causes of extinctions are fully compatible with the multiverse. Using only one of the sources of heat only alters these numbers by a few percent, whichever we take. The distribution of observers including both sources of heat is displayed in Figure 4.

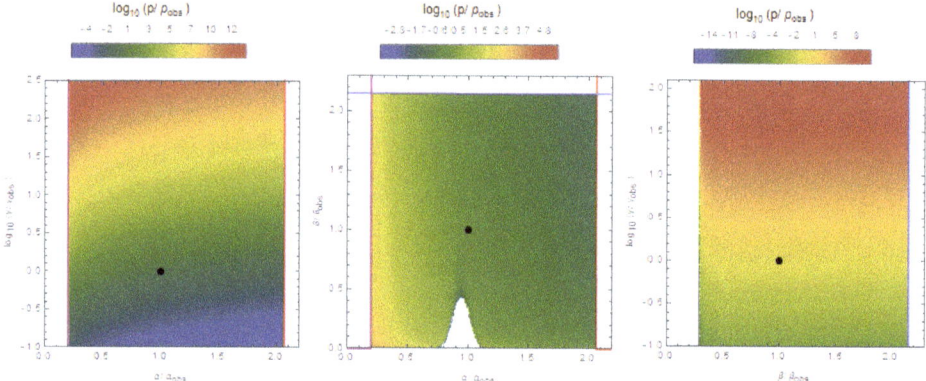

Figure 4. Distribution of observers taking glaciations and sea level rise into account. The small white regions for small β and α close to our value have such high geologic activity that an extinction would occur every 10 Myr.

4.2. Volcanoes

A similar consideration can be made for the rate of supervolcanic eruptions. By these, we mean those rare events that are powerful enough to cause a mass extinction. Volcanic eruptions have been implicated as a causal factor for several of Earth's mass extinctions, most notably the end Permian, which was argued to be a result of the Siberian eruptions, the largest known volcanic event that occurred on land [10].

Volcanoes of all sizes are constantly erupting on Earth, and most are small enough to only affect their immediate vicinity. Some larger ones are capable of exerting a noticeable influence on the whole Earth, but these are correspondingly rare. These large events are qualitatively different in nature from the smaller explosive eruptions, and are most likely a result of a convective instability in the

---

[6] This Gaussian approximation can be improved upon by treating the sum as an integral and using the saddle point approximation: this yields $f_{rad} = 2.7 \, e^{\hat{\ell}} \left(1 + \ell(1-x) - e^{\ell(1-x)}\right)$, with $x = 1/\sqrt{144\alpha}$ and $\ell = 52.72 + \log(t_{rec}/\hat{t}_{decay})$, where the hatted quantity inside the log evaluates to 1 for our values. Using this more accurate expression does not affect our results at all.

Earth's mantle [44]. This results in mantle plumes, which are enormous billows originating in the lower mantle, and emplace millions of cubic kilometers of lava on the surface in geologically short periods of time. While these eruptions inject cooling aerosols into the upper atmosphere, these are short lived and are not currently thought to be the major cause of extinction [45]. More likely is the large amount of carbon dioxide, which for the end Permian extinction injected as much as 10 times preindustrial levels, causing an increase in temperature, acidity, and hypoxia [46].

The size and frequency of these events is then dictated by the mantle physics, as outlined in [47]. There, they found the timescale governing both the evolution and periodicity of convective plumes to be

$$t_{conv} \sim \left(\frac{n \nu A_\oplus}{g \alpha_t Q} \right)^{1/2} \tag{34}$$

Here, $\alpha_t \sim 0.02/T_{melt}$ is the coefficient of thermal expansion, ν is the viscosity of the mantle, and g is the gravitational acceleration. The only quantity in this expression that requires an explanation of any detail is the viscosity: an expression for this was given in [48] based on the diffusion creep model, where defects in the rock structure such as vacancies migrate in response to stresses. They find

$$\nu = \frac{10 \, T \, d^2}{D_0 \, m_a} e^{\Delta H/T} \tag{35}$$

where d is the typical spacing and D_0 is the diffusion rate of holes. The exponential term involving the binding energy varies by over three orders of magnitude throughout the mantle, but the overall normalization is dictated by a very large constant offset reflecting the difficulty of deformation. This offset, which is typical of any quantity with Arrhenius type temperature dependence, is difficult to derive without a detailed model of the microscopic physics, but should be relatively independent of the physical constants. Then the final thing to note is that the diffusion constant is given by $D_0 \propto T d^2$, so that $\nu \sim 10^{25}/m_p$. Our final expression for the rate is

$$\Gamma_{vol} = 3.3 \times 10^{-15} \frac{m_p^{15/8} Q^{1/2}}{\alpha^{3/4} m_e^{3/8} M_{pl}^{3/2}} \tag{36}$$

This has been normalized to $1/(90 \, \text{Myr})$.

We can also check the typical plume volume, and compare this to that required to have a significant impact on the atmosphere. Again from [47], the characteristic size of a plume is given by

$$\lambda_{plume} \sim 31 \left(\frac{n \, \kappa_{heat}^2 \, \nu \, A_\oplus}{g \, \alpha_t \, Q} \right)^{1/4} \tag{37}$$

This corresponds to around 100 km. In [49], it was shown that all plumes are within a narrow range around this value. Here, κ_{heat} is the thermal diffusivity, which was expressed in terms of fundamental constants as $\kappa_{heat} = 2/(m_e^{1/4} m_p^{3/4})$ in [3]. In terms of constants, the volume of the basalt flow is

$$V_{vol} = 1.3 \times 10^{23} \frac{\alpha^{9/8} m_e^{3/16} M_{pl}^{9/4}}{m_p^{21/16} Q^{3/4}} \tag{38}$$

Most of this will consist of lava which, although devastating for local ecosystems, would not have much impact on the global biosphere. We can estimate how much associated carbon dioxide gas is released by noting that for the end Permian eruption, 2000 km^3 basalt flow corresponded to 12 Gt C [50]. The total weight was 10^4 Gt, so for the total amount of carbon we use $M_C \sim 10^{-3} \rho V_{vol}$. This can be compared to the amount needed to significantly warm the climate, which is set by the optical depth becoming appreciable: $M_{\text{warm}} \sim 44 m_p A_\oplus / \sigma_T$. Then the ratio of these two masses is given by

$$\frac{M_C}{M_{\text{warm}}} = 7.9 \times 10^{19} \frac{\alpha^{57/8} m_e^{43/16} M_{pl}^{1/4}}{m_p^{23/16} Q^{3/4}} \qquad (39)$$

If this quantity is less than around 10% of its observed value, then these supervolcanoes will not cause a mass extinction. Though this criteria is somewhat crude, it only affects the overall probabilities by several percent, and so will not be included in our analysis.

If volcanoes are taken to be the only cause of mass extinctions, then the probabilities of observing our values of the constants are

$$\mathbb{P}(\alpha_{obs}) = 0.19, \quad \mathbb{P}(\beta_{obs}) = 0.43, \quad \mathbb{P}(\gamma_{obs}) = 0.30 \qquad (40)$$

From here, it can be seen that volcanoes have very little impact on these values, so that this mechanism for mass extinctions is compatible with the multiverse. The distribution of observers is shown in Figure 5.

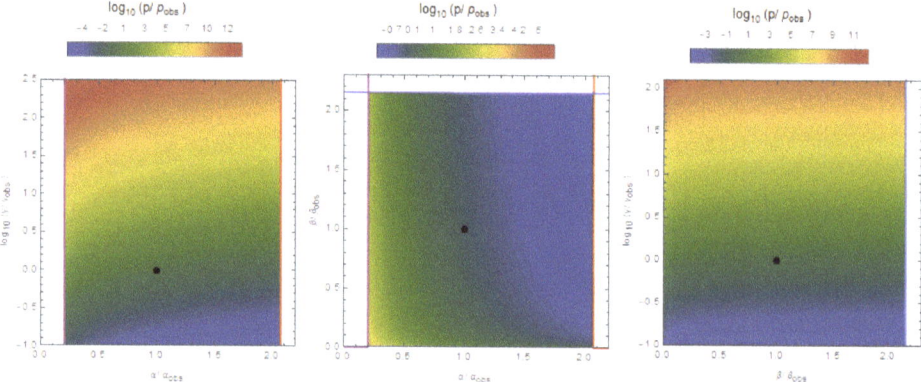

Figure 5. Distribution of observers with volcanic extinctions included.

5. Gamma Ray Bursts

5.1. Gamma Ray Bursts and Extinctions

Stellar explosions, in the form of supernovae, are essential for the creation and redistribution of heavy elements within our universe. However, this mechanism also gives rise to much more powerful events as well, which are orders of magnitude stronger than a typical supernova, and accompanied by an initial pulse of high energy gamma ray radiation, called a gamma ray burst (GRB). While these bursts only last several seconds to minutes in duration, they are so extreme that an event that occurs a substantial fraction of the galaxy away may be capable of throwing the ecosystem of a fragile environment such as our planet into complete disarray [51,52].

GRBs are not all identical, but instead come with a distribution of properties. Their luminosity follows a broken power law that rises for small energies, peaks at the value $L_0 = 10^{52.5}$ erg/s, and then decreases for luminosities beyond this value [53]. The initial pulse of gamma rays would only be for around ten seconds, though the spread in duration is large as well. At a deadly distance, this imparts less than an order of magnitude more than the natural fluence (time integrated energy flux) provided by the sun in visible wavelengths, so animals outside at the time would scarcely notice a major difference (if it happened during the day). However, the effects do not end with the subsidence of the initial burst.

Because most of the energy is delivered in the form of high energy photons, as they pass through the atmosphere they will ionize a large number of atoms. Each of these ions will then convert atmospheric nitrogen N_2 into nitrous oxide molecules NO_x [54]. It is this significant buildup of 'noxious gases' that ultimately leads to effects that can last for months to years. Nitrous oxide molecules react with ozone through the reaction $NO_x + O_3 \rightarrow NO_{x+1} + O_2$, causing its depletion over a sustained period. For the fluence of 100 kJ/m^2, this results in a depletion of the ozone by 50% [55]. Ozone is an excellent UV absorber, and on Earth acts as a shield against otherwise harmful solar radiation. A depletion by this amount will lead to an increase in UV flux by a factor of 3. This increased level of radiation would lead to an enhanced mutation rate among all cells exposed. This would merely lead to an enhanced cancer rate among multicellular organisms like ourselves, but would devastate microscopic organisms such as algae and plankton. Any communities exposed to such harsh radiation may collapse which, through an immense trophic cascade, could trigger the collapse of the entire ecosystem that relies on these organisms for food.

Not every ecosystem would be affected by this, however. Deep sea and benthic (ocean floor) communities are shielded from this radiation by the UV absorption of water. Any ecosystem that is ultimately reliant on the sun, however, will experience this effect, and since the sun is the largest free energy reservoir on our planet, it is naturally responsible for the majority of the complexity we observe. Even among terrestrial and surface ocean communities, atmospheric circulation simulations [54] indicate that the enhancement of ultraviolet light is confined to mid latitudes, leaving the poles relatively untouched. There they also vary the time of day, month, incident latitude, fluence of the event, and atmospheric composition. These variables have some effects on the overall atmospheric response, but none important enough to alter the qualitative conclusion of allowable flux. In [56] the duration of the burst was also shown to have no effect, as each photon acts relatively independently of the others, on a timescale that is long compared to the ionization time, and short compared to the chemical buildup time. The energy of the photons also only marginally affects the final result, since they are all very much above the ionization threshold. Thus, the quantity of primary relevance is the total fluence: only if enough photons to appreciably deplete the ozone layer are incident will drastic effects occur. The value of the fluence required to significantly perturb the ecosystem is 100 kJ/m^2 [55]. For a GRB at the peak value of the luminosity distribution, this corresponds to it being situated a distance 2 kpc away.

In fact, such an event may have been responsible for one of the several mass extinction events that the Earth went through in the past half a billion years. The Ordovician mass extinction, which was the second worst in terms of genera that went extinct, seems particularly compatible with a GRB trigger [8]. Firstly, in was marked by a sudden glaciation in an otherwise climatically stable period, which may have been caused by a GRB. Of the aquatic phyla affected, those that spent more time near the surface of the water seem to have been selected against. The Ordovician extinction is also unique in that afterwards the planet was repopulated by 'high latitude survivors'. There is also some indication that the repopulation happened on land before the water communities recovered. This would be consistent with a resultant nitrate rain that would have been the outcome of a GRB, which would act as fertilizer for land plants but further suppress aquatic communities. However, the cause of the Ordovician extinction remains unproven. It will ultimately be possible to conclusively link a GRB with this event based on the complete record of environmental effects the moon keeps in its regolith [57]. However, such a test of this hypothesis remains outside the foreseeable future.

In the following, we estimate the rate of deadly GRB bursts, and how this depends on the fundamental constants.

5.2. GRB Rate

Data from the SWIFT survey has been used [53] to estimate the rate of gamma ray bursts, arriving at the cosmic average of $1.3^{+0.6}_{-0.7}$/Gpc^3/yr. The number density of galaxies has then been used to estimate that within the Milky Way, one GRB occurs about every 10^7 years. This is shorter than the typical time between mass extinctions, but most GRBs that occur even within our galaxy would be outside the sphere of influence necessary to affect life.

There are several reasons why such a simple extrapolation of the cosmic rate may be too naive, which is why this number is treated as uncertain in much of the literature. The biggest complication is the dependence of the rate on the metallicity of the environment. GRBs roughly track the star formation rate (SFR) [58], since they are the result of very massive stars with lifetimes of only several million years. However, it was noticed in [59] that the observed GRBs show a clear preference for low metallicity environments, which can be explained by noting that several of the observed GRBs at that time were associated with type 1c supernova events. These are supernovae explosions whose hydrogen and helium envelopes are absent, indicating that the star remained well mixed enough throughout its lifetime that the outer envelopes continued to participate in nuclear fusion (as well as the final bang). This can only happen if the star was rapidly rotating up until its death, and, since the presence of metals enhances the rate of angular momentum loss, GRBs can only occur in regions below a certain threshold metallicity. This is observed to be about $Z \sim 0.1 Z_\odot$.

It was then argued that since this threshold metallicity is below the lower range of typical metallicities within our Milky Way, the rate in our galaxy should be suppressed relative to the naive extrapolation from the cosmic rate. However, it was pointed out in [60] that this is incompatible with the fact that an object that looks like a GRB remnant has been detected within our galaxy that is only ten thousand years old [61]. In fact, our galaxy is continually replenished with low metallicity gas from infalling high velocity clouds coming from the outskirts of the galactic disk [62]. These low metallicity mergers, while only about 2% of our galaxy, lead to regions of greatly enhanced star formation because the collision results in regions of locally denser gas. Because of this, [63] concludes that the naive extrapolation of galactic GRB rate is likely an *underestimate*.

To proceed, we outline a simple model that takes the GRB rate to be directly proportional to the star formation rate The fraction of stars which become GRBs, $f_{\text{grb}} \sim 10^{-7}$, will be assumed independent of physical constants at this juncture, as the theoretical underpinning of this number is not on solid enough footing to track the dependence of the constants. Star formation follows the Kennicutt-Schmidt law, $\psi_{\text{sfr}} \propto \rho^{1.4}$ [64], with quite a tight correlation over a wide range of scales, as far as astrophysical observations go. A simplistic account for this scaling can be understood in terms of a model where the star formation rate is proportional to the density divided by the free-fall time of the gas, which scales as $t_{\text{ff}} \sim (G\rho)^{-1/2}$, yielding $\psi_{\text{sfr}} = \epsilon_{\text{sfr}} G^{1/2} \rho^{3/2}$, though this neglects the many intricacies accompanying the star formation process. We note that this law seems to break down below the kiloparsec scale, but since the deadly distance is larger than this breakdown, we effectively average over regions of any smaller size than this. We model the galactic disk as having uniform density and radius r_{gal}, and relatively thin height $h_{\text{gal}} \sim r_{\text{gal}}/10$, so that $\rho = 10 M_{\text{gal}}/(\pi r_{\text{gal}}^3)$. Then the rate of deadly GRBs can be expressed as

$$\Gamma_{\text{grb}} = \frac{f_{\text{grb}} \epsilon_{\text{sfr}} M_{\text{gal}}}{t_{\text{ff}} M_\star} f_{\text{vol}}\left(\frac{r_\dagger}{r_{\text{gal}}}\right) \qquad (41)$$

The fraction of GRBs which are deadly to Earth, f_{vol}, takes into account that the Milky Way is larger than the deadly distance $r_\dagger$, and so only a fraction of GRBs will be dangerous. This distance is $r_\dagger \sim 2$ kpc, several times smaller than the radius of our galaxy, and several times larger than its height. On this scale the galaxy is effectively two dimensional, but we introduce the (slightly crude) ramp function, which holds more generically:

$$f_{\text{vol}}(x) = \begin{cases} \frac{40}{3} x^3 & x < \frac{3}{40} \\ x^2 & \frac{3}{40} < x < 1 \\ 1 & 1 < x \end{cases} \qquad (42)$$

For our values, the fraction of the galaxy that can affect our planet is $(r_\dagger/r_{\rm disk})^2 \sim 0.03$, leading to a lethal event every few 100 Myr [65]. This is compatible with the expectation that there has been one GRB driven extinction since the emergence of complex life.

Before moving on, we note several simplifications we have used in this expression: firstly, we did not take into account that the density of the galaxy is a function of the radius, leading to a potential effect that the interior parts are expected to experience a higher rate of GRBs [66]. Additionally, we do not take into account the exponential decrease in star formation with time, arising from the gas reserves becoming depleted. With this effect, even if a galaxy initially has a GRB rate high enough to disrupt its habitable systems, the rate will ultimately drop to a low enough value that any habitable stars that are still present will be capable of supporting complex life, an observation used in [67] to argue that the universe may have only recently become habitable.

Lastly, we have focused exclusively on smaller, more numerous GRBs from within our galaxy, though more powerful GRBs from nearby dwarf galaxies are potentially relevant as well. These were the focus in [68], where they calculated the extinction rate as a function of the cosmological constant Λ. Extinctions from this type increase for smaller values of Λ because hierarchical structure formation continues for longer. This effect may actually favor larger values of Λ, as then galaxies would be more isolated, leading a decreased extinction rate.

5.3. What Sets $r_\dagger$?

We now calculate the distance to which a gamma ray burst can affect a planetary atmosphere. In our universe this is 2 kpc, which is set by the requirement that the number of high energy photons rivals the amount of ozone in the planet's atmosphere. We track the dependence of both these numbers on the fundamental physics parameters in turn.

Though in actuality highly complex, the physics of gamma ray bursts can be distilled down to a very simple picture, known as the fireshell model [69]: energy densities, in the form of magnetic fields, build up in the environment of a collapsing star. Because of the near total participation in these fully convective systems, this process continues until electrons are capable of being pair produced in this environment. The subsequent annihilation produces photons in the 100 keV-MeV range, which subsequently escape the system and propagate through the universe in a highly collimated beam. If the fraction of the stellar energy converted into photons through this process is denoted by $\epsilon_{\rm grb}$, then the number will be

$$N_{\rm grb} = \epsilon_{\rm grb} \frac{1}{\beta \gamma^3} \quad (43)$$

The number of incident photons on a planet a distance r away will then be related to this quantity through $4\pi\alpha r^2 N_{\rm hits} = A_\oplus N_{\rm grb}$, where $\alpha \sim 10^{-2}$ is the opening angle of the jet (which depends very weakly on the underlying physics [70]). The blast will be lethal when $N_{\rm hits} \sim N_{\rm ozone}$, and so this leads to a lethal distance of

$$r_\dagger = 5\, R_\oplus \sqrt{\frac{N_{\rm grb}}{N_{\rm ozone}}} \quad (44)$$

To proceed, we need the total number of ozone molecules in the atmosphere. Thankfully, this is not as environmentally dependent as one might at first suppose. Recall that ozone is produced when a UV photon breaks apart an O_2 molecule, producing two Os that can then react with an ambient O_2 to produce an O_3 (e.g., [71]). The energy needed for this first reaction is associated with photons of wavelength shorter than 242 nm [72]. Additionally, it is relatively easy for ozone to be photodissociated, which occurs with photons of wavelength less than 1100 nm. This process proceeds until sufficient ozone has built up that these photons are effectively screened from the lower atmosphere, preventing the ozone from further destruction. This makes the column density of ozone simply equal to the inverse of its cross section, $n^c = 1/\sigma \sim 7 \times 10^{18}/{\rm cm}^2$. The cross section in this wavelength region is purely geometric, being far away from any enhancements that come from resonances affecting shorter wavelengths. That this simple picture is correct is evidenced by the fact that the ozone layer is practically constant no matter what the oxygen content of the atmosphere is over almost four orders of magnitude [71], once it reaches sufficient density to saturate this criterion. From here, it is straightforward to find the total ozone in the atmosphere by multiplying by the surface area of the planet. Then, the final expression for the lethal distance is

$$r_\dagger = 6.5 \frac{M_{pl}^{3/2}}{\alpha\, m_e^{3/2}\, m_p} \qquad (45)$$

The coefficient has been set to match the observed value in our universe. Using expressions for the galaxy mass and radius from the appendix in [2], the extinction rate becomes

$$\Gamma_{\rm grb} = 9.1 \times 10^{-14} \frac{Q^{3/2} m_p^2}{M_{pl}} f_{\rm vol} \left(\frac{3.6 \times 10^4\, \kappa^{3/2}}{Q^{1/2}\, \alpha\, \beta^{3/2}\, \gamma^{1/2}} \right) \qquad (46)$$

We have included the cosmological density parameter $\kappa = 10^{-16}$ and amplitude of fluctuations $Q = 1.8 \times 10^{-5}$ for completeness, but these will be held fixed in our analysis. The distribution of observers with this rate is plotted in Figure 6.

The probabilities if GRBs are the only cause of mass extinction are

$$\mathbb{P}(\alpha_{obs}) = 0.32, \quad \mathbb{P}(\beta_{obs}) = 0.23, \quad \mathbb{P}(\gamma_{obs}) = 0.04 \qquad (47)$$

Of the four possible causes of extinctions we consider, this gives the lowest probability values of all.

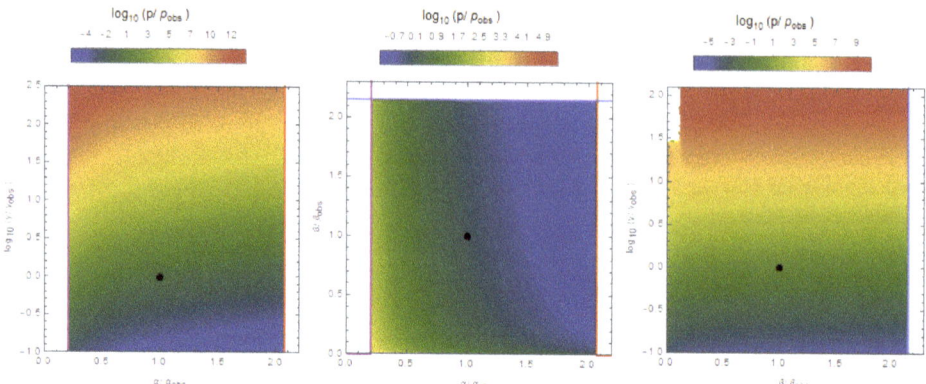

Figure 6. Distribution of observers with gamma ray burst-induced extinctions.

6. Discussion

6.1. Multiple Causes

Up to this point, we have derived expressions for the rates of various purported extinction causing processes, with an aim to be able to extrapolate these rates to different values of the physical constants α, β and γ. When we computed the probabilities of observing our values for the fiducial values $t_{\rm rec} = 10$ Myr, $\Gamma_\dagger = 1/(90\,{\rm Myr})$, we found that these effects did not exert as large an influence on these quantities as the factors we discussed in our previous work, but some of the processes had a larger impact on these values than others. In particular, while glaciations and volcanoes had almost no effect on the probabilities, comets decreased $\mathbb{P}(\beta_{\rm obs})$ by a factor of 5, and GRBs decreased $\mathbb{P}(\gamma_{\rm obs})$ by a factor of 8. Here, we extend this previous analysis to include the effects of multiple extinction causes simultaneously. Additionally, we consider the effects of using the alternative mass extinction models, and the parameterizations of the recovery timescale, discussed in Section 2.

In Table 1 we present values for the setback model, given by Equation (2). The most interesting thing to note is that including GRBs as a factor uniformly makes the probability of observing our strength of gravity below 10%. The worst case is when GRBs are taken to be the only cause of mass extinctions, and several of the combinations are only just below 10%, but it is certainly fair to say that our position in this universe is better explained without including this hazard. The prediction we can derive from this is that technosignatures should not be more prevalent in GRB-quiet regions of the galaxy. Also of note is that taking the recovery time to be set by the length of the year usually makes $\mathbb{P}(\gamma_{obs})$ about 2 times lower, but $\mathbb{P}(\beta_{obs})$ several times higher. The dependence of generation time on the length of the year may in principle be detectable remotely too, for example by looking at seasonal variations of biosignature gases. Since year length shows extreme variation from planet to planet, this effect could be quite noticeable. The probabilities with only comets, for instance, are increased with this choice, but the effect is too mild to make any predictions based on this.

Table 1. Probabilities for various extinction hypotheses discussed in the text. Here the reset model is used for the two choices of the recovery time and including all possible combinations of extinction processes. In all of the above, the recovery time is set to be 10 Myr and the total extinction rate is $(90 \text{ Myr})^{-1}$.

	Setback Model					
t_{rec}	Molecular			Year		
	$\mathbb{P}(\alpha_{obs})$	$\mathbb{P}(\beta_{obs})$	$\mathbb{P}(\gamma_{obs})$	$\mathbb{P}(\alpha_{obs})$	$\mathbb{P}(\beta_{obs})$	$\mathbb{P}(\gamma_{obs})$
comets	0.125	0.0809	0.287	0.169	0.217	0.117
grbs	0.323	0.233	0.042	0.241	0.277	0.019
volcanoes	0.193	0.426	0.303	0.192	0.425	0.293
glaciations	0.193	0.423	0.304	0.193	0.426	0.305
comets+grbs	0.208	0.0761	0.0757	0.207	0.242	0.0284
comets+glaciations	0.123	0.116	0.267	0.169	0.25	0.154
comets+volcanoes	0.124	0.115	0.265	0.169	0.25	0.148
grbs+glaciations	0.299	0.253	0.0527	0.233	0.289	0.0309
grbs+volcanoes	0.299	0.253	0.052	0.232	0.289	0.0301
glaciations+volcanoes	0.193	0.423	0.303	0.193	0.426	0.299
comets+grbs+glac	0.198	0.1	0.0905	0.207	0.255	0.037
comets+grbs+vol	0.198	0.0998	0.0895	0.206	0.256	0.036
comets+glac+vol	0.125	0.146	0.262	0.169	0.274	0.171
grbs+glac+vol	0.283	0.262	0.0697	0.228	0.297	0.0387
all	0.193	0.123	0.0995	0.205	0.266	0.043

Table 2 displays the same results for the reset model, from Equation (4). The probabilities are broadly similar with those of the setback model, since their functional dependences are so similar. Table 3 reports on the intermediate disturbance model, Equation (5): from here one can see that the problems that plague the other two models are largely absent in this one, the majority of the probabilities being either above 10%, or nearly so. From this, we conclude that this model of mass extinctions fares the best with the multiverse hypothesis.

Table 2. Probabilities for different combinations of extinction processes and recovery times with the reset model. Here the noogenesis timescale is set to be 100 Myr.

	Reset Model					
t_{noo}	Molecular			Year		
	$\mathbb{P}(\alpha_{obs})$	$\mathbb{P}(\beta_{obs})$	$\mathbb{P}(\gamma_{obs})$	$\mathbb{P}(\alpha_{obs})$	$\mathbb{P}(\beta_{obs})$	$\mathbb{P}(\gamma_{obs})$
comets	0.105	0.0717	0.251	0.168	0.26	0.136
grbs	0.294	0.243	0.0332	0.226	0.293	0.0194
volcanoes	0.19	0.436	0.317	0.188	0.432	0.31
glaciations	0.192	0.43	0.312	0.19	0.43	0.31
comets+grbs	0.179	0.0804	0.0617	0.202	0.273	0.0351
comets+glaciations	0.114	0.132	0.248	0.167	0.288	0.174
comets+volcanoes	0.114	0.131	0.247	0.168	0.289	0.168
grbs+glaciations	0.265	0.262	0.0506	0.22	0.308	0.038
grbs+volcanoes	0.265	0.262	0.0497	0.22	0.309	0.0368
glaciations+volcanoes	0.191	0.431	0.313	0.189	0.431	0.31
comets+grbs+glac	0.175	0.119	0.07	0.201	0.288	0.0428
comets+grbs+vol	0.175	0.119	0.0692	0.201	0.288	0.042
comets+glac+vol	0.12	0.189	0.257	0.167	0.303	0.19
grbs+glac+vol	0.25	0.271	0.0616	0.217	0.317	0.0467
all	0.173	0.145	0.0874	0.2	0.297	0.0504

Table 3. Probabilities for different combinations of extinction processes and recovery times with the intermediate disturbance hypothesis model. Here the disturbance timescale is set to be 10 Myr.

	Intermediate Disturbance Hypothesis Model					
t_{dist}	Molecular			Year		
	$\mathbb{P}(\alpha_{obs})$	$\mathbb{P}(\beta_{obs})$	$\mathbb{P}(\gamma_{obs})$	$\mathbb{P}(\alpha_{obs})$	$\mathbb{P}(\beta_{obs})$	$\mathbb{P}(\gamma_{obs})$
comets	0.106	0.119	0.251	0.151	0.462	0.4
grbs	0.2	0.329	0.0984	0.129	0.384	0.0963
volcanoes	0.128	0.356	0.368	0.156	0.404	0.262
glaciations	0.131	0.323	0.337	0.148	0.466	0.455
comets+grbs	0.165	0.135	0.0818	0.18	0.453	0.124
comets+glaciations	0.113	0.205	0.256	0.144	0.498	0.454
comets+volcanoes	0.113	0.202	0.261	0.147	0.496	0.489
grbs+glaciations	0.185	0.353	0.113	0.142	0.414	0.146
grbs+volcanoes	0.185	0.351	0.113	0.141	0.423	0.162
glaciations+volcanoes	0.129	0.345	0.359	0.153	0.435	0.4
comets+grbs+glac	0.16	0.18	0.132	0.175	0.46	0.158
comets+grbs+vol	0.16	0.179	0.131	0.179	0.472	0.166
comets+glac+vol	0.123	0.281	0.274	0.147	0.438	0.489
grbs+glac+vol	0.167	0.349	0.178	0.138	0.443	0.186
all	0.159	0.227	0.154	0.172	0.482	0.187

Additionally, we report on the effects of varying the extinction rate and recovery time: since the values of both t_{rec} and $\Gamma_†$ are uncertain, we report the largest value of $\Gamma_† t_{rec}$ for which all values of the probabilities are greater than 10%. This is shown in Table 4. In this table, we have actually fixed $t_{rec} = 0.1 t_{noo} = t_{dist} = 10$ Myr, and report the smallest values for $1/\Gamma_†$ for which all probabilities are greater than 0.1, in Myr. Since for the first two columns, taking the rate of mass extinctions to be very small recovers the scenario where they can be neglected entirely, this minimal value is guaranteed to exist. In the intermediate disturbance hypothesis case, this is not guaranteed, as this scenario favors values for which $\Gamma \sim 1/t_{dist}$, but nevertheless this value does exist for all cases we consider. Though our results are phrased in terms of a fixed recovery time and letting the rate vary, all quantities are only dependent on the product of these, so that if the value x is reported in the table, the actual restriction is $\Gamma_† t_{rec} < 10/x$. If one prefers to hold the extinction rate fixed at the observed value and contemplate varying the recovery time, one has $t_{noo} > 900/x$ Myr. Though in this case t_{noo} also appears in the exponent of the expression for f_{int}, the effects of varying this are comparatively small.

Several features can be extracted from this table. Firstly, for many combinations, the smallest allowable extinction time is less than the observed estimate of 90 Myr. For glaciations and volcanoes in particular, the extinction timescale could be 10 Myr (the smallest allowable by our formalism), and still we would be in this universe. The reason for this is that universes with lower extinction rates were bad real estate anyway, being disfavored regions of the probability distribution for other reasons. Combining multiple processes always results in roughly an average of the lone extinction times. Though we made the simplification that, when multiple effects are presented, each individual rate is taken to be the same, interpolating to more general mixtures can be made by using this observation.

Table 4. The smallest extinction interval Γ_f^{-1} for which all values are above 10%. All values in Myr.

	Setback		Reset		IDH	
t_{rec}	mol	Year	mol	Year	mol	Year
comets	140	71	132	55	70	12
grbs	424	1061	422	813	96	95
volcanoes	10	10	10	10	10	10
glaciations	10	10	10	10	10	10
comets+grbs	160	640	201	473	102	68
comets+glaciations	74	40	69	32	39	10
comets+volcanoes	75	45	70	36	40	13
grbs+glaciations	216	537	215	412	46	58
grbs+volcanoes	218	540	217	415	50	52
glaciations+volcanoes	10	10	10	10	10	10
comets+grbs+glac	114	430	138	319	69	52
comets+grbs+vol	117	433	139	321	69	47
comets+glac+vol	53	33	49	27	29	10
grbs+glac+vol	148	364	148	280	33	41
all	91	328	107	243	53	41

One simplification we have made in our analysis is that the rates of all these processes were treated as constant throughout the universe. In fact, they all are most likely environmental: the Oort clouds of stars born in large clusters can be severely depleted [73], galaxies and regions of galaxies with lower star formation rate will have less GRBs, and planets with less internal heat will have less glaciations and volcanoes. If there truly is a selection pressure for quiet environments, this intra-universe variability will surely play a role. Our analysis has been wholly complementary to this line of reasoning; it serves to investigate how much the coincidence of these timescales can be explained by multiverse reasoning, but not necessarily how much must. Considering both these selection effects in unison is worth further investigation.

6.2. Why Are We in This Universe?

The answer to this question can depend on a great number of factors. In a multiverse context, the probability of being in a particular universe is directly proportional to the number of observers in that universe. The trouble is, there is a large number of things this could depend on, covering a range of scales, from the subatomic to the cosmic. In order to begin to address this question, it is necessary to get a rough idea for what the most important factors are in controlling the total number of intelligent beings in our universe. In this initial series of four papers, we have surveyed a host of different potential controlling processes, and are now in a position to gauge the relative importance of each. The discussion, representing an initial attempt to make progress on this question, has been necessarily inchoate. Very many of the factors involved have only been crudely represented in this analysis, and a great many more have been omitted altogether. Nevertheless, several key insights have been gained, and additionally, this framework can be used as a scaffolding to incorporate arbitrarily sophisticated criteria for the creation and development of complex life. At the close of this first attempt, let us reflect on the generic lessons that have been learned.

The tools for estimating the total number of observers in the universe have been around for decades in the form of the Drake equation. Furthermore, even though we may not have a good idea on the absolute magnitude for some of the factors, often it is possible to determine how each will depend on the physical constants, given a criteria for habitability. Of these, some factors were more sensitive to the laws of physics, and to the assumptions about what life needs that were put in. To recapitulate our results: the number of habitable stars in the universe is the backbone of this computation, and this factor exerts a pressure to live in universes with stronger gravity that must be overcome by one of the other factors in order to provide a consistent picture. The fraction of stars that have planets, on the other hand, was relatively insensitive to the laws of physics, and to the assumptions about planet formation and galactic evolution that we made. Likewise, the properties of planets is likely not a key factor for determining which universe we live in. By far the most important factor was found to be the fraction of planets that develop life. This was most sensitive to the assumptions made, and led to the largest number of predictions for the distribution of life throughout our universe. The fraction of planets that develop intelligence can be similarly constraining, as found with the few different models we explored in [3]. The follow-up we performed here, detailing the purported stymieing effects of mass extinctions, does not play as large a role, but the effects can still be nontrivial.

There is one final factor which we have not discussed, which is the average number of observers per civilization. In principle this may lead to drastic changes in our conclusions, if this depends sensitively on physics. However, we refrain from incorporating this into our analysis at the present moment, largely because it is hard to say anything concrete about this factor without veering into the realm of wild speculation. One thing that may be noted is that it is a perfectly consistent prescription to neglect this factor altogether, as in [74]. The viewpoint here is that "it takes a village to raise a question": that is, that the consciousness you enjoy is not wholly your own, but is in part inherited from the whole history of society. By shifting the selection pressure onto the civilization rather than the individual, this sidesteps this complication completely. While controversial (see [75,76]), this is the de facto stance we have adopted in these papers.

While the exploration of these topics has sometimes resulted in discovering that some factors lead to more predictions than others, it was necessary to establish which of these factors were the most important early, in order to guide the direction of future research. From this, our recommendation would be to look most closely at the properties of stars and the factors that influence the origin of life, as these have generated the most predictions so far. This is not to say that the others are not expected to yield any interesting results at all: indeed, the nature of the task at hand is that any criterion, however innocuous seeming, may be found to be the absolute key driving factor for why we arose in this universe. Thankfully (for these purposes), we have a few decades ahead of us before we start to measure a robust number of exoatmospheres, so there is potentially ample time to sort out the key influences in this proposal.

It should be stressed that a substantial fraction of habitability criteria are incompatible with the multiverse. Then, if the multiverse scenario is true, this leads to the prediction that future surveys will determine these criteria are wrong. In addition, while each individual instance of prediction is a far cry from proving that other universes exist, by now we have accumulated a list of them: taken together, the ultimate case for the multiverse can be made far stronger. The more factors we incorporate, the more predictions we will be able to make, and the stronger our case will be. Though this method will never be able to do better than an indirect inference of the existence of other realms forever outside our reach, in the absence of a direct way of observing them, it represents the best path forward.

Through the course of our analysis we found the additional complication that some habitability criteria are only compatible with the multiverse if others are simultaneously employed. This was the case, for instance, when the requirement of plate tectonics was found incompatible on its own, but consistent when the tidal locking condition was included. This conditional interdependence prevents us from testing each criterion in complete isolation, instead necessitating a check of its compatibility with all previously considered hypotheses. This leads to an exponential proliferation of different choices, which is already becoming overly cumbersome to enumerate, test and report on.

Additionally, attention so far has been restricted to just three physical constants, which do a fair job of determining the character of the macroscopic world. However, for full consistency, about 10 of the parameters of the standard models of particle physics and cosmology must be incorporated. Furthermore, more attention needs to be paid to local environmental variations within the universe, which to this point has crudely been represented only by stellar mass. Expanding the calculation in these ways will significantly increase computational costs, but promises to extend our predictive power. Of equal importance will be to identify further criteria for what life needs and how these processes likely depend on physics to incorporate into this analysis. The more criteria are put into the system, the more predictions will be returned, and the stronger the case either for or against the multiverse will be.

7. Conclusions

In this work, we have investigated the influence of mass extinctions on the probability of our presence in this universe, within the multiverse context. On the whole, we find that this factor is not as important as the others in the Drake equation, as explored in other papers of this series, and so we do not expect the extinction rate to be the determining factor for why we live here. However, depending on the assumptions we made about the relative importance of the various extinction mechanisms we consider and their overall effect, more can be said. Firstly, taking mass extinctions to be solely caused by gamma ray bursts led to an uncomfortably small probability for observing our strength of gravity, signaling that this assumption should be wrong. In contrast, taking comets to be the only cause had a much smaller impact on the probabilities, and the geologic influences had almost none at all. Combining various processes usually tempers the effects each would exert individually. Various models for extinction effects were explored, and while the setback and reset model were broadly similar, the intermediate disturbance hypothesis was very forgiving, resulting in probabilities almost always compatible with our existence here. Lastly, taking the recovery time to be dictated by the year length rather than the molecular timescale changed some of the probability values by as much as a factor of a few, but not drastically. So, while certain specific assumptions are incompatible with the multiverse hypothesis, for the most part we can expect the selective influence exerted by extinction events to be minimal.

Funding: This research received no external funding.

Acknowledgments: I would like to thank Fred Adams, Gary Bernstein, RJ Graham, Mario Livio, Aki Roberge, and Alex Vilenkin for useful discussions.

Conflicts of Interest: The author declares no conflict of interest.

References

1. Sandora, M. Multiverse Predictions for Habitability I: The Number of Stars and Their Properties. *arXiv* **2019**, arXiv:1901.04614.
2. Sandora, M. Multiverse Predictions for Habitability II: Number of Habitable Planets. *arXiv* **2019**, arXiv:1902.06784.
3. Sandora, M. Multiverse Predictions for Habitability III: Fraction of Planets That Develop Life. *arXiv* **2019**, arXiv:1903.06283 .
4. Hallam, A.; Wignall, P.B. *Mass Extinctions and Their Aftermath*; Oxford University Press: Oxford, UK, 1997.
5. Alvarez, L.W.; Alvarez, W.; Asaro, F.; Michel, H.V. Extraterrestrial cause for the Cretaceous-Tertiary extinction. *Science* **1980**, *208*, 1095–1108. [CrossRef] [PubMed]

6. Schulte, P.; Alegret, L.; Arenillas, I.; Arz, J.A.; Barton, P.J.; Bown, P.R.; Bralower, T.J.; Christeson, G.L.; Claeys, P.; Cockell, C.S.; et al. The Chicxulub asteroid impact and mass extinction at the Cretaceous-Paleogene boundary. *Science* **2010**, *327*, 1214–1218. [CrossRef]
7. Berry, W.B.; Boucot, A.J. Glacio-eustatic control of Late Ordovician–Early Silurian platform sedimentation and faunal changes. *Geol. Soc. Am. Bull.* **1973**, *84*, 275–284. [CrossRef]
8. Melott, A.L.; Lieberman, B.S.; Laird, C.M.; Martin, L.D.; Medvedev, M.V.; Thomas, B.C.; Cannizzo, J.; Gehrels, N.; Jackman, C. Did a gamma-ray burst initiate the late Ordovician mass extinction? *Int. J. Astrobiol.* **2004**, *3*, 55–61. [CrossRef]
9. Algeo, T.J.; Scheckler, S.E. Terrestrial-marine teleconnections in the Devonian: Links between the evolution of land plants, weathering processes, and marine anoxic events. *Philos. Trans. R. Soc. Lond. Ser. B Biol. Sci.* **1998**, *353*, 113–130. [CrossRef]
10. Kamo, S.L.; Czamanske, G.K.; Amelin, Y.; Fedorenko, V.A.; Davis, D.; Trofimov, V. Rapid eruption of Siberian flood-volcanic rocks and evidence for coincidence with the Permian–Triassic boundary and mass extinction at 251 Ma. *Earth Planet. Sci. Lett.* **2003**, *214*, 75–91. [CrossRef]
11. Rothman, D.H.; Fournier, G.P.; French, K.L.; Alm, E.J.; Boyle, E.A.; Cao, C.; Summons, R.E. Methanogenic burst in the end-Permian carbon cycle. *Proc. Natl. Acad. Sci. USA* **2014**, *111*, 5462–5467. [CrossRef]
12. Embry, A.F. Triassic Sea-Level Changes: Evidence from the Canadian Arctic Archipelago. 1988. Available online: http://archives.datapages.com/data/sepm_sp/SP42/Triassic_Sea-Level_Changes.htm (accessed on 16 July 2019).
13. Barnosky, A.D.; Matzke, N.; Tomiya, S.; Wogan, G.O.; Swartz, B.; Quental, T.B.; Marshall, C.; McGuire, J.L.; Lindsey, E.L.; Maguire, K.C.; et al. Has the Earth's sixth mass extinction already arrived? *Nature* **2011**, *471*, 51. [CrossRef] [PubMed]
14. Gorban, A.N.; Pokidysheva, L.I.; Smirnova, E.V.; Tyukina, T.A. Law of the minimum paradoxes. *Bull. Math. Biol.* **2011**, *73*, 2013–2044. [CrossRef] [PubMed]
15. Connell, J.H. Diversity in tropical rain forests and coral reefs. *Science* **1978**, *199*, 1302–1310. [CrossRef] [PubMed]
16. Randall Hughes, A.; Byrnes, J.E.; Kimbro, D.L.; Stachowicz, J.J. Reciprocal relationships and potential feedbacks between biodiversity and disturbance. *Ecol. Lett.* **2007**, *10*, 849–864. [CrossRef] [PubMed]
17. Miller, A.D.; Roxburgh, S.H.; Shea, K. How frequency and intensity shape diversity—Disturbance relationships. *Proc. Natl. Acad. Sci. USA* **2011**, *108*, 5643–5648. [CrossRef] [PubMed]
18. Press, W.H.; Lightman, A.P. Dependence of macrophysical phenomena on the values of the fundamental constants. *Philos. Trans. R. Soc. Lond. Ser. A* **1983**, *310*, 323–334. [CrossRef]
19. Lightman, A.P. A fundamental determination of the planetary day and year. *Am. J. Phys.* **1984**, *52*, 211–214. [CrossRef]
20. Horner, J.; Jones, B. Jupiter: Friend or foe? An answer. *Astron. Geophys.* **2010**, *51*, 6–16. [CrossRef]
21. Fernández, J.; Ip, W.H. Statistical and evolutionary aspects of cometary orbits. In *International Astronomical Union Colloquium*; Cambridge University Press: Cambridge, UK, 1989; Volume 116, pp. 487–535.
22. Dones, L.; Weissman, P.R.; Levison, H.F.; Duncan, M.J. Oort cloud formation and dynamics. In *Star Formation in the Interstellar Medium: In Honor of David Hollenbach*; ASP: San Francisco, CA, USA, 2004; Volume 323, p. 371.
23. Heisler, J.; Tremaine, S. The influence of the galactic tidal field on the Oort comet cloud. *Icarus* **1986**, *65*, 13–26. [CrossRef]
24. Gaidos, E.J. Paleodynamics: Solar System formation and the early environment of the Sun. *Icarus* **1995**, *114*, 258–268. [CrossRef]
25. Brasser, R.; Duncan, M.; Levison, H. Embedded star clusters and the formation of the Oort Cloud: II. The effect of the primordial Solar nebula. *Icarus* **2007**, *191*, 413–433. [CrossRef]
26. Heisler, J.; Tremaine, S.; Alcock, C. The frequency and intensity of comet showers from the Oort cloud. *Icarus* **1987**, *70*, 269–288. [CrossRef]
27. Byl, J. Galactic perturbations on nearly-parabolic cometary orbits. *Moon Planets* **1983**, *29*, 121–137. [CrossRef]
28. Meech, K.; Hainaut, O.; Marsden, B. Comet nucleus size distributions from HST and Keck telescopes. *Icarus* **2004**, *170*, 463–491. [CrossRef]

29. Safronov, V. Ejection of Bodies from the Solar System in the Course of the Accumulation of the Giant Planets and the Formation of the Cometary Cloud. In *Symposium-International Astronomical Union*; Cambridge University Press: Cambridge, UK, 1972; Volume 45, pp. 329–334.
30. Tremaine, S. The distribution of comets around stars. *Planets Around Pulsars* **1993**, *36*, 335–344.
31. Bland, P.A.; Artemieva, N.A. The rate of small impacts on Earth. *Meteorit. Planet. Sci.* **2006**, *41*, 607–631. [CrossRef]
32. Toon, O.B.; Zahnle, K.; Morrison, D.; Turco, R.P.; Covey, C. Environmental perturbations caused by the impacts of asteroids and comets. *Rev. Geophys.* **1997**, *35*, 41–78. [CrossRef]
33. Pope, K.O. Impact dust not the cause of the Cretaceous-Tertiary mass extinction. *Geology* **2002**, *30*, 99–102. [CrossRef]
34. Belcher, C.M.; Collinson, M.E.; Sweet, A.R.; Hildebrand, A.R.; Scott, A.C. Fireball passes and nothing burns—The role of thermal radiation in the Cretaceous-Tertiary event: Evidence from the charcoal record of North America. *Geology* **2003**, *31*, 1061–1064. [CrossRef]
35. Alegret, L.; Thomas, E.; Lohmann, K.C. End-Cretaceous marine mass extinction not caused by productivity collapse. *Proc. Natl. Acad. Sci. USA* **2012**, *109*, 728–732. [CrossRef]
36. Twitchett, R.J. The palaeoclimatology, palaeoecology and palaeoenvironmental analysis of mass extinction events. *Palaeogeogr. Palaeoclimatol. Palaeoecol.* **2006**, *232*, 190–213. [CrossRef]
37. Brugger, J.; Feulner, G.; Petri, S. Baby, it's cold outside: Climate model simulations of the effects of the asteroid impact at the end of the Cretaceous. *Geophys. Res. Lett.* **2017**, *44*, 419–427. [CrossRef]
38. Melosh, H.J. *Impact Cratering: A Geologic Process*; Research Supported by NASA; Oxford University Press: New York, NY, USA, 1989; Volume 253, p.11.
39. Manins, P. Cloud heights and stratospheric injections resulting from a thermonuclear war. *Atmos. Environ. (1967)* **1985**, *19*, 1245–1255. [CrossRef]
40. Kennett, J.P. Cenozoic evolution of Antarctic glaciation, the circum-Antarctic Ocean, and their impact on global paleoceanography. *J. Geophys. Res.* **1977**, *82*, 3843–3860. [CrossRef]
41. Konhauser, K.O. *Introduction to Geomicrobiology*; John Wiley & Sons: Hoboken, NJ, USA, 2009.
42. Gando, A.; Gando, Y.; Ichimura, K.; Ikeda, H.; Inoue, K.; Kibe, Y.; Kishimoto, Y.; Koga, M.; Minekawa, Y.; Mitsui, T.; et al. Partial radiogenic heat model for Earth revealed by geoneutrino measurements. *Nat. Geosci.* **2011**, *4*, 647.
43. Sandora, M. The fine structure constant and habitable planets. *J. Cosmol. Astropart. Phys.* **2016**, *8*, 048. [CrossRef]
44. Saunders, A.D. Large igneous provinces: Origin and environmental consequences. *Elements* **2005**, *1*, 259–263. [CrossRef]
45. Schmidt, A.; Skeffington, R.A.; Thordarson, T.; Self, S.; Forster, P.M.; Rap, A.; Ridgwell, A.; Fowler, D.; Wilson, M.; Mann, G.W.; et al. Selective environmental stress from sulphur emitted by continental flood basalt eruptions. *Nat. Geosci.* **2016**, *9*, 77. [CrossRef]
46. Benton, M.J. Hyperthermal-driven mass extinctions: Killing models during the Permian—Triassic mass extinction. *Philos. Trans. R. Soc. A Math. Phys. Eng. Sci.* **2018**, *376*, 20170076. [CrossRef]
47. Tan, K.K.; Thorpe, R.B.; Zhao, Z. On predicting mantle mushroom plumes. *Geosci. Front.* **2011**, *2*, 223–235. [CrossRef]
48. McKenzie, D.P. The viscosity of the lower mantle. *J. Geophys. Res.* **1966**, *71*, 3995–4010. [CrossRef]
49. Tan, K.K.; Thorpe, R.B. The onset of convection driven by buoyancy caused by various modes of transient heat conduction: Part II. The sizes of plumes. *Chem. Eng. Sci.* **1999**, *54*, 239–244. [CrossRef]
50. Saunders, A.; Reichow, M. The Siberian Traps and the End-Permian mass extinction: A critical review. *Chin. Sci. Bull.* **2009**, *54*, 20–37. [CrossRef]
51. Thorsett, S. Terrestrial implications of cosmological gamma-ray burst models. *arXiv* **1995**, arXiv:10.1086/187858.
52. Scalo, J.; Wheeler, J.C. Astrophysical and astrobiological implications of gamma-ray burst properties. *Astrophys. J.* **2002**, *566*, 723. [CrossRef]
53. Wanderman, D.; Piran, T. The luminosity function and the rate of Swift's gamma-ray bursts. *Mon. Not. R. Astron. Soc.* **2010**, *406*, 1944–1958. [CrossRef]

54. Thomas, B.C.; Melott, A.L.; Jackman, C.H.; Laird, C.M.; Medvedev, M.V.; Stolarski, R.S.; Gehrels, N.; Cannizzo, J.K.; Hogan, D.P.; Ejzak, L.M. Gamma-ray bursts and the Earth: Exploration of atmospheric, biological, climatic, and biogeochemical effects. *Astrophys. J.* **2005**, *634*, 509. [CrossRef]
55. Thomas, B.C.; Jackman, C.H.; Melott, A.L.; Laird, C.M.; Stolarski, R.S.; Gehrels, N.; Cannizzo, J.K.; Hogan, D.P. Terrestrial ozone depletion due to a Milky Way gamma-ray burst. *Astrophys. J. Lett.* **2005**, *622*, L153. [CrossRef]
56. Ejzak, L.M.; Melott, A.L.; Medvedev, M.V.; Thomas, B.C. Terrestrial consequences of spectral and temporal variability in ionizing photon events. *Astrophys. J.* **2007**, *654*, 373. [CrossRef]
57. Crawford, I.A. The moon as a recorder of nearby supernovae. *arXiv* **2016**, arXiv:1608.03926.
58. Bromm, V.; Loeb, A. The expected redshift distribution of gamma-ray bursts. *Astrophys. J.* **2002**, *575*, 111. [CrossRef]
59. Stanek, K.Z.; Gnedin, O.; Beacom, J.; Gould, A.; Johnson, J.; Kollmeier, J.; Modjaz, M.; Pinsonneault, M.; Pogge, R.; Weinberg, D. Protecting life in the Milky Way: Metals keep the GRBs away. *Acta Astron.* **2006**, *56*, 333–345.
60. Melott, A.L. Comment on: Protecting Life in the Milky Way: Metals Keep the GRBs Away by Stanek et al. *arXiv* **2006**, arXiv:astro-ph/0604440.
61. Atoyan, A.; Buckley, J.; Krawczynski, H. A Gamma-Ray Burst Remnant in Our Galaxy: HESS J1303–1631. *Astrophys. J. Lett.* **2006**, *642*, L153. [CrossRef]
62. Putman, M.; Peek, J.; Joung, M. Gaseous galaxy halos. *Annu. Rev. Astron. Astrophys.* **2012**, *50*, 491–529. [CrossRef]
63. Jimenez, R.; Piran, T. Reconciling the gamma-ray burst rate and star formation histories. *Astrophys. J.* **2013**, *773*, 126. [CrossRef]
64. Kennicutt, R.C. The global Schmidt law in star-forming galaxies. *Astrophys. J.* **1998**, *498*, 541. [CrossRef]
65. Gowanlock, M.G. Astrobiological effects of gamma-ray bursts in the Milky Way galaxy. *Astrophys. J.* **2016**, *832*, 38. [CrossRef]
66. Gowanlock, M.G.; Morrison, I.S. The Habitability of our Evolving Galaxy. *arXiv* **2018**, arXiv:1802.07036.
67. Annis, J. An astrophysical explanation for the great silence. *arXiv* **1999**, arXiv:astro-ph/9901322.
68. Piran, T.; Jimenez, R.; Cuesta, A.J.; Simpson, F.; Verde, L. Cosmic explosions, life in the universe, and the cosmological constant. *Phys. Rev. Lett.* **2016**, *116*, 081301. [CrossRef] [PubMed]
69. Ruffini, R.; Rueda, J.; Muccino, M.; Aimuratov, Y.; Becerra, L.; Bianco, C.; Kovacevic, M.; Moradi, R.; Oliveira, F.; Pisani, G.; et al. On the classification of GRBs and their occurrence rates. *Astrophys. J.* **2016**, *832*, 136. [CrossRef]
70. Goldstein, A.; Connaughton, V.; Briggs, M.S.; Burns, E. Estimating long grb jet opening angles and rest-frame energetics. *Astrophys. J.* **2016**, *818*, 18. [CrossRef]
71. Ratner, M.I.; Walker, J.C. Atmospheric ozone and the history of life. *J. Atmos. Sci.* **1972**, *29*, 803–808. [CrossRef]
72. Craig, R.A. *The Upper Atmosphere: Meteorology and Physics*; Elsevier: Amsterdam, the Netherlands, 2016.
73. Brasser, R.; Morbidelli, A. Oort cloud and Scattered Disc formation during a late dynamical instability in the Solar System. *Icarus* **2013**, *225*, 40–49. [CrossRef]
74. Vilenkin, A. Predictions from Quantum Cosmology. *Phys. Rev. Lett.* **1995**, *74*, 846–849. [CrossRef] [PubMed]
75. Bostrom, N. *Anthropic bias: Observation Selection Effects in Science and Philosophy*; Routledge: New York, NY, USA, 2013.
76. Olum, K.D. Conflict between anthropic reasoning and observation. *Analysis* **2004**, *64*, 1–8. [CrossRef]

 © 2019 by the author. Licensee MDPI, Basel, Switzerland. This article is an open access article distributed under the terms and conditions of the Creative Commons Attribution (CC BY) license (http://creativecommons.org/licenses/by/4.0/).

MDPI
St. Alban-Anlage 66
4052 Basel
Switzerland
Tel. +41 61 683 77 34
Fax +41 61 302 89 18
www.mdpi.com

Universe Editorial Office
E-mail: universe@mdpi.com
www.mdpi.com/journal/universe

www.ingramcontent.com/pod-product-compliance
Lightning Source LLC
LaVergne TN
LVHW070429100526
838202LV00014B/1555